ANNALS *of* THE NEW YORK ACADEMY OF SCIENCES

The New York Academy of Sciences
7 World Trade Center
250 Greenwich Street, 40th Floor
New York, NY 10007-2157

annals@nyas.org
www.nyas.org/annals

The New York Academy of Sciences

Published by Blackwell Publishing
On behalf of the New York Academy of Sciences

Boston, Massachusetts
2010

ANNALS *of* THE NEW YORK ACADEMY OF SCIENCES

VOLUME 1214

ISSUE

The Year in Human and Medical Genetics

New Trends in Mendelian Genetics

ISSUE EDITORS

Jean-Laurent Casanova, Laurent Abel, and Stanislas Lyonnet

Supported in part by a grant from the Life Technologies Foundation.

TABLE OF CONTENTS

Ann. N.Y. Acad. Sci. ISSN 0077-8923

ANNALS OF THE NEW YORK ACADEMY OF SCIENCES
Issue: *The Year in Human and Medical Genetics*

The role played by natural selection on Mendelian traits in humans

Lluis Quintana-Murci[1,2] and Luis B. Barreiro[3]

[1]Institut Pasteur, Human Evolutionary Genetics, Department of Genomes and Genetics, Paris, France. [2]Centre National de la Recherche Scientifique, Paris, France. [3]Department of Human Genetics, University of Chicago, Chicago, Illinois

Address for correspondence: Lluis Quintana-Murci, UP Human Evolutionary Genetics, Institut Pasteur, 25, rue Dr Roux F-75015 Paris, France. quintana@pasteur.fr

Detecting whether and how natural selection has targeted regions of the human genome represents a complementary strategy for identifying functionally important loci and variants involved in disease resistance and adaptation to the environment. In contrast with most complex diseases or traits, the genetic architecture of most Mendelian traits is relatively well established. Most mutations associated with Mendelian disease-related traits are highly penetrant and kept at low population frequencies because of the effects of purifying selection. However, this is not always the case. Here, we review several examples of Mendelian mutations—associated with various disease conditions or other traits of anthropological interest—that have increased in frequency in the human population as a result of past positive selection. These examples clearly illustrate the value of a population genetics approach to unravel the biological mechanisms that have been central to our past and present survival against the selective pressures imposed by diseases and other environmental factors.

Keywords: natural selection; Mendelian traits; malaria; noroviruses; lactase persistence; cystic fibrosis

Introduction

The ultimate challenge of modern human genetics is to identify the genetic basis of human diseases and other phenotypes of interest. Determining how genetic variants influence the susceptibility or severity of human disease and the occurrence of other traits will provide new insights into the mechanisms underlying disease etiology and phenotypic variation. In turn, this could lead to the development of new therapeutics and novel genetic screening tools to identify individuals at increased risk. During the last few decades, there has been much success in identifying genes responsible for a wide range of Mendelian diseases or traits. However, the identification of loci involved in complex traits has proved to be more complicated. Mendelian and complex traits can be considered as opposite extremes of a continuous spectrum of relationships between genotypes and phenotypes.[1,2] Traits in which single genotypic changes result in large, often discrete phenotypic differences that follow Mendelian rules of inheritance in families are referred to as "Mendelian traits." However, the segregation of phenotypic traits in a "strict" Mendelian fashion is the exception, rather than the rule. First, many Mendelian traits are not caused by a single, unique genetic variant and feature extremely high levels of allelic heterogeneity. For example, although the cystic fibrosis (CF) locus harbors a single variant that accounts for about 70% of all disease alleles, many additional disease alleles with various phenotypic effects have been identified.[3] In addition, various modifier factors can influence the penetrance or severity of a Mendelian trait; these include genetic modifiers (i.e., modifier genes or variation in regulatory elements) and nongenetic modifiers linked to lifestyle and environmental factors (age, gender, alcohol consumption, etc.). Second, most phenotypic differences in morphology, physiology, and disease are quantitative rather than discrete, and display a continuous, almost normal phenotypic distribution.[1] These so-called complex traits are caused by a number of genetic and environmental

doi: 10.1111/j.1749-6632.2010.05856.x

factors and therefore do not follow simple Mendelian rules of inheritance. Because the line between Mendelian and complex traits is fairly fuzzy, a better understanding of the causes and evolutionary forces that shape Mendelian traits should also inform the more elusive genetic causes of complex traits.[4,5]

Mendelian diseases are usually rare and in most cases result from highly penetrant, deleterious alleles that segregate in families.[1,2] Linkage methods for mapping Mendelian traits are well established and have identified the molecular causes of hundreds of diseases. In most cases, the mutations underlying Mendelian diseases are found at very low frequencies in natural populations ($\ll$1%). These low frequencies are most likely determined by mutation–selection balance, that is, a balance between the input of new mutations and the effects of purifying selection that gradually remove them.[1,6] Mutation–selection balance predicts high allelic heterogeneity and low disease prevalence, which are both common features of Mendelian diseases. However, highly penetrant alleles associated with a disease condition can occasionally reach high frequencies in the general population, challenging the dogma that Mendelian diseases are clinical rarities. In this context, the field of population genetics has proven to be a useful complementary strategy in highlighting functionally important loci and variants associated with disease resistance and human adaptation throughout evolution. By analyzing the patterns of genetic variability within natural populations, empirical population genetics seeks to understand which evolutionary forces have shaped such diversity. Specifically, identifying genes that have been targeted by natural selection, in its different forms and intensities, should highlight the biological mechanisms that have been critical for our past and present survival. In the context of disease, genetic variants that influence susceptibility to disease are likely to affect the fitness of the organism, unless the disease concerned begins later in the life, after reproductive age. There is, therefore, an intimate relationship between disease and selection that can be exploited for the better understanding of disease susceptibility in contemporary human populations.[6–13]

Here, we will review how natural selection has influenced the frequency of some Mendelian (or quasi-Mendelian) mutations associated with different diseases and other human phenotypic traits. We start with a description of the different forms of natural selection and the distinctive molecular signature they leave in the targeted genomic region. We then discuss several examples of how natural selection can maintain, or even favor the presence of, mutations in the population that are associated with various disease conditions. We also discuss the selective advantage afforded by a number of traits not directly related to disease, such as lactase persistence or bitter-taste sensitivity. Finally, we conclude by highlighting the utility of assessing the form and intensity of natural selection targeting the human genome as a vehicle for increasing our knowledge of the genetic factors that underlie Mendelian and complex diseases.

Natural selection and genome diversity

Over 40 years ago, Motoo Kimura suggested that most genetic variation is selectively neutral, so that the frequency dynamics of such mutations in the population is primarily determined by chance, a process known as genetic drift.[14] An alternative view is that a significant proportion of genetic variation does affect fitness—the ability of an organism to survive and reproduce—and will therefore be subject to the effects of natural selection. Even though the relative contributions of genetic drift and natural selection in human evolution are still being debated, there is little doubt that both forces had a significant role in shaping human genome diversity.[11–13,15] A variety of forms of natural selection can take place that generate different molecular "signatures" in the patterns of genetic variation (Fig. 1).

Negative or purifying selection refers to the selective removal of deleterious alleles from a population. In humans, 38–75% of all new amino-acid altering mutations are estimated to be affected by moderate to strong negative selection.[10,16] Under conditions of strong negative selection, the frequent removal of deleterious variants will also lead to the occasional removal of neutral-linked variation, resulting in decreased levels of variation around the selected locus (Fig. 1A). This selective regime is by far the most common form of selection responsible for the removal (or maintenance at low frequencies) of mutations associated with severe Mendelian disorders.[17,18] Conversely, *positive or directional selection* refers to selection that favors advantageous mutations. Most population genetic

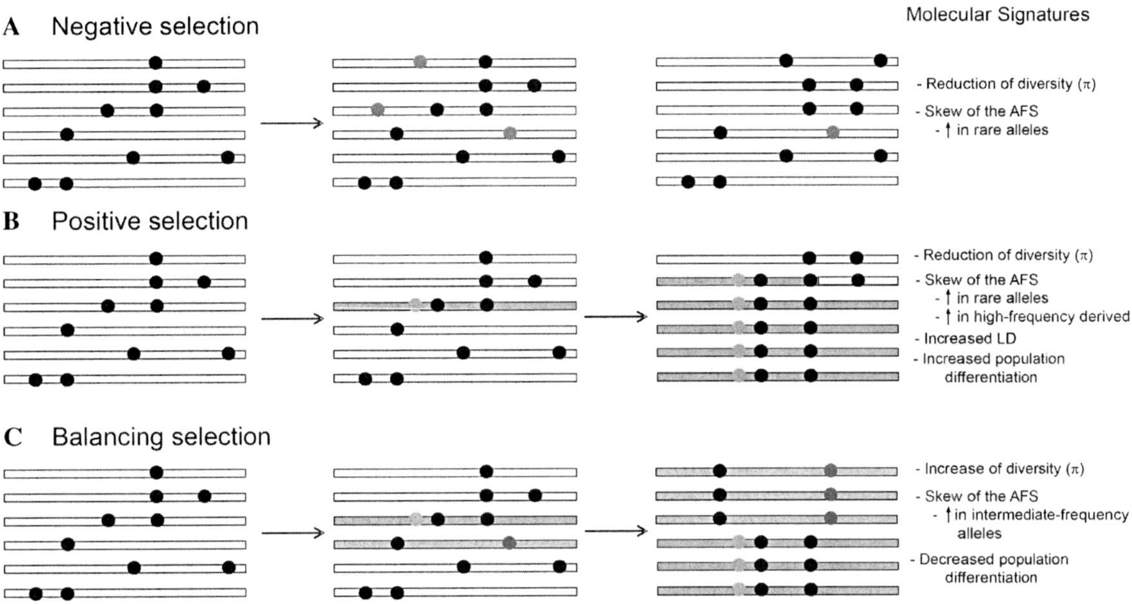

Figure 1. The molecular signatures of natural selection. (A) On the left, patterns of neutral polymorphism (denoted as black circles) are shown for a sample of six haplotypes. When deleterious mutations arise (indicated by the red circle), they can either disappear from the population (if extremely deleterious) or be retained at low population frequencies (if weakly deleterious). The main consequence will be a reduction of diversity and an increase of rare alleles in the case of weak negative selection. (B) On the left, patterns of neutral polymorphism are shown. A new advantageous mutation arises (indicated by the green circle) on a specific haplotype (mid-panel highlighted in orange). As the advantageous allele increases in frequency, it drags along linked neutral polymorphisms. On the right, an incomplete sweep is represented, such as the advantageous allele that has not yet reached fixation. The expected molecular signatures of this selective event are also described. Note that the relative magnitude of these signatures of positive selection depend on many factors, such as when the advantageous allele arose, the strength of selection, or whether the sweep is ongoing or has reached already reached fixation. (C) On the left, patterns of neutral polymorphism are shown. In the middle panel, two advantageous alleles arise (circles highlighted in green and blue). If they confer an advantage at the heterozygote state, they will increase toward intermediate frequencies in the population, "dragging" with them all linked neutral polymorphisms. An alternative form of balancing selection is frequency-dependent selection. In this case, all rare variants are advantaged with respect to common variants. In both situations, there is an increase of the overall levels of diversity of the regions. We also expect to observe an excess of intermediate frequency allele in the regions targeted by balancing selection.

approaches to detect regions of the genome targeted by positive selection rely on the fact that a beneficial allele will increase to a high frequency within the population at a rate that is much faster than that of a neutrally evolving allele. As an advantageous allele is driven quickly to an intermediate or high frequency, only a few recombinant events will take place and, consequently, the neutral variation linked to the selected variant will also increase in frequency within the population. This process if often referred to as a *partial or incomplete selective sweep* and results, among other things, in a transitory increase in the strength of linkage disequilibrium (LD) (i.e., haplotype homozygosity) associated with the selected allele.[13,19] This signature can be identified by means of LD-based tests, such as the relative extended homozygosity test (rEHH),[19] the integrated haplotype score test (iHS),[20] or the linkage disequilibrium decay test (LDD).[21] If positive selection acts for a long enough period of time, the favored allele can reach fixation within the population. In this case, known as a *complete selective sweep*, most of the neutral variation linked to the selected variant will also become fixed, leading to an overall reduction of genetic diversity surrounding the selected site (Fig. 1B). Additional signatures of a complete selective sweep include a skew in the distribution of allele frequencies toward rarer alleles and an excess of high-frequency derived variants. These signals can be detected by classical neutrality tests, such as Tajima's *D* and Fay and Wu's *H*, or by more recent (and powerful) methods, such as composite likelihood approaches (e.g., CLR test) (for an extensive review on statistical tests of neutrality, see Refs. 22–24).

Positive selection can act at the level of the entire human species (i.e., a mutation that is advantageous in all human populations regardless of the varying environmental contexts) or within specific population(s). When positive selection occurs in only a subset of populations, the frequency of the selected variant may differ across populations to a greater extent than that predicted by chance. This will increase the genetic differentiation between populations. The levels of genetic differentiation between populations are commonly quantified by the statistic F_{ST}.[25,26] One interesting property of the F_{ST} statistic is that, unlike classical sequence-based neutrality tests, it can be SNP-specific, and can thus directly identify the genetic variant(s) under selection. In humans, genome-wide analysis of population differentiation levels between different human groups shows that positive selection has primarily targeted amino acid altering mutations and ensured the regional adaptation of human populations.[27]

A final form of natural selection that can also shape genetic diversity is *balancing selection.* Balancing selection is a general type of selective regime in which two, or multiple alleles, at a given locus are maintained in the population.[28,29] The principal molecular signature left by balancing selection is an increase of genetic diversity (Fig. 1C), because polymorphisms are maintained in the population for a longer period of time than expected under neutrality.[28] There are several mechanisms by which balancing selection works to maintain polymorphism, including heterozygote advantage (or overdominance) and frequency-dependent selection. *Heterozygote advantage* refers to a type of selection in which individuals who are heterozygous at a particular locus have a greater fitness than homozygous individuals (e.g., HbS [sickle-cell] variant). Frequency-dependent selection occurs when the fitness of a phenotype is dependent on its frequency relative to other phenotypes in a given population. In negative frequency-dependent selection, the fitness of a phenotype decreases as it becomes more common. Contrary to what is expected under a selective sweep, balancing selection is expected to lead to an excess of intermediate frequency variants and a depletion of rare alleles.[28] In humans, genome scans of balancing selection suggest that balancing selection is less pervasive than positive selection. However, it is a key force affecting the evolution of a number of genes in humans, particularly genes associated with immune functions.[30]

Natural selection imposed by malaria parasites

The contemporary increased population frequencies of some disease-related mutations may result from the action of past selection. Evolutionary selection by malaria is remarkable in this respect. Sickle-cell disease, thalassemia, glucose-6-phosphate dehydrogenase deficiency, and other erythrocyte defects are the most common Mendelian disorders of humankind.[31,32] In some parts of the world, the prevalence of these conditions is actually the rule rather than the exception. Malaria appears to be the evolutionary selective force that accounts for the high prevalence of these Mendelian disorders around the globe.[31,32] This idea was first proposed more than 60 years ago by John B. S. Haldane, who hypothesized that red-blood-cell disorders (thalassemia) could protect from malaria,[33] and gained later experimental support with Anthony C. Allison, whose work showed that heterozygotes for the sickle-cell variant had some protection against malaria.[34,35] Despite the use of antimalarial drugs and other control measures, the malaria death toll remains one of the leading causes of child mortality. It is therefore not surprising that malaria remains the strongest selective pressure acting on the human genome.[32]

Hemoglobinopathies and resistance to malaria

The hemoglobinopathies are inherited single-gene disorders that, in most cases, are inherited as autosomal codominant traits.[31] Hemoglobinopathies result from structural abnormalities in α-globin (encoded by *HBA1* and *HBA2*) or β-globin (*HBB*), which together makes up the tetrameric protein backbone of hemoglobin. In contrast, thalassemias, which are the most common Mendelian diseases in humans, usually result from the defective production of α- or β-globin chains, often through regulatory mutations.[31] Balancing selection at the β-globin gene, HBB, is probably the most emblematic case of natural selection maintaining high population frequencies of "deleterious" mutations underlying hemoglobinopathies. HBB presents three different common amino-acid altering variants, each conferring different levels of resistance against malaria: HbS or "sickle cell trait" (6 Glu → Val at codon 6), HbC (136 Glu → Lys at codon 6) and HbE (Glu → Lys at codon 26).[31] Although these three mutations segregate in a Mendelian fashion with the phenotype, the clinical outcomes can vary substantially due to a combination of genetic modifiers and environmental factors.[36]

Of the HBB variants, the worldwide frequency and functional effects of the HbS allele are probably the most extensively characterized. HbS displays the highest frequencies in equatorial Africa, up to 30% in Cameroon, Guinea, Uganda, and Kenya, as well as in parts of the Middle East, Saudi Arabia, and India.[31,37] HbS homozygotes have sickle-cell disease, a debilitating and often fatal disorder caused by red-cell deformities. However, the heterozygous state (HbAS) is usually not associated with any clinical abnormality and confers ~10-fold increase in protection from life-threatening forms of malaria.[34,38,39] Malaria selection undoubtedly explains why HbS has reached high frequencies in some populations, representing the most convincing case of heterozygote advantage reported so far.[32,34] Like HbS, HbC is primarily observed in Africa, although its distribution is virtually restricted to West African groups.[31] Both the heterozygote and homozygote states of HbC have been shown to confer strong protection against severe falciparum malaria,[40,41] with a substantially greater protective effect for homozygotes.[41] In contrast with HbS homozygotes, HbC homozygotes present a much milder phenotype than sickle-cell disease, displaying a relatively mild hemolytic anemia.[42] This mild deleterious effect of HbC in homozygosity has led some researchers to propose that in the absence of malaria control, HbC would in the long-term naturally replace HbS in central West Africa as a less costly evolutionary measure to combat malaria.[41]

In Asia, HbE is the commonest mutation, with carrier rates of up to 50% in some regions from northern India to China and South East Asia.[31,37,43] HbE accounts for the second most frequent form of abnormal hemoglobin in the world. Like HbC, homozygotes are not lethal and generally have symptomless anemia. HbE heterozygotes appear to enjoy some protection against *Plasmodium falciparum* malaria, as supported by both epidemiological data[44,45] and experimental evidence showing that erythrocytes from HbE heterozygous individuals are more resistant to invasion by *P. falciparum*.[46] It cannot be denied that population migrations and genetic drift have, to varying extents, also contributed to the increased frequencies of these three *HBB* mutations. However, their rapid increase in frequency due to positive selection attests that the selective pressure imposed by malaria is the main cause underlying the prevalence of hemoglobinopathies worldwide.[31,47,48]

The trade-offs of G6PD deficiency worldwide

Glucose-6-phosphate dehydrogenase (G6PD) deficiency is an X-linked recessive hereditary disease characterized by abnormally low levels of G6PD, a metabolic enzyme involved in the pentose phosphate pathway that is especially important in red blood cell metabolism.[49] G6PD deficiency is the commonest human enzymopathy, affecting ~400 million people worldwide. When triggered by certain foods (e.g., fava beans), drugs, or infection, G6PD deficiency can lead to a number of hemopathologies.[49–51] There are more than 100 different *G6PD* genetic variants, most of which are amino-acid altering mutations leading to enzyme deficiency.[51] The most commonly found are the $G6PD^{A}$ and $G6PD^{A-}$ in sub-Saharan Africa; $G6PD^{Med}$ in Southern Europe, the Middle East, and India; and $G6PD^{Mahidol}$ in South East Asia.[51] The distribution of G6PD deficiency is highly correlated with the distribution of current or past malaria endemicity.[52] This implies that the frequency of the mutations leading to G6PD deficiency might have been driven by the selective pressure imposed by malaria parasites. Concordantly, deficiencies in G6PD enzyme activity have been shown to correlate with protection against severe malaria in Nigerian children.[53] Furthermore, a large case-control study of more than 2,000 African children demonstrated that the deficiency allele $G6PD^{A-}$ is associated with a ~50% reduction in risk of severe malaria for female heterozygotes and male hemizygotes.[50] These observations led to the generally accepted hypothesis that the G6PD deficiency trait has been maintained in the population as it confers reduced risk from infection by the *Plasmodium* parasite. This hypothesis is supported by several population genetics studies, showing that different G6PD deficiency variants have been independently targeted by recent positive selection.[19,54–56] In Africa and the Mediterranean basin, it has been shown that the rapid increase in frequency of the deficiency alleles $G6PD^{A-}$ and $G6PD^{Med}$ is due to a recent positive selection within the last 10,000 years.[19,55,56] This supports the hypothesis that malaria has had a major impact on humans only since the introduction of agriculture, which is also consistent with the signs of expansion of *P. falciparum* in Africa in recent times (<10,000 years ago).[57]

Most clinical, epidemiological, and evolutionary studies of the relationships between G6PD deficiency and malaria protection have centered on *P. falciparum* malaria and the African continent. A recent study has focused instead on G6PD deficiency in South East Asia, where the $G6PD^{Mahidol}$ mutation dominates.[54] $G6PD^{Mahidol}$ is present in the populations of greater Southeast Asia including mainland China and leads to a reduction in G6PD activity by up to 32% in healthy individuals. This study explored the evolutionary patterns associated with $G6PD^{Mahidol}$ and its epidemiological significance in Thailand. Recent positive selection targeting the $G6PD^{Mahidol}$ variant over the past 1,500 years was demonstrated, and this coincides with the time at which rice started to be extensively cultured in Thailand.[54] In addition, the selective advantage conferred by this mutation is among the strongest detected so far in the human genome, including the malaria-protective $G6PD^{A-}$ mutation,[55] some β-globin variants,[58,59] as well as lactase persistence.[60] The same study also showed that $G6PD^{Mahidol}$ is associated with reduced *P. vivax*, but not *P. falciparum*, parasite density in humans.[54] These findings provide evidence that *P. vivax* has been a driving force behind the strong selective advantage conferred by the $G6PD^{Mahidol}$ mutation and supports the notion that *P. vivax* malaria has historically had a considerable impact on human health, at least in Southeast Asia.

The evolutionary trade-off illustrated by the G6PD case, where a deficient allele is adaptive for one phenotype but deleterious for another, could be exacerbated in particular circumstances. Environmental changes related to temporarily or spatially varying pathogen presence or changes in lifestyle (e.g., transition from hunter-gathering to farming, industrialization, etc.) may be situations in which alleles that represented a selective advantage in the past becomes harmful in present conditions (e.g., the pathogen exerting pressure is no longer present).[61] For example, the high incidence of autoimmune and inflammatory disorders in modern societies could reflect, at least partially, the consequences of past selection for increased immune responses to combat infection.[8] Additional clinical, epidemiological, and evolutionary studies are needed to test this seducing hypothesis.

When less is more: the case of DARC deficiency

Loss-of-function mutations can also be a target of natural selection, particularly when they are associated with a phenotype that substantially increases individual's fitness.[62,63] For example, when a pathogen uses host immune receptors as a mechanism of cell entry and survival, mutations that inactivate those receptors are likely to represent a selective advantage for the host. Positive selection targeting loss-of-function mutations has been documented for *CCR5*,[64] *caspase 12*,[65] *SERPINA2*,[66] and some leukocyte Ig-like receptor genes.[67] In the context of malaria, the Duffy antigen receptor for chemokines gene (*DARC* or *FY*) represents a paradigmatic case of positive selection acting on a loss-of-function mutation. DARC is a chemokine receptor expressed in various cell types and is used by *P. vivax* to enter erythrocytes.[68] The null mutation FY*0, due to a SNP in a regulatory element, disrupts protein expression in erythroid cells.[69] This loss-of-function mutation completely prevents *P. vivax* from invading erythrocytes, and imparts complete resistance to *P. vivax* infection in FY*0 homozygote individuals.[69] The FY*O allele is present at very high frequencies in most sub-Saharan African populations, being at or near fixation in some of them, for example, East Africans, 95%; Pygmies, 100%; West Africans, 100%, whereas it is virtually absent outside of Africa[70] (Fig. 2).

The association of the FY*O allele with *P. vivax* malaria resistance led to the hypothesis that the almost-fixation of this null allele in much of Africa is due to a conferred selective advantage against *P. vivax* malaria. Population genetic studies support this notion; the levels of population differentiation (F_{ST}) observed for the FY*O allele is among the highest reported in humans. The *DARC* gene also displays other features (e.g., excess of high-frequency derived variants in Africa) that strongly suggest positive

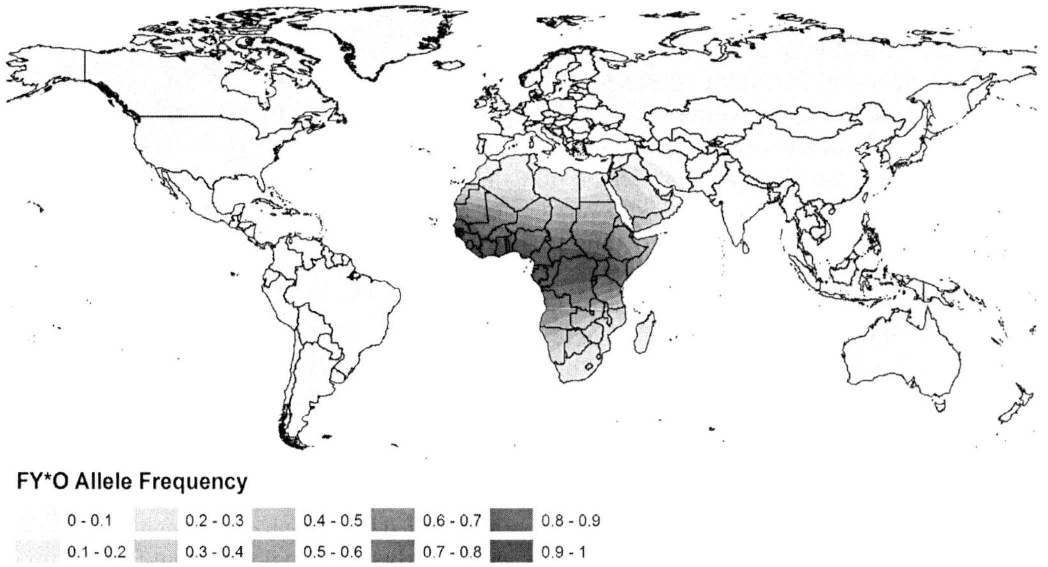

Figure 2. Worldwide distribution of the DARC FY*O allele. The FY*O allele, which confers resistance to *P. vivax* malaria, is found at very high frequency in African populations, but is virtually absent outside Africa. The frequencies of the FY*O allele were retrieved from the Alfred database (http://alfred.med.yale.edu/).

selection.[71,72] The highest frequencies of the FY*O allele in Africa are in regions where *P. vivax* malaria is virtually absent,[73] suggesting that *P. vivax* malaria has been eradicated *via* human adaptation to the parasite (mediated, at least partially, by the resistance afforded by the FY*O allele).

There is another, completely independent DARC null mutation (FY*A^{null}) among Papua New Guineans. This mutation also decreases protein expression and is associated with reduced *P. vivax* invasion efficiency,[74,75] further demonstrating the protective effect of DARC deficiency against *P. vivax* invasion. However, recent data suggest that *P. vivax* can also use DARC-independent pathways to invade human erythrocytes.[76] This casts doubts upon the complete Mendelian protection against *P. vivax* malaria afforded by DARC null alleles. Because various pathogens can utilize DARC and other cell surface receptors can be exploited by various pathogens to enter host cells, other pathogens may have also contributed to the increased frequency of the FY*O allele in Africa.[71–73,77]

Sideeffects of ancient selection: the case of CCR5 and HIV

Besides the examples of natural selection exerted by the malaria parasite, the identification of the genuine selective agent responsible for the patterns of selection observed is usually a daunting task. This is further complicated by several independent selective pressures (e.g., different pathogens) acting simultaneously on the same genes and that these pressures can vary in form, intensity, time, and space. An example of such a conundrum is the *CCR5* gene, which encodes a chemokine receptor that, with CD4, serves as an entry port on macrophages and monocytes for primary human immunodeficiency virus (HIV)-1 strains.[78,79] The importance of CCR5 variation in the pathogenesis of AIDS was first apparent from two observations. First, heterozygote individuals for a 32-bp deletion in the ORF of *CCR5* (i.e., CCR5-Δ32 preventing the functional expression of the CCR5) displayed reduced susceptibility to infection and delayed onset of AIDS. Second, homozygous carriers were resistant to HIV infection.[64,80] Subsequent studies confirmed that coding polymorphisms causing reduced or absent cell-surface expression of CCR5 are associated with decreased susceptibility to HIV-1 infection.[81] The CCR5-Δ32 deletion is found in Europe and western Asia at frequencies ranging from 4% to 14%, whereas it is absent elsewhere.[82,83] Population genetic studies have suggested that the deletion arose within the last 1,000 years and reached its present high frequency across Europe as a result of strong positive selection.[82,84,85] This is based on the patterns of long-range linkage disequilibrium at the CCR5-Δ32 locus. Another study has suggested that such increased frequencies of the CCR5-Δ32 can also be explained under neutral evolution.[86] In parallel, balancing selection has been documented as shaping variation at the 5′-*cis*-regulatory region of *CCR5*,[87] which has also been associated with varying susceptibility to HIV-1 infection and time to developing AIDS.[88–90]

The recent emergence of the HIV-1 epidemics in humans (only ~30 years ago) suggests that the resistance afforded by CCR5 variation results from adaptive changes to older pathogens.[83,91] The coincidence of the estimated age of the CCR5-Δ32 deletion with the bubonic plague, which killed ~30% of Europe's population between 1,346 and 1,352, led to the theory that plague was responsible for the selective rise of the CCR5-Δ32 allele.[92] However, a study using a model combining population genetics with disease dynamics showed that plague did not generate sufficient selection to drive the CCR5-Δ32 deletion to a frequency of 10%.[93] Different lines of evidence suggest instead that smallpox is a more plausible candidate for selective pressure on the CCR5-Δ32 deletion.[83,91] Smallpox caused significant mortality in Europe during both epidemic and interepidemic periods until its recent eradication.[94] In addition, the cumulative number of deaths from smallpox during the last 700 years was greater than that from plague, with smallpox disproportionately affecting younger people. A population-genetic model that formally incorporates all of these factors has shown that smallpox could indeed have posed sufficient selection to account for the increased frequency of the CCR5-Δ32 allele in Europe.[93]

FUT2 deficiency confers protection against norovirus

Noroviruses, also known as Norwalk-like viruses, are responsible for the majority of nonbacterial gastroenteritis outbreaks around the world.[95] Human resistance to noroviruses is associated with mutations in the *FUT2* gene, which encodes an α(1,2)fucotransferase and is responsible for the synthesis of ABO histo-blood group antigens in saliva and mucosa.[96,97] About 20% of individuals in most populations fail to secrete ABO antigens in body fluids because they are homozygous for a nonfunctional copy of *FUT2*.[98] These individuals are referred to as nonsecretor (se^-) individuals. Individuals carrying at least one functional copy of *FUT2* secrete ABO antigens and are called secretors (se^+). Several mutations in *FUT2* lead to *FUT2* deficiency and, consequently, the nonsecretor phenotype. However, there are two common polymorphisms that account for most of the worldwide nonsecretor status. The first is a stop codon at position 143, which explains over 95% of the nonsecretor phenotype in Europeans and Africans. The second is a missense mutation at codon 129, which is the most frequent cause of the nonsecretor phenotype among South and East Asians.[98–100]

Several population genetic studies suggest that the high-prevalence of the nonsecretor phenotype around the globe is a result of balancing selection.[99–102] The evidence for balancing selection is particularly strong in populations from Europe, the Middle East, and sub-Saharan Africa, where both the secretor and nonsecretor haplotypes are found at intermediate population frequencies.[99–102] The patterns of balancing selection are thought to be at least partly due to the increased resistance to noroviruses afforded by se^- individuals. Indeed, a study of volunteers infected with Norwalk virus revealed that se^- individuals do not develop symptoms of infection after challenge, regardless of the dose of inoculum used.[96] The complete Mendelian resistance to symptomatic norovirus infections enjoyed by se^- individuals was later confirmed in multiple studies.[97,103,104] For example, Thorven *et al.*[97] showed that after a series of nosocomial and sporadic norovirus outbreaks only se^+ individuals developed symptoms of infection. All se^- individuals appeared to be fully protected against any of the symptoms characteristic of gastroenteritis infection. However, only ~20% of the se^+ individuals develop symptoms after infection with norovirus,[96] suggesting that other mechanisms, genetic or nongenetic, might also contribute to disease pathogenesis.

Although an increased resistance to norovirus infection appears to be a likely explanation for the high worldwide frequency of the nonsecretor phenotype, this alone does not explain the patterns of balancing selection observed in *FUT2*. Indeed, if increased resistance to noroviruses was the only driving force shaping the evolution of *FUT2*, the nonsecretor phenotype would be expected to rapidly reach fixation worldwide. Instead, secretor and nonsecretor haplotypes are maintained at intermediate frequencies throughout the world, suggesting that secretor individuals also afford some sort of selective advantage over nonsecretors. Consistent with this notion, secretor individuals have lower susceptibility to urinary tract and candida infection, and increased protection against Neisseria meningitis and Streptococcus.[105,106] Thus, the postulated long-term maintenance by balancing selection of both secretor and nonsecretor haplotypes may be because secretor and nonsecretor individuals will experience advantages or disadvantages depending on temporal and spatially varying pathogenic pressures.

The controversial case of cystic fibrosis

CF is the most common life-threatening autosomal recessive disorder in populations of European descent.[107] CF is estimated to affect as many as 1 in 2,000–3,000 Caucasian newborns.[107] All CF cases are explained by mutations in the CF transmembrane conductance regulatory gene (*CFTR*).[108] The most common of these mutant alleles, ΔF508, is a three base pair in-frame deletion that results in a loss of structural stability and subsequent proteosome-mediated degradation of the CFTR protein.[108,109] Although over 1,500 mutations in *CFTR* have been shown to lead to CF, the ΔF508 mutation alone accounts for 66% of all cases worldwide, and 83% of cases in northern European populations.[107] Multiple age estimates place the origin of the ΔF508 mutation in the pre-Neolithic era (>10,000 years ago), suggesting that this is an ancient mutation in human history.[110,111]

Despite the lethal nature of CF, 1 in 26 individuals of European descent are heterozygous carriers of a CF mutation, mostly the ΔF508 allele.[107,112] It has been speculated that the relatively high frequency of CF lethal mutations in some populations could be a result of natural selection, if the heterozygous carriers had a fitness advantage. In accordance with such hypothesis, several studies suggest that parents of children with CF had increased fertility compared with noncarrier parents.[113,114] However, these studies may be biased because, at the time, carriers could only be identified after having a child with CF. Indeed, a later reexamination of the data with an ascertainment correction revealed that the previously observed fertility advantages tended to disappear.[115,116] A more recent population-based study has also failed to show any association between CF carrier status and overall fertility.[117] Nevertheless, all of these studies have been performed in relatively small sample sizes, which afford a low power to detect any eventual association. The hypothesis that CF carriers have some sort of fertility advantage can therefore not be completely discarded and future population-based studies with a large number of carrier parents are now necessary to resourcefully test this hypothesis. That said, if there is any selective advantage, this advantage is likely to be small.

Increased resistance to infectious diseases has also been postulated as an explanation for the high frequency of the ΔF508 allele among European individuals.[118–120] Several mice studies have provided support for this hypothesis. These show that heterozygous mice with only one functional copy of *CFTR* have increased protection against the dehydrating action of *Vibrio cholerae* and *Salmonella typhi* enterotoxins.[118–120] Under these conditions, although ΔF508 homozygous individuals would develop CF, heterozygous individuals would enjoy a fitness advantage because of a greater resistance to bacterial toxins, which induce diarrhea in normal homozygotes.[121] Even though this is an attractive evolutionary model explaining the high frequency of the ΔF508 allele (or other CF mutations) in humans, there is no evidence that CF carriers have increased protection against intestinal bacterial toxins as seen in mice[121] and therefore, this hypothesis remains largely speculative. Taken together, the evolutionary data presently available provide limited evidence that the high frequency of CF mutations is the result of natural selection and rather supports the notion that genetic drift might be the primary force accounting for the worldwide distribution of CF variants.

Natural selection and Mendelian traits in health

Besides the abovementioned examples of Mendelian disease-related traits being targeted by natural selection, there are other cases of selection acting on Mendelian, or quasi-Mendelian, traits that are not (directly) related to any disease status. Traits such as lactase persistence, skin color, or bitter-taste sensation have been convincingly shown to evolve adaptively in humans, highlighting how phenotypic variation in these traits has provided varying selective advantages in human populations.[77]

The extraordinary advantage of the lactase persistence trait

The ability to digest milk in adulthood, a trait known as lactase persistence or lactose tolerance, remains one of the most paradigmatic cases of recent, strong natural selection acting on the human genome. As in most mammals, the human ability to digest milk rapidly declines after weaning, owing to a decrease in lactase levels in the small intestine.[122] However, many human populations worldwide, particularly those that have traditionally based their mode of subsistence on pastoralism, can continue to drink milk into adulthood without problems, because lactase expression persists in the gut.[122] The prevalence of lactase persistence is very high in North European populations and their descendants (>90% in some populations), decreases gradually toward southern Europe and the Middle east (~50%), is low in nonpastoralists Asian (~1%) and African populations (5–20%), while present at high frequencies in African pastoralist populations (50–90%).[122] Today, the genetic bases of lactase persistence are well known.[123] It is inherited as a dominant Mendelian trait in Europeans, with a mutation (C/T-13910) in the promoter of LCT (the gene encoding lactase). This mutation has a 100% association with the lactase persistence trait in Finnish populations[123] and a ~86–98% association in other European populations.[124] Furthermore, *in vitro* studies demonstrate that the −13910T variant functions as a *cis*-element, increasing gene transcription.[125,126]

In Europeans, the lactase-persistence −13910T allele lies on a long-range haplotype that largely exceeds its expected length under neutrality.[20,127,128] This clear signature of recent positive selection explains the very high frequencies of the lactase persistence allele in European populations, particularly in northern Europe. There is a strong correlation between the geographic distribution of the −13910T allele and historical milk consumption and the estimated age of expansion of this allele is between 5,000 and 12,000 years ago.[129] This suggests that lactase persistence emerged in response to the cultural innovation of dairying associated with an agricultural lifestyle. Some East African populations with a high prevalence of the lactase persistence trait do not present with the European −13910T allele.[130] A different mutation in the LCT promoter region (−14010C), which also increases LCT expression, has been shown to be positively selected over the last 7,000 years among East African pastoralists.[131] This provides a fascinating example of convergent adaptation within our species. It also demonstrates that the cultural trait of milk consumption conferred such a strong selective advantage in terms of human survival that lactase persistence alleles increased in frequency independently in several parts of the world.

Bitter-taste sensitivity: independent evolution in humans and chimpanzees

Variation in taste sensitivity to the bitter compound phenylthiocarbamide (PTC) is another example of a benign trait that segregates in a Mendelian fashion in humans. Variation in PTC taste sensitivity was first shown in 1931 when Fox discovered that some people found crystallized PTC to be bitter while others did not.[132] The molecular basis of this trait was identified 73 years later by linkage and association genetic analyses.[133,134] These studies revealed that variation at the *TAS2R38* locus accounts for 50–80% of the phenotypic variance in PTC sensitivity. Most of the variance is explained by the presence of a "taster" and a "nontaster" haplotype, which differ by three amino acid mutations. Thus, although PTC sensitivity behaves as a Mendelian trait, it displays allelic heterogeneity.

Although PTC is a synthetic compound, the ability to taste it is strongly correlated with the ability to taste bitter compounds present in food.[135,136] Variation in PTC sensitivity has therefore appealed to evolutionary biologists for a long time. Only recently, Wooding *et al.* have formally evaluated the action of natural selection at the *TAS2R38* locus.[137] On the basis of a resequencing survey of *TAS2R38* in different human populations, they identified distinct molecular "signatures" suggesting that the worldwide distribution of "taster" and "nontaster" haplotypes results from the action of balancing selection. The patterns of variation at *TAS2R38* are characterized by unexpectedly high levels of nucleotide diversity, an excess of intermediate frequency alleles, and low levels of differentiation between populations.[137] The "taster" and "nontaster" haplotypes are found at intermediate frequencies in most parts of the world (Fig. 3A). This led to the theory that taster/nontaster heterozygous individuals would have a selective advantage over the homozygotes for either allele, due to their ability to sense a larger number of toxins.[137] However, this selective scenario assumes that the

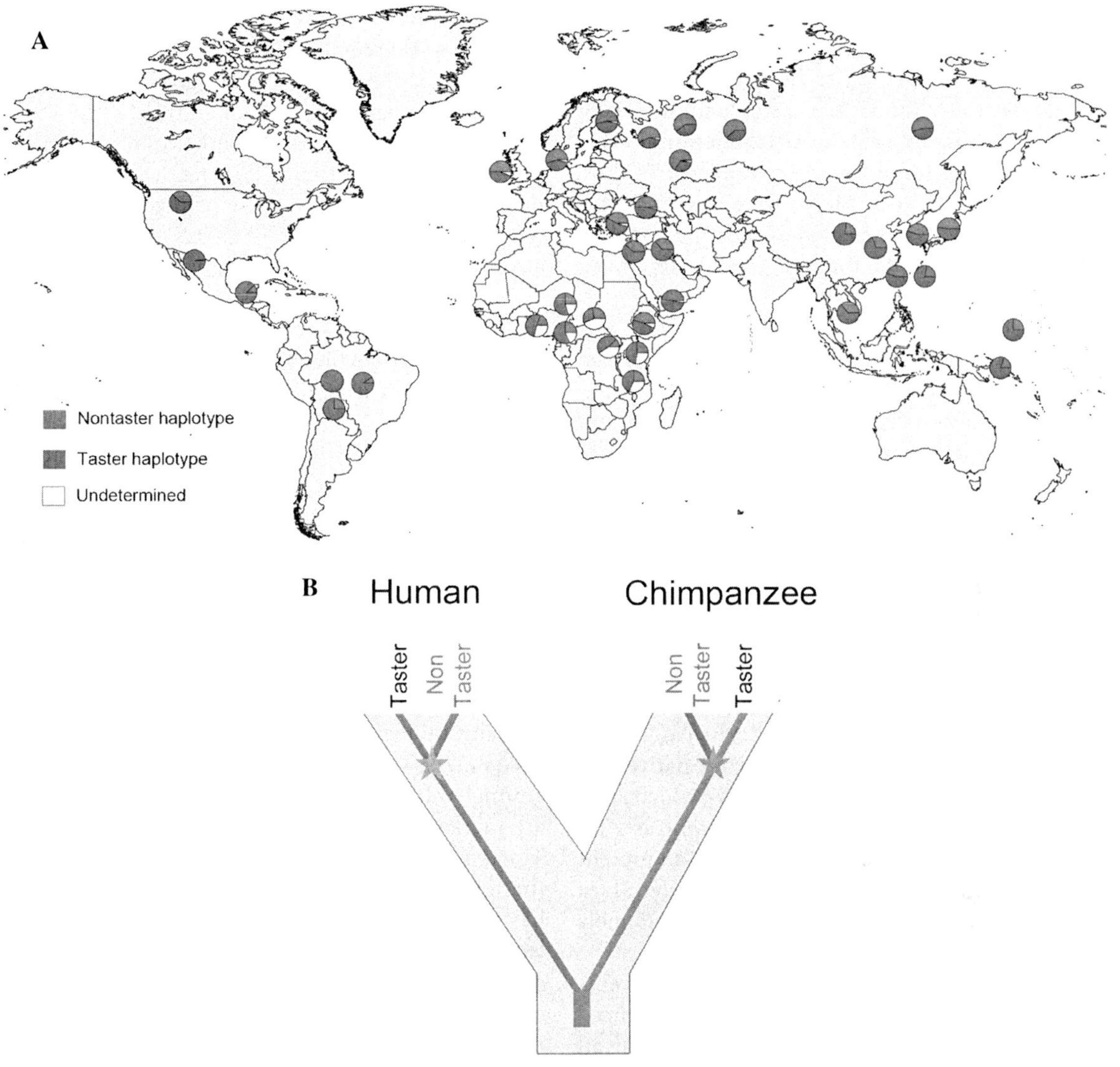

Figure 3. Worldwide distribution of taster and nontaster haplotypes. (A) Frequency of taster and nontaster haplotypes in 36 populations. Haplotype frequencies were retrieved from the Alfred database (http://alfred.med.yale.edu/). *Undetermined* refers to recombinant haplotypes for the three amino acid substitutions that define the PTC sensitivity phenotype. (B) Illustration of the independent origin of taster and nontaster phenotypes in humans and chimpanzees, after the human–chimpanzee divergence.

nontaster PTC allele is also functional and confers sensitivity to a set of compounds other than PTC. Although, the specific ligand(s) for the PTC nontaster allele have yet to be identified, the fruit of the *Antidesma bunius* plant taste bitter to PTC nontasters, but sweet to PTC tasters.[138] This raises the possibility that the *Antidesma bunius* fruit contains such a ligand.

Variation in PTC sensitivity also occurs in chimpanzees. This phenomenon was first described by Fisher *et al.*, who argued for a single origin of PTC taster and nontaster alleles prior to the evolutionary divergence of humans and chimpanzees.[139] In other words, the taster and nontaster haplotypes would have been maintained by balancing selection separately in both species for more than 6 million years. However, sequencing of *TAS2R38* in 86 captive chimpanzees rejects Fisher *et al.*'s longstanding hypothesis. Just like humans, chimpanzees also harbor two common haplotypes at the *TAS2R38* locus, which are strongly associated with PTC sensitivity.[140] However, whereas the human taster and nontaster haplotypes differ by three amino acid mutations, the chimpanzee taster and nontaster haplotypes differ by a single nucleotide variant that is unlike any of those found in humans.[140] Thus, although humans and chimpanzees share variable taste sensitivity to bitter compounds mediated by PTC receptor variants, the molecular basis of this variation has evolved independently at least twice, over the course of hominid evolution (Fig. 3B). Signatures of natural selection have also been described on other human bitter taste receptors,[141,142] suggesting that taste-perception, particularly to bitter compounds, has been an important evolutionary pressure during human evolution.

Genetic adaptation to high altitude

One of the most challenging environments inhabited by humans are the high-altitude plateaus of Tibet and the Andes, where oxygen concentrations are about 40% lower than those of sea level. Populations residing in high-altitude zones possess unique suites of physiological traits.[143–145] When ascending to high altitudes, nonadapted lowland natives increase the production of red blood cells, which leads to increase Hb concentrations.[146] In contrast, Tibetans living at elevations of 3,500–4,500 m present a hematological profile similar to that expected at sea level,[143] suggesting that they have adapted uniquely to their extreme high-altitude conditions.

Recently, the genetic basis of such adaptive events has started to be deciphered. Yi *et al.* have sequenced the exomes of 50 residents of the Tibetan plateau and compared the frequency of the variants identified against those observed in a sample of Han Chinese.[147] Strikingly, the genetic variants showing the strongest differentiation between Tibetans and Han Chinese were enriched among genes classified as involved in "response to hypoxia."[147] The strongest signal of selection was observed for the endothelial Per-Arnt-Sim (PAS) domain protein 1 (*EPAS1*) gene (also known as hypoxia-inducible factor 2a, HIF-1a), which encodes the oxygen-sensitive subunit of the HIF-2 transcription factor and plays an important role in regulating erythropoiesis.[148] They identified a variant in *EPAS1* with a derived allele at 87% frequency in Tibetans but only 9% in the Han. Interestingly, the "Tibetan" *EPAS1* allele is associated with significantly lower erythrocyte quantities and low Hb levels.[147,149] Simonson *et al.* have also searched for Tibetan-specific signatures of selection at genes whose products are likely involved in adaptation to high altitude.[150] In addition to *EPAS1*, they identified other putative targets of selection, *EGLN1* and *PPRARA*, which are also involved in the HIF oxygen-signaling pathway. The advantageous haplotypes of *EGLN1* and *PPRARA* also showed a significantly negative correlation with Hb concentrations.[150] This suggests that the additive effects of variation in multiple genes involved in the HIF oxygen-signaling pathway might be the ultimate reason accounting for the low Hb concentrations observed in Tibetans. It remains, however, to be determined whether low-Hb concentration represents the phenotypic target of selection in Tibetans, or whether changes in Hb concentration reflect instead a side effect of HIF-mediated adaptation to hypoxia conditions.

In contrast to Tibetans, Andean populations show elevated Hb concentration, which indicates that Andean and Tibetan populations have taken different evolutionary routes to adapt to high altitude[144]; yet the same biological pathways might be involved. Indeed, *EGLN1* has also been targeted by selection among Andeans although the selected haplotypes are markedly different from those selected among Tibetans.[151] Taken together, these studies support the notion that the HIF oxygen signaling pathway has been subject to strong and recent positive selection in high-altitude populations. Future surveys aiming to characterize the entire genome diversity of different highland populations might lead to a more complete portray of the genetic changes having contributed to the adaptation of high-altitude populations to their unique environment and lifestyle.

Conclusion

The last few years have seen a sudden increase of population genetics studies aiming to understand the effects of natural selection in the shaping of human genome diversity. These have been bolstered by the advent of massive genome-wide surveys of population genetic variation based on genotyping and resequencing data (e.g., International HapMap Project, 1000 Genomes Project, etc.). The identification of genomic regions that seem to have been targeted by different forms of natural selection has improved our knowledge on how past selection has impacted on medically related traits and various phenotypes of anthropological interest. For example, a recent study at the genome-wide scale has shown that Mendelian disease genes are under stronger purifying selection than complex disease genes, especially when the disease-causing alleles are dominant.[17] However, as shown by the different examples discussed in this review, mutations associated with Mendelian traits can, under particular circumstances, be the object of positive selection. For several of these, the advantages they confer in human survival are clearly much stronger than the associated deleterious effect (e.g., HbS sickle cell allele). In the era where next-generation sequencing technologies are becoming more and more affordable, it is now possible to carry out genetic surveys based on complete human genome sequences from diverse human populations coupled with precise definition of phenotypes, disease-related or not. These studies will reveal new genotype–phenotype combinations—including variants located in regions of the genome that have long been ignored because of their assumed nonfunctional role (e.g., noncoding DNA)—that have been selected for throughout human history. In turn, this will increase our understanding of the physiological mechanisms that have played a crucial role in the adaption of our species and its survival against disease over time.

Acknowledgments

We warmly thank F. G. Sperone for generating the GIS maps in Figures 2 and 3. This work was supported by the Institut Pasteur, the Centre National de la Recherche Scientifique, the Fondation pour la Recherche Médicale, the EPFL-Debiopharm Life Sciences Award, and an Agence Nationale de la Recherche research grant (ANR-08-MIEN-009–01) to L.Q.-M. L.B.B. is supported by the Human Frontiers Science Foundation (LT6572009-L).

Conflicts of interest

The authors declare no conflicts of interest.

References

1. Pritchard, J.K. & N.J. Cox. 2002. The allelic architecture of human disease genes: common disease-common variant—or not? *Hum. Mol. Genet.* **11:** 2417–2423.
2. Zwick, M.E., D.J. Cutler & A. Chakravarti. 2000. Patterns of genetic variation in Mendelian and complex traits. *Annu. Rev. Genomics Hum. Genet.* **1:** 387–407.
3. Zielenski, J. & L.C. Tsui. 1995. Cystic fibrosis: genotypic and phenotypic variations. *Annu. Rev. Genet.* **29:** 777–807.
4. Badano, J.L. & N. Katsanis. 2002. Beyond Mendel: an evolving view of human genetic disease transmission. *Nat. Rev. Genet.* **3:** 779–789.
5. Botstein, D. & N. Risch. 2003. Discovering genotypes underlying human phenotypes: past successes for Mendelian disease, future approaches for complex disease. *Nat. Genet.* **33**(Suppl.): 228–237.
6. Di Rienzo, A. 2006. Population genetics models of common diseases. *Curr. Opin. Genet. Dev.* **16:** 630–636.
7. Akey, J.M. 2009. Constructing genomic maps of positive selection in humans: where do we go from here? *Genome Res.* **19:** 711–722.
8. Barreiro, L.B. & L. Quintana-Murci. 2010. From evolutionary genetics to human immunology: how selection shapes host defence genes. *Nat. Rev. Genet.* **11:** 17–30.
9. Di Rienzo, A. & R.R. Hudson. 2005. An evolutionary framework for common diseases: the ancestral-susceptibility model. *Trends Genet.* **21:** 596–601.
10. Kryukov, G.V., L.A. Pennacchio & S.R. Sunyaev. 2007. Most rare missense alleles are deleterious in humans: implications for complex disease and association studies. *Am. J. Hum. Genet.* **80:** 727–739.
11. Nielsen, R., I. Hellmann, M. Hubisz, *et al.* 2007. Recent and ongoing selection in the human genome. *Nat. Rev. Genet.* **8:** 857–868.
12. Pritchard, J.K., J.K. Pickrell & G. Coop. 2010. The genetics of human adaptation: hard sweeps, soft sweeps, and polygenic adaptation. *Curr. Biol.* **20:** R208–R215.
13. Sabeti, P.C., S.F. Schaffner, B. Fry, *et al.* 2006. Positive natural selection in the human lineage. *Science* **312:** 1614–1620.
14. Kimura, M. 1968. Evolutionary rate at the molecular level. *Nature* **217:** 624–626.
15. Coop, G., J.K. Pickrell, J. Novembre, *et al.* 2009. The role of geography in human adaptation. *PLoS Genet.* **5:** e1000500.
16. Eyre-Walker, A. & P.D. Keightley. 1999. High genomic deleterious mutation rates in hominids. *Nature* **397:** 344–347.
17. Blekhman, R., O. Man, L. Herrmann, *et al.* 2008. Natural selection on genes that underlie human disease susceptibility. *Curr. Biol.* **18:** 883–889.
18. Bustamante, C.D., A. Fledel-Alon, S. Williamson, *et al.* 2005. Natural selection on protein-coding genes in the human genome. *Nature* **437:** 1153–1157.
19. Sabeti, P.C., D.E. Reich, J.M. Higgins, *et al.* 2002. Detecting recent positive selection in the human genome from haplotype structure. *Nature* **419:** 832–837.
20. Voight, B.F., S. Kudaravalli, X. Wen, *et al.* 2006. A map of recent positive selection in the human genome. *PLoS Biol.* **4:** e72.
21. Wang, E.T., G. Kodama, P. Baldi, *et al.* 2006. Global landscape of recent inferred Darwinian selection for Homo sapiens. *Proc. Natl. Acad. Sci. USA* **103:** 135–140.
22. Kreitman, M. 2000. Methods to detect selection in populations with applications to the human. *Annu. Rev. Genomics Hum. Genet.* **1:** 539–559.
23. Nielsen, R. 2005. Molecular signatures of natural selection. *Annu. Rev. Genet.* **39:** 197–218.

24. Thornton, K.R., J.D. Jensen, C. Becquet, *et al.* 2007. Progress and prospects in mapping recent selection in the genome. *Heredity* **98:** 340–348.
25. Cavalli-Sforza, L.L. 1966. Population structure and human evolution. *Proc. R. Soc. Lond. B Biol. Sci.* **164:** 362–379.
26. Lewontin, R.C. & J. Krakauer. 1973. Distribution of gene frequency as a test of the theory of the selective neutrality of polymorphisms. *Genetics* **74:** 175–195.
27. Barreiro, L.B., G. Laval, H. Quach, *et al.* 2008. Natural selection has driven population differentiation in modern humans. *Nat. Genet.* **40:** 340–345.
28. Charlesworth, D. 2006. Balancing selection and its effects on sequences in nearby genome regions. *PLoS Genet.* **2:** e64.
29. Hurst, L.D. 2009. Fundamental concepts in genetics: genetics and the understanding of selection. *Nat. Rev. Genet.* **10:** 83–93.
30. Andres, A.M., M.J. Hubisz, A. Indap, *et al.* 2009. Targets of balancing selection in the human genome. *Mol. Biol. Evol.* **26:** 2755–2764.
31. Flint, J., R.M. Harding, A.J. Boyce, *et al.* 1998. The population genetics of the haemoglobinopathies. *Baillieres Clin. Haematol.* **11:** 1–51.
32. Kwiatkowski, D.P. 2005. How malaria has affected the human genome and what human genetics can teach us about malaria. *Am. J. Hum. Genet.* **77:** 171–192.
33. Haldane, J.B.S. 1949. Disease and evolution. *Ric. Sci.* **19**(Suppl. A): 68–76.
34. Allison, A.C. 1954. Protection afforded by sickle-cell trait against subtertian malareal infection. *Br. Med. J.* **1:** 290–294.
35. Allison, A.C. 2004. Two lessons from the interface of genetics and medicine. *Genetics* **166:** 1591–1599.
36. Driss, A., K.O. Asare, J.M. Hibbert, *et al.* 2009. Sickle cell disease in the post genomic era: a monogenic disease with a polygenic phenotype. *Genomics Insights* **2009:** 23–48.
37. Livingstone, F.B. 1985. *Frequencies of Haemoglobin Variants.* Oxford University Press. Oxford.
38. Ackerman, H., S. Usen, M. Jallow, *et al.* 2005. A comparison of case-control and family-based association methods: the example of sickle-cell and malaria. *Ann. Hum. Genet.* **69:** 559–565.
39. Hill, A.V., C.E. Allsopp, D. Kwiatkowski, *et al.* 1991. Common west African HLA antigens are associated with protection from severe malaria. *Nature* **352:** 595–600.
40. Agarwal, A., A. Guindo, Y. Cissoko, *et al.* 2000. Hemoglobin C associated with protection from severe malaria in the Dogon of Mali, a West African population with a low prevalence of hemoglobin S. *Blood* **96:** 2358–2363.
41. Modiano, D., G. Luoni, B.S. Sirima, *et al.* 2001. Haemoglobin C protects against clinical *Plasmodium falciparum* malaria. *Nature* **414:** 305–308.
42. Diallo, D.A., O.K. Doumbo, A. Dicko, *et al.* 2004. A comparison of anemia in hemoglobin C and normal hemoglobin A children with *Plasmodium falciparum* malaria. *Acta Trop.* **90:** 295–299.
43. Fucharoen, S. & P. Winichagoon. 1987. Hemoglobinopathies in Southeast Asia. *Hemoglobin* **11:** 65–88.
44. Hutagalung, R., P. Wilairatana, S. Looareesuwan, *et al.* 1999. Influence of hemoglobin E trait on the severity of Falciparum malaria. *J. Infect. Dis.* **179:** 283–286.
45. Hutagalung, R., P. Wilairatana, S. Looareesuwan, *et al.* 2000. Influence of hemoglobin E trait on the antimalarial effect of artemisinin derivatives. *J. Infect. Dis.* **181:** 1513–1516.
46. Chotivanich, K., R. Udomsangpetch, K. Pattanapanyasat, *et al.* 2002. Hemoglobin E: a balanced polymorphism protective against high parasitemias and thus severe *P. falciparum* malaria. *Blood* **100:** 1172–1176.
47. Currat, M., G. Trabuchet, D. Rees, *et al.* 2002. Molecular analysis of the beta-globin gene cluster in the Niokholo Mandenka population reveals a recent origin of the beta(S) Senegal mutation. *Am. J. Hum. Genet.* **70:** 207–223.
48. Ohashi, J., I. Naka, J. Patarapotikul, *et al.* 2004. Extended linkage disequilibrium surrounding the hemoglobin E variant due to malarial selection. *Am. J. Hum. Genet.* **74:** 1198–1208.
49. Beutler, E. 1994. G6PD deficiency. *Blood* **84:** 3613–3636.
50. Ruwende, C. & A. Hill. 1998. Glucose-6-phosphate dehydrogenase deficiency and malaria. *J. Mol. Med.* **76:** 581–588.
51. Vulliamy, T., P. Mason & L. Luzzatto. 1992. The molecular basis of glucose-6-phosphate dehydrogenase deficiency. *Trends Genet.* **8:** 138–143.
52. Ganczakowski, M., M. Town, D.K. Bowden, *et al.* 1995. Multiple glucose 6-phosphate dehydrogenase-deficient variants correlate with malaria endemicity in the Vanuatu archipelago (southwestern Pacific). *Am. J. Hum. Genet.* **56:** 294–301.
53. Gilles, H.M., K.A. Fletcher, R.G. Hendrickse, *et al.* 1967. Glucose-6-phosphate-dehydrogenase deficiency, sickling, and malaria in African children in South Western Nigeria. *Lancet* **1:** 138–140.
54. Louicharoen, C., E. Patin, R. Paul, *et al.* 2009. Positively selected G6PD-Mahidol mutation reduces *Plasmodium vivax* density in Southeast Asians. *Science* **326:** 1546–1549.
55. Saunders, M.A., M.F. Hammer & M.W. Nachman. 2002. Nucleotide variability at G6pd and the signature of malarial selection in humans. *Genetics* **162:** 1849–1861.
56. Tishkoff, S.A., R. Varkonyi, N. Cahinhinan, *et al.* 2001. Haplotype diversity and linkage disequilibrium at human G6PD: recent origin of alleles that confer malarial resistance. *Science* **293:** 455–462.
57. Joy, D.A., X. Feng, J. Mu, *et al.* 2003. Early origin and recent expansion of *Plasmodium falciparum*. *Science* **300:** 318–321.
58. Allison, A.C. 1956. The sickle-cell and haemoglobin C genes in some African populations. *Ann. Hum. Genet.* **21:** 67–89.
59. Wood, E.T., D.A. Stover, M. Slatkin, *et al.* 2005. The beta-globin recombinational hotspot reduces the effects of strong selection around HbC, a recently arisen mutation providing resistance to malaria. *Am. J. Hum. Genet.* **77:** 637–642.
60. Enattah, N.S., T.G. Jensen, M. Nielsen, *et al.* 2008. Independent introduction of two lactase-persistence alleles into human populations reflects different history of adaptation to milk culture. *Am. J. Hum. Genet.* **82:** 57–72.

61. Wolfe, N.D., C.P. Dunavan & J. Diamond. 2007. Origins of major human infectious diseases. *Nature* **447:** 279–283.
62. Yngvadottir, B., Y. Xue, S. Searle, *et al.* 2009. A genome-wide survey of the prevalence and evolutionary forces acting on human nonsense SNPs. *Am. J. Hum. Genet.* **84:** 224–234.
63. Olson, M.V. 1999. When less is more: gene loss as an engine of evolutionary change. *Am. J. Hum. Genet.* **64:** 18–23.
64. Arenzana-Seisdedos, F. & M. Parmentier. 2006. Genetics of resistance to HIV infection: role of co-receptors and co-receptor ligands. *Semin. Immunol.* **18:** 387–403.
65. Xue, Y., A. Daly, B. Yngvadottir, *et al.* 2006. Spread of an inactive form of caspase-12 in humans is due to recent positive selection. *Am. J. Hum. Genet.* **78:** 659–670.
66. Seixas, S., G. Suriano, F. Carvalho, *et al.* 2007. Sequence diversity at the proximal 14q32.1 SERPIN subcluster: evidence for natural selection favoring the pseudogenization of SERPINA2. *Mol. Biol. Evol.* **24:** 587–598.
67. Hirayasu, K., J. Ohashi, H. Tanaka, *et al.* 2008. Evidence for natural selection on leukocyte immunoglobulin-like receptors for HLA class I in Northeast Asians. *Am. J. Hum. Genet.* **82:** 1075–1083.
68. Chitnis, C.E. & L.H. Miller. 1994. Identification of the erythrocyte binding domains of *Plasmodium vivax* and *Plasmodium knowlesi* proteins involved in erythrocyte invasion. *J. Exp. Med.* **180:** 497–506.
69. Tournamille, C., Y. Colin, J.P. Cartron, *et al.* 1995. Disruption of a GATA motif in the Duffy gene promoter abolishes erythroid gene expression in Duffy-negative individuals. *Nat. Genet.* **10:** 224–228.
70. Cavalli-Sforza, L.L., P. Menozzi & A. Piazza. 1994. *The History and Geography of Human Genes.* Princeton University Press. Princeton, NJ.
71. Hamblin, M.T. & A. Di Rienzo. 2000. Detection of the signature of natural selection in humans: evidence from the Duffy blood group locus. *Am. J. Hum. Genet.* **66:** 1669–1679.
72. Hamblin, M.T., E.E. Thompson & A. Di Rienzo. 2002. Complex signatures of natural selection at the Duffy blood group locus. *Am. J. Hum. Genet.* **70:** 369–383.
73. Livingstone, F.B. 1984. The Duffy blood groups, vivax malaria, and malaria selection in human populations: a review. *Hum. Biol.* **56:** 413–425.
74. Michon, P., I. Woolley, E.M. Wood, *et al.* 2001. Duffy-null promoter heterozygosity reduces DARC expression and abrogates adhesion of the *P. vivax* ligand required for blood-stage infection. *FEBS Lett.* **495:** 111–114.
75. Zimmerman, P.A., I. Woolley, G.L. Masinde, *et al.* 1999. Emergence of FY* A(null) in a Plasmodium vivax-endemic region of Papua New Guinea. *Proc. Natl. Acad. Sci. USA* **96:** 13973–13977.
76. Menard, D., C. Barnadas, C. Bouchier, *et al.* 2010. *Plasmodium vivax* clinical malaria is commonly observed in Duffy-negative Malagasy people. *Proc. Natl. Acad. Sci. USA* **107:** 5967–5971.
77. Harris, E.E. & D. Meyer. 2006. The molecular signature of selection underlying human adaptations. *Am. J. Phys. Anthropol.* **43**(Suppl.): 89–130.
78. Alkhatib, G., C. Combadiere, C.C. Broder, *et al.* 1996. CC CKR5: a RANTES, MIP-1alpha, MIP-1beta receptor as a fusion cofactor for macrophage-tropic HIV-1. *Science* **272:** 1955–1958.
79. Dragic, T., V. Litwin, G.P. Allaway, *et al.* 1996. HIV-1 entry into CD4+ cells is mediated by the chemokine receptor CC-CKR-5. *Nature* **381:** 667–673.
80. Dean, M., M. Carrington, C. Winkler, *et al.* 1996. Genetic restriction of HIV-1 infection and progression to AIDS by a deletion allele of the CKR5 structural gene. Hemophilia Growth and Development Study, Multicenter AIDS Cohort Study, Multicenter Hemophilia Cohort Study, San Francisco City Cohort, ALIVE Study. *Science* **273:** 1856–1862.
81. O'brien, S.J. & J.P. Moore. 2000. The effect of genetic variation in chemokines and their receptors on HIV transmission and progression to AIDS. *Immunol. Rev.* **177:** 99–111.
82. Stephens, J.C., D.E. Reich, D.B. Goldstein, *et al.* 1998. Dating the origin of the CCR5-Delta32 AIDS-resistance allele by the coalescence of haplotypes. *Am. J. Hum. Genet.* **62:** 1507–1515.
83. Galvani, A.P. & J. Novembre. 2005. The evolutionary history of the CCR5-Delta32 HIV-resistance mutation. *Microbes Infect.* **7:** 302–309.
84. Libert, F., P. Cochaux, G. Beckman, *et al.* 1998. The deltaccr5 mutation conferring protection against HIV-1 in Caucasian populations has a single and recent origin in Northeastern Europe. *Hum. Mol. Genet.* **7:** 399–406.
85. Novembre, J., A.P. Galvani & M. Slatkin. 2005. The geographic spread of the CCR5 Delta32 HIV-resistance allele. *PLoS Biol.* **3:** e339.
86. Sabeti, P.C., E. Walsh, S.F. Schaffner, *et al.* 2005. The case for selection at CCR5-Delta32. *PLoS Biol.* **3:** e378.
87. Bamshad, M.J., S. Mummidi, E. Gonzalez, *et al.* 2002. A strong signature of balancing selection in the 5' cis-regulatory region of CCR5. *Proc. Natl. Acad. Sci. USA* **99:** 10539–10544.
88. Gonzalez, E., M. Bamshad, N. Sato, *et al.* 1999. Race-specific HIV-1 disease-modifying effects associated with CCR5 haplotypes. *Proc. Natl. Acad. Sci. USA* **96:** 12004–12009.
89. Gonzalez, E., R. Dhanda, M. Bamshad, *et al.* 2001. Global survey of genetic variation in CCR5, RANTES, and MIP-1alpha: impact on the epidemiology of the HIV-1 pandemic. *Proc. Natl. Acad. Sci. USA* **98:** 5199–5204.
90. Martin, M.P., M. Dean, M.W. Smith, *et al.* 1998. Genetic acceleration of AIDS progression by a promoter variant of CCR5. *Science* **282:** 1907–1911.
91. Lalani, A.S., J. Masters, W. Zeng, *et al.* 1999. Use of chemokine receptors by poxviruses. *Science* **286:** 1968–1971.
92. Mcevedy, C. 1988. The bubonic plague. *Sci. Am.* **258:** 118–123.
93. Galvani, A.P. & M. Slatkin. 2003. Evaluating plague and smallpox as historical selective pressures for the CCR5-Delta 32 HIV-resistance allele. *Proc. Natl. Acad. Sci. USA* **100:** 15276–15279.
94. Hopkins, D.R. 2002. *The Greatest Killer in History: Smallpox.* The University of Chicago Press. Chicago and London.

95. Donaldson, E.F., L.C. Lindesmith, A.D. Lobue, *et al.* 2010. Viral shape-shifting: norovirus evasion of the human immune system. *Nat. Rev. Microbiol.* **8:** 231–241.
96. Lindesmith, L., C. Moe, S. Marionneau, *et al.* 2003. Human susceptibility and resistance to Norwalk virus infection. *Nat. Med.* **9:** 548–553.
97. Thorven, M., A. Grahn, K.O. Hedlund, *et al.* 2005. A homozygous nonsense mutation (428G–>A) in the human secretor (FUT2) gene provides resistance to symptomatic norovirus (GGII) infections. *J. Virol.* **79:** 15351–15355.
98. Kelly, R.J., S. Rouquier, D. Giorgi, *et al.* 1995. Sequence and expression of a candidate for the human secretor blood group alpha(1,2)fucosyltransferase gene (FUT2). Homozygosity for an enzyme-inactivating nonsense mutation commonly correlates with the non-secretor phenotype. *J. Biol. Chem.* **270:** 4640–4649.
99. Ferrer-Admetlla, A., M. Sikora, H. Laayouni, *et al.* 2009. A natural history of FUT2 polymorphism in humans. *Mol. Biol. Evol.* **26:** 1993–2003.
100. Koda, Y., H. Tachida, H. Pang, *et al.* 2001. Contrasting patterns of polymorphisms at the ABO-secretor gene (FUT2) and plasma alpha(1,3)fucosyltransferase gene (FUT6) in human populations. *Genetics* **158:** 747–756.
101. Fumagalli, M., R. Cagliani, U. Pozzoli, *et al.* 2009. Widespread balancing selection and pathogen-driven selection at blood group antigen genes. *Genome Res.* **19:** 199–212.
102. Walsh, E.C., P. Sabeti, H.B. Hutcheson, *et al.* 2006. Searching for signals of evolutionary selection in 168 genes related to immune function. *Hum. Genet.* **119:** 92–102.
103. Kindberg, E., B. Akerlind, C. Johnsen, *et al.* 2007. Host genetic resistance to symptomatic norovirus (GGII.4) infections in Denmark. *J. Clin. Microbiol.* **45:** 2720–2722.
104. Tan, M., M. Jin, H. Xie, *et al.* 2008. Outbreak studies of a GII-3 and a GII-4 norovirus revealed an association between HBGA phenotypes and viral infection. *J. Med. Virol.* **80:** 1296–1301.
105. Blackwell, C.C., D.M. Weir, V.S. James, *et al.* 1990. Secretor status, smoking and carriage of Neisseria meningitidis. *Epidemiol. Infect.* **104:** 203–209.
106. Haverkorn, M.J. & W.R. Goslings. 1969. Streptococci, ABO blood groups, and secretor status. *Am. J. Hum. Genet.* **21:** 360–375.
107. Bobadilla, J.L., M. Macek Jr., J.P. Fine, *et al.* 2002. Cystic fibrosis: a worldwide analysis of CFTR mutations—correlation with incidence data and application to screening. *Hum. Mutat.* **19:** 575–606.
108. Consortium, C.F.G.A. 1994. Population variation of common cystic fibrosis mutations. The Cystic Fibrosis Genetic Analysis Consortium. *Hum. Mutat.* **4:** 167–177.
109. Denning, G.M., M.P. Anderson, J.F. Amara, *et al.* 1992. Processing of mutant cystic fibrosis transmembrane conductance regulator is temperature-sensitive. *Nature* **358:** 761–764.
110. Morral, N., J. Bertranpetit, X. Estivill, *et al.* 1994. The origin of the major cystic fibrosis mutation (delta F508) in European populations. *Nat. Genet.* **7:** 169–175.
111. Wiuf, C. 2001. Do delta F508 heterozygotes have a selective advantage? *Genet. Res.* **78:** 41–47.
112. Welsh, M., L. Tsui, T. Boat, *et al.* 1995. The metabolic and molecular bases of inherited disease. *Cystic Fibrosis.* 17th ed.: 3799–3876. New York: McGraw-Hill.
113. Kitzis, A., J.C. Chomel, J.C. Kaplan, *et al.* 1988. Unusual segregation of cystic fibrosis allele to males. *Nature* **333:** 215.
114. Knudson, A.G. Jr., L. Wayne & W. Y. Hallett. 1967. On the selective advantage of cystic fibrosis heterozygotes. *Am. J. Hum. Genet.* **19:** 388–392.
115. Jorde, L.B. & G.M. Lathrop. 1988. A test of the heterozygote-advantage hypothesis in cystic fibrosis carriers. *Am. J. Hum. Genet.* **42:** 808–815.
116. Tummler, B., A. Aschendorf, T. Darnedde, *et al.* 1988. Segregation of cystic fibrosis allele. *Nature* **334:** 110.
117. Gallego Romero, I. & C. Ober. 2008. CFTR mutations and reproductive outcomes in a population isolate. *Hum. Genet.* **122:** 583–588.
118. Gabriel, S.E., K.N. Brigman, B.H. Koller, *et al.* 1994. Cystic fibrosis heterozygote resistance to cholera toxin in the cystic fibrosis mouse model. *Science* **266:** 107–109.
119. Pier, G.B., M. Grout, T. Zaidi, *et al.* 1998. Salmonella typhi uses CFTR to enter intestinal epithelial cells. *Nature* **393:** 79–82.
120. Van De Vosse, E., S. Ali, A.W. De Visser, *et al.* 2005. Susceptibility to typhoid fever is associated with a polymorphism in the cystic fibrosis transmembrane conductance regulator (CFTR). *Hum. Genet.* **118:** 138–140.
121. Hogenauer, C., C.A. Santa Ana, J.L. Porter, *et al.* 2000. Active intestinal chloride secretion in human carriers of cystic fibrosis mutations: an evaluation of the hypothesis that heterozygotes have subnormal active intestinal chloride secretion. *Am. J. Hum. Genet.* **67:** 1422–1427.
122. Swallow, D.M. 2003. Genetics of lactase persistence and lactose intolerance. *Annu. Rev. Genet.* **37:** 197–219.
123. Enattah, N.S., T. Sahi, E. Savilahti, *et al.* 2002. Identification of a variant associated with adult-type hypolactasia. *Nat. Genet.* **30:** 233–237.
124. Poulter, M., E. Hollox, C.B. Harvey, *et al.* 2003. The causal element for the lactase persistence/non-persistence polymorphism is located in a 1 Mb region of linkage disequilibrium in Europeans. *Ann. Hum. Genet.* **67:** 298–311.
125. Lewinsky, R.H., T.G. Jensen, J. Moller, *et al.* 2005. T-13910 DNA variant associated with lactase persistence interacts with Oct-1 and stimulates lactase promoter activity in vitro. *Hum. Mol. Genet.* **14:** 3945–3953.
126. Olds, L.C. & E. Sibley. 2003. Lactase persistence DNA variant enhances lactase promoter activity in vitro: functional role as a cis regulatory element. *Hum. Mol. Genet.* **12:** 2333–2340.
127. Bersaglieri, T., P.C. Sabeti, N. Patterson, *et al.* 2004. Genetic signatures of strong recent positive selection at the lactase gene. *Am. J. Hum. Genet.* **74:** 1111–1120.
128. Nielsen, R., C. Bustamante, A.G. Clark, *et al.* 2005. A scan for positively selected genes in the genomes of humans and chimpanzees. *PLoS Biol.* **3:** e170.
129. Beja-Pereira, A., G. Luikart, P.R. England, *et al.* 2003. Gene-culture coevolution between cattle milk protein genes and human lactase genes. *Nat. Genet.* **35:** 311–313.

130. Ingram, C.J., M.F. Elamin, C.A. Mulcare, *et al.* 2007. A novel polymorphism associated with lactose tolerance in Africa: multiple causes for lactase persistence? *Hum. Genet.* **120:** 779–788.
131. Tishkoff, S.A., F.A. Reed, A. Ranciaro, *et al.* 2007. Convergent adaptation of human lactase persistence in Africa and Europe. *Nat. Genet.* **39:** 31–40.
132. Fox, A.L. 1932. The relationship between chemical constitution and taste. *Proc. Natl. Acad. Sci. USA* **18:** 115–120.
133. Drayna, D., H. Coon, U.K. Kim, *et al.* 2003. Genetic analysis of a complex trait in the Utah Genetic Reference Project: a major locus for PTC taste ability on chromosome 7q and a secondary locus on chromosome 16p. *Hum. Genet.* **112:** 567–572.
134. Kim, U.K., E. Jorgenson, H. Coon, *et al.* 2003. Positional cloning of the human quantitative trait locus underlying taste sensitivity to phenylthiocarbamide. *Science* **299:** 1221–1225.
135. Drewnowski, A. & C.L. Rock. 1995. The influence of genetic taste markers on food acceptance. *Am. J. Clin. Nutr.* **62:** 506–511.
136. Hall, M.J., L.M. Bartoshuk, W.S. Cain, *et al.* 1975. PTC taste blindness and the taste of caffeine. *Nature* **253:** 442–443.
137. Wooding, S., U.K. Kim, M.J. Bamshad, *et al.* 2004. Natural selection and molecular evolution in PTC, a bitter-taste receptor gene. *Am. J. Hum. Genet.* **74:** 637–646.
138. Henkin, R.I. & W.T. Gillis. 1977. Divergent taste responsiveness to fruit of the tree Antidesma bunius. *Nature* **265:** 536–537.
139. Fisher, R., E. Ford & J. Huxley. 1939. Taste-testing the anthropoid apes. *Nature* **144:** 750.
140. Wooding, S., B. Bufe, C. Grassi, *et al.* 2006. Independent evolution of bitter-taste sensitivity in humans and chimpanzees. *Nature* **440:** 930–934.
141. Kim, U., S. Wooding, D. Ricci, *et al.* 2005. Worldwide haplotype diversity and coding sequence variation at human bitter taste receptor loci. *Hum. Mutat.* **26:** 199–204.
142. Soranzo, N., B. Bufe, P.C. Sabeti, *et al.* 2005. Positive selection on a high-sensitivity allele of the human bitter-taste receptor TAS2R16. *Curr. Biol.* **15:** 1257–1265.
143. Beall, C.M. 2007. Detecting natural selection in high-altitude human populations. *Respir. Physiol. Neurobiol.* **158:** 161–171.
144. Beall, C.M. 2007. Two routes to functional adaptation: Tibetan and Andean high-altitude natives. *Proc. Natl. Acad. Sci. USA* **104**(Suppl 1): 8655–8660.
145. Beall, C.M., G.M. Brittenham, K.P. Strohl, *et al.* 1998. Hemoglobin concentration of high-altitude Tibetans and Bolivian Aymara. *Am. J. Phys. Anthropol.* **106:** 385–400.
146. Wu, T., X. Wang, C. Wei, *et al.* 2005. Hemoglobin levels in Qinghai-Tibet: different effects of gender for Tibetans vs. Han. *J. Appl. Physiol.* **98:** 598–604.
147. Yi, X., Y. Liang, E. Huerta-Sanchez, *et al.* 2010. Sequencing of 50 human exomes reveals adaptation to high altitude. *Science* **329:** 75–78.
148. Rankin, E.B., M.P. Biju, Q. Liu, *et al.* 2007. Hypoxia-inducible factor-2 (HIF-2) regulates hepatic erythropoietin in vivo. *J. Clin. Invest.* **117:** 1068–1077.
149. Beall, C.M., G.L. Cavalleri, L. Deng, *et al.* 2010. Natural selection on EPAS1 (HIF2alpha) associated with low hemoglobin concentration in Tibetan highlanders. *Proc. Natl. Acad. Sci. USA* **107:** 11459–11464.
150. Simonson, T.S., Y. Yang, C.D. Huff, *et al.* 2010. Genetic evidence for high-altitude adaptation in Tibet. *Science* **329:** 72–75.
151. Bigham, A., M. Bauchet, D. Pinto, *et al.* 2010. Identifying signatures of natural selection in Tibetan and Andean populations using dense genome scan data. *PLoS Genet.* **6:** e1001116.

Ann. N.Y. Acad. Sci. ISSN 0077-8923

ANNALS OF THE NEW YORK ACADEMY OF SCIENCES
Issue: *The Year in Human and Medical Genetics*

Life-threatening infectious diseases of childhood: single-gene inborn errors of immunity?

Alexandre Alcaïs,[1,2] Lluis Quintana-Murci,[3] David S. Thaler,[4] Erwin Schurr,[5] Laurent Abel,[1,2] and Jean-Laurent Casanova[1,2]

[1]Laboratory of Human Genetics of Infectious Diseases, Necker Branch, Institut National de la Santé et de la Recherche Médicale, University Paris Descartes, Paris, France. [2]St. Giles Laboratory of Human Genetics of Infectious Diseases, Rockefeller Branch, The Rockefeller University, New York, New York. [3]Institut Pasteur, Human Evolutionary Genetics, Centre National de la Recherche Scientifique, Paris, France. [4]Howard Hughes Medical Institute, Department of Microbiology and Immunology, Albert Einstein College of Medicine, Bronx, New York. [5]McGill Centre for the Study of Host Resistance & Departments of Medicine and Human Genetics, McGill University, Montréal, Québec, Canada

Address for correspondence: Jean-Laurent Casanova, M.D., Ph.D., St Giles Laboratory of Human Genetics of Infectious Diseases, Rockefeller Branch, The Rockefeller University, 1230 York Avenue, New York, NY 10065. jean-laurent.casanova@rockefeller.edu

The hypothesis that inborn errors of immunity underlie infectious diseases is gaining experimental support. However, the apparent modes of inheritance of predisposition or resistance differ considerably among diseases and among studies. A coherent genetic architecture of infectious diseases is lacking. We suggest here that life-threatening infectious diseases in childhood, occurring in the course of primary infection, result mostly from individually rare but collectively diverse single-gene variations of variable clinical penetrance, whereas the genetic component of predisposition to secondary or reactivation infections in adults is more complex. This model is consistent with (i) the high incidence of most infectious diseases in early childhood, followed by a steady decline; (ii) theoretical modeling of the impact of monogenic or polygenic predisposition on the incidence distribution of infectious diseases before reproductive age; (iii) available molecular evidence from both monogenic and complex genetics of infectious diseases in children and adults; (iv) current knowledge of immunity to primary and secondary or latent infections; (v) the state of the art in the clinical genetics of noninfectious pediatric and adult diseases; and (vi) evolutionary data for the genes underlying single-gene and complex disease risk. With the recent advent of new-generation deep resequencing, this model of single-gene variations underlying severe pediatric infectious diseases is experimentally testable.

Keywords: immunity; inborn errors; infectious diseases

Background

Clinical heterogeneity among infected patients was known to the microbiologists of the late 19th century. Nevertheless, it was not until Charles Nicolle's discovery of asymptomatic infections between 1911 and 1917 that the question of the mechanisms underlying interindividual, interfamilial, and interpopulation clinical variability among infected individuals became a fundamental challenge in the field of infectious diseases.[1–4] If microbes are necessary but not sufficient for the manifestation of infectious diseases, what other factors are involved in their pathogenesis? Infectious diseases are commonly thought to be well understood and often proposed as textbook examples of pure environmental diseases. However, nearly 150 years after Pasteur's microbial theory of disease and 100 years after Nicolle's discovery of asymptomatic infections, it is striking that the actual pathogenesis of most infectious diseases remains unknown for the vast majority of patients. Environmental factors, whether microbial (the pathogen) or nonmicrobial (ecological, including other microbes), and host factors, whether genetic (germline-encoded) or nongenetic (somatic and epigenetic, e.g., acquired immunity), all make potential contributions of various sizes to this complex process.[5] The respective merits of the

doi: 10.1111/j.1749-6632.2010.05834.x

microbial, ecological, immunological, and genetic theories of infectious diseases will not be discussed here. However, poverty, for example, favors infectious diseases,[6] largely because of its impact on environmental and host nongenetic factors, but only a small fraction of the poor fall victim to any specific infectious agent. In this context, that is, within groups in which other factors are mostly equivalent, host genetic background is progressively emerging as a key determinant.[7–9] Building on the continual progress made in the forward genetics of infection in the mouse model since the 1920s,[10] the field of human genetics of infectious diseases aims to identify the genes and alleles rendering certain individuals, families, and populations susceptible or resistant to past and present infectious diseases.[11]

The common view of the genetic basis of infectious diseases reflects the theoretical and experimental work of two different schools of thought that historically tackled this problem separately.[12] Since the early 1950s, an increasing number of Garrodian clinicians-scientists, pediatricians-immunologists, in particular, have described and deciphered an expanding group of over 200 monogenic inborn errors of immunity, designated as primary immunodeficiencies (PIDs).[13–16] Over the same period, Galtonian population geneticists, particularly those interested in common tropical diseases, have studied infectious diseases from a human genetics standpoint.[17–20] These investigations have led to the widespread acknowledgment that rare infections with weakly virulent ("opportunistic") microbes and/or multiple, recurrent infections in a single patient result from rare monogenic PIDs (rare infectious phenotypes, rare alleles; multiple infectious diseases in the same patient, one morbid gene).[13–16] Conversely, common infections with more virulent microbes in otherwise normally resistant patients are often thought to result from the polygenic inheritance of common susceptibility alleles (common infectious phenotypes, common alleles; one infectious disease per patient, multiple genes).[17–20] This paradigm is consistent with the view that rare inherited disorders are caused by rare alleles, whereas common diseases are favored by common alleles.[21] It is also supported by the identification of rare deleterious mutations as the molecular genetic basis of numerous Mendelian PIDs, and by the large number of association studies examining the contribution of common polymorphisms to the complex genetic basis of many common infectious diseases.[7]

However, the genetic architecture described is unsatisfactory in at least four ways. First, human infectious diseases cannot be rigorously broken down into rare or common infections, as annual incidence differs tremendously between diseases (resulting in an almost continuous range from less than 10^{-6} to more than 0.1). Likewise, the incidence of any given disease varies with the geographic region and historical period considered. Second, patients cannot easily be subdivided into those vulnerable to a single or to many microbes, as there are various patterns of susceptibility between these two extremes. Third, rare infections are not necessarily associated with broad susceptibility, just as common infections are not necessarily linked to narrow vulnerability, as illustrated by rare children with single life-threatening infections in developed countries and the large number of children with multiple infections in developing countries. Fourth, despite the spectacular progress made in the field of PIDs in the last 50 years, most complex genetics studies have failed to identify a causal link between the proposed genetic risk factors and biological function or clinical use. The recent discovery of rare monogenic PIDs and common major genes, predisposing patients and populations, respectively, to a single infectious disease (one gene, one infectious disease), has bridged the gap between monogenic and complex predispositions to infection.[9,12] These findings have triggered a move toward unification in the field of human genetics of infectious diseases, hitherto separated into two separate branches.

Hypothesis

This edifice is, however, fragile, due to the large differences in apparent modes of inheritance between diseases and between studies. A predictive architecture of genetic predisposition to infectious diseases is currently lacking in this field. We suggest here that age at disease onset is the key factor determining whether genetic predisposition is monogenic or complex, regardless of the incidence of the infection considered and the overall resistance of the patient to other infections. We also suggest that, in populations widely exposed to the causal microbes from birth, life-threatening infectious diseases in children mostly (but not invariably) result from collections of diverse single-gene variants affecting immunity

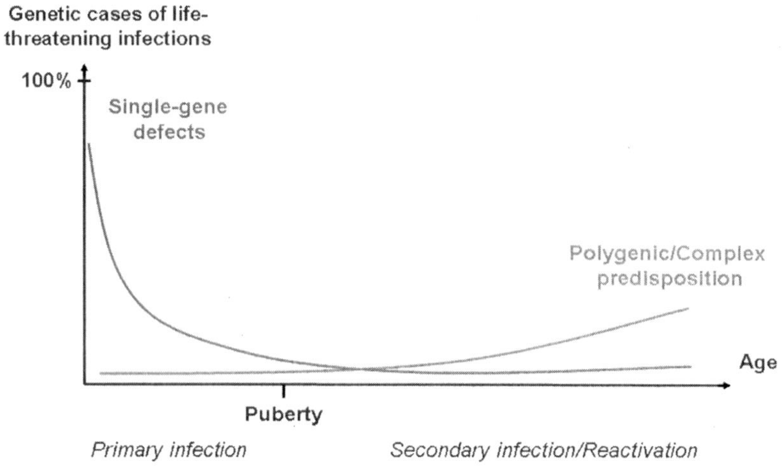

Figure 1. Schematic representation of a hypothetical, age-dependent, human genetic architecture of infectious diseases. We suggest that single-gene human variants make an important contribution to the determinism of life-threatening infectious diseases of childhood, in the course of primary infection. By contrast, predisposition to severe infectious diseases in adults, in the course of microbial reactivation from latency or secondary infection, is less influenced by germline human genetic variations, resulting in a more complex and, perhaps, polygenic contribution. Somatic and epigenetic processes are likely to play a greater role with advanced age.

to primary infection—mostly rare variants displaying incomplete clinical penetrance. By contrast, the corresponding infectious diseases in adults, particularly the elderly, are thought to be less influenced by germline variations and to reflect more complex predisposition affecting immunity to secondary or latent infection, including polygenic and monogenic variations (Fig. 1 and see Text Box). This model may apply to most infectious diseases, regardless of the proportion of cases in the population—ranging from exceedingly rare cases, in which the monogenic and complex predisposition in the patients would be investigated, to very frequent cases, in which attention would instead be focused on monogenic and complex resistance in asymptomatic individuals. According to this model, genetic susceptibility to invariably benign infections, if such susceptibility exists, would obey more complex rules of inheritance, and emerging pathogens may reveal the single-gene component of susceptibility and resistance in patients of all ages.

What do we specifically mean by "single-gene variations conferring a predisposition to severe pediatric infectious diseases"? These variants do not often define fully penetrant Mendelian traits, because the familial segregation of infectious diseases is only rarely Mendelian. In addition, they are predicted to be rare, typically with a frequency of less than 1% in the population. Indeed, we prefer to focus on the frequency of the at-risk genotype, a more relevant entity in this context, particularly for recessive traits. This frequency can be denoted as q_V (Fig. 2). We consider these rare single-gene variants to display a continuous spectrum of effects. At one end of this spectrum, there are "classical Mendelian" variants: very rare ($q_V < 10^{-4}$), with a high penetrance (denoted as F_R) (>0.5) and a very high relative risk (RR) (>1,000). At the other end of the spectrum, there are less rare variants (q_V in the range of 0.001–0.01) with a substantial RR (e.g., an RR value of 10) making a potentially large contribution to mortality rates, depending on the medical environment. The example of severe childhood malaria is discussed in a specific section later. In all infectious diseases, there are, of course, a very large number of intermediate situations, as illustrated in Figure 2, including, for example, quite rare variants (10^{-4}) in very rare disorders (10^{-5}) with a high RR (1,000) but very low penetrance (0.01).

However, these estimates probably vary geographically and historically, and with changes in the microbial and medical environment. Before improvements in hygiene and the advent of vaccination and antibiotics, 50% of the population used to die before the age of 15 years, the vast majority succumbing to infectious diseases.[8,22] This observation highlights a critical but misunderstood aspect of the natural history of genetic predisposition to

The organisms-as-machines analogy

In this analogy, living organisms are considered as machines. This analogy is imperfect but assists reflection. The automobile is a machine with which we are all familiar. Consider an automobile and its lifetime as analogous to the life span of a multicellular differentiated organism. If a new car breaks down in the first 1,000 miles, the breakdown is probably caused by the failure of single component or an incorrect interaction between two components. Furthermore, it is likely that the broken component was defective in its original manufacture. The defect in manufacturing probably occurred before the component was assembled into the car. The defect and its consequences are exclusively restricted to a single component. If a junction between components failed, this connection was probably assembled in a defective manner. By contrast, consider a car that breaks down after having been driven 100,000 miles. A single component may still be responsible for the final breakdown; however, the context of the rest of the car is involved in a way that was not true for the failure in the first 1,000 miles. The component that has failed was probably manufactured well and its failure results from two factors with an intuitively clear meaning that nonetheless requires careful consideration: (1) "Wear and tear" (W&T) resulting from normal machine function and (2) "Cumulative environmental effects" (CEE). These two factors predominate beyond 100,000 miles, but have almost no effect before 1,000 miles. More work is required to define W&T and CEE more precisely in this context, but a few key points can already be made. These two factors are akin to the physical concept of increasing entropy in the system. W&T and CEE are distributed diffusely throughout the system. In the context of W&T and CEE, each discrete component becomes more likely to fail. The degrading effects concern both the individual components and the connections between all the individual components. A part that was originally manufactured with a minor defect, perhaps well within the tolerable limits, may be differentially vulnerable to W&T. In an engineering context, parts that do not fit together perfectly are likely to become sites of vibrations, such as shimmy and tramp. This analogy can be articulated in precise genetic terms (Thaler *et al.* in preparation). In the vocabulary of genetics, the model predicts that individual genes become increasingly pleiotropic as the organism ages. Thus, phenotypes controlled by a single gene in the young gradually come under the influence of many genes in adult and aged organisms.

infectious diseases: the dynamic nature of the effect of a given genetic variant (Fig. 3). Nowadays, only a small fraction of all genetic variants conferring a predisposition to potentially fatal infectious diseases confer a detectable phenotype in developed countries, thanks largely to considerable medical progress (hygiene, vaccination, drug-based treatments for infection, and supportive care, in particular).[8,22] By contrast, fewer than 200 years ago, the penetrance of these variants was higher, simply because their effects were not yet masked by medicine. It also seems likely that the range of allelic frequencies of such morbid variants was initially larger than that observed today, with even relatively frequent morbid alleles ($q_V > 0.01$), the penetrance of which decreased with the recent conquests of medicine. This idea of dynamic penetrance should be borne in mind when reflecting on the genetic architecture of infectious diseases. Indirect evidence for the existence of rare variants underlying pediatric infectious diseases is provided by the lack of compelling evidence favoring the alternative, "polygenic" model attributing an essential role in life-threatening pediatric infections to "common variants."[18,19] This model is increasingly being called into question, even for late-onset nonlethal diseases.[23–28] Furthermore, several lines of evidence are consistent with rare variants making an important contribution, as reviewed later.

Supporting evidence

Epidemiologic evidence: U or L incidence distribution pattern of severe infectious diseases

Infectious diseases, whether endemic or epidemic, have historically been among the major killers of humans, particularly before recent improvements in hygiene and the advent of vaccines and antibiotics.[8,22] About 60% of deaths in mid-19th century England were due to infectious diseases, and

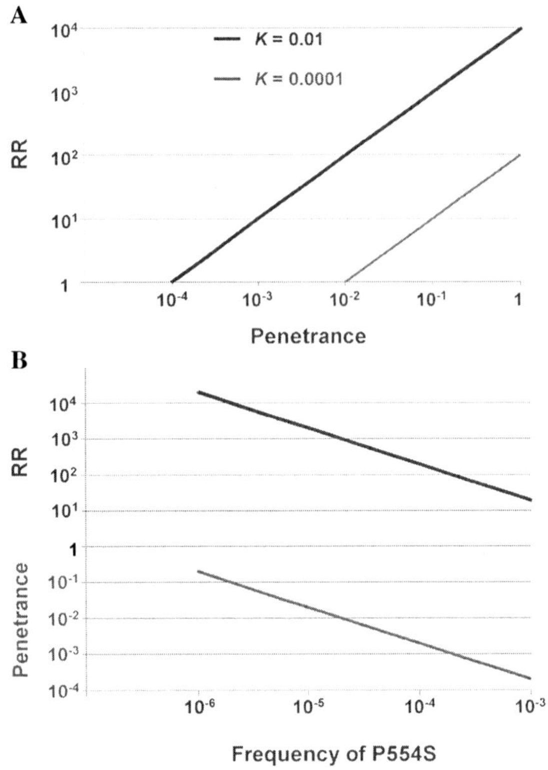

Figure 2. Illustration of the concept of rare single-gene variants. Our general hypothesis is that, in young children, life-threatening infectious diseases are mostly controlled by a collection of rare single-gene variants with strong effects on the occurrence of the disease/mortality (see also Fig. 3). Based on the classical definition of polymorphism frequencies, we will consider *rare* to be defined as a frequency of the risk variant of less than 0.01. Note that, for variants with a recessive effect, this frequency of 0.01 applies to the frequency of the homozygous genotype at risk, that is, the frequency of the risk variant itself may be as high as 0.1. We will denote as q_V the frequency of subjects carrying the at-risk genotype for a specific variant V. The effect of the variant can be measured in terms of penetrance (denoted as F_R)—that is, the probability of a subject carrying the genotype at risk actually having the disease—and/or in terms of relative risk (RR) —F_R divided by the probability of having the disease, for a subject who is wild type at the risk locus, denoted by F_{WT}. F_{WT} is often estimated as the general prevalence of the disease, denoted by K, which is a reasonable approximation of F_{WT}, provided that the risk variant accounts for a low proportion of total cases. Assuming that this is the case, when K is fixed, there is a straightforward correspondence between RR and F_R, with RR being equal to F_R/K (A). In addition, the maximum value of F_R depends directly on q_V, as it should verify the following condition, $q_V \times F_R \leq K$. Thus, penetrance is complete ($F_R = 1$) only when $q_V \leq K$; otherwise, max $F_R = K/q_V$. As an example, we will consider the *TLR3* P554S dominant mutation we recently identified as conferring a predisposition to herpes simplex encephalitis (HSE) (B).[34] We do not know q_V for *TLR3* P554S, but it is likely to be $<10^{-3}$, as P554S was not found in more than 2,000 control chromosomes. We also assume, for HSE, that $K = 10^{-5}$ and that the proportion of HSE cases due to *TLR3* P554S is 0.02 (we found two patients heterozygous for P554S in a sample of approximately 100 HSE patients sequenced for TLR3). From q_V, we can therefore calculate F_R as $0.02 \times K/q_V$ (we know that penetrance is incomplete, as some *TLR3* P554S carriers in the patient's family are healthy). Figure B presents the *TLR3* P554S F_R and RR values associated with HSE for q_V values ranging from 10^{-6} to 10^{-3}. This results in penetrance and RR values decreasing from 0.2 to 0.0002 and from 20,000 to 20, respectively.

this proportion was probably even higher in previous centuries. The impact of infectious diseases before diagnosis became widely available at the start of the 20th century can be estimated from overall mortality curves. In 1690, mortality rates were very high in the first year of life (~28%), sharply decreasing thereafter until early adulthood, and then progressively increasing again in adults over the age of 30 years (U pattern, Fig. 4A). From the late 19th century onwards, it is possible to measure the impact of infectious diseases by means of disease-specific mortality curves. Tuberculosis is probably the only major infectious disease for which reliable data were collected in the 19th century. This disease displays a very similar U pattern, with a golden age of low mortality rates around puberty (Fig. 4A).[29] Life expectancy has now increased considerably, principally due to progress in the control of infectious diseases, itself resulting from the development of hygiene, vaccines, and drugs for treating infections. The WHO has reported specific mortality rates for infectious diseases in several countries, including the United States, Brazil, and Egypt (Fig. 4B). Again, these three curves follow a very similar pattern to those for the 17th and 19th centuries, although the absolute values differ between studies. Mortality rates are always much higher for subjects under the age of 1 year and over the age of 44 years than in subjects between the ages of 5 and 24 years, regardless of whether the data originate from a country with a low (United States) or high (Brazil and Egypt) infectious mortality rate. Finally, worldwide age-dependent mortality curves for the two main WHO-defined causes of infection-related mortality (infectious and parasitic diseases, and respiratory infections) are still U-shaped, as are curves for mortality due to diarrheal diseases, the leading cause of mortality in the "infectious and parasitic diseases" category

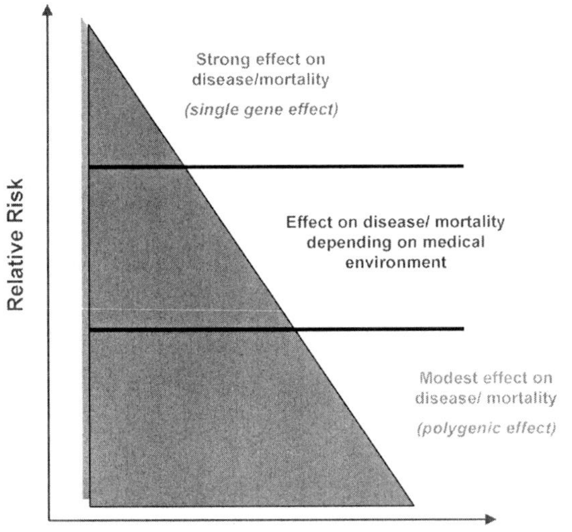

Figure 3. The relative risk as a function of allelic frequency is dependent on both the microbial and medical environment and, therefore, on the geographic region and historical period considered. Before improvements in hygiene and the advent of vaccination and antibiotics, 50% of the population used to die before the age of 15 years, and it is clear that the vast majority of these individuals died from infectious diseases. The impact of any given genetic variant may therefore change with time. Nowadays, in developed countries, only a small fraction of all genetic variants intrinsically predisposing to death from a particular infectious disease confer a detectable phenotype (the upper part of the triangle; strong effect), because of considerable medical progress, including the development of preventive strategies such as improved hygiene and vaccines (preventing disease) and prompt drug-based treatments and supportive care (preventing death from the disease). By contrast, 200 years ago, these predisposing variants had a much higher penetrance, simply because their effects were not masked by preventive and therapeutic measures. The range of allelic frequencies of these morbid variants was probably much larger, and there were probably many frequent alleles (the middle part of the triangle; effect depending on medical environment), the penetrance of which decreased and became negligible with medical progress. Some of these alleles may make an oligo- or polygenic contribution to susceptibility to other infectious diseases (i.e., only individuals carrying a large number of susceptibility alleles at variant loci will display a detectable phenotype, perhaps different from that associated with such variants centuries ago). Finally, a group of common variants intrinsically have little or no impact on the risk of developing the disease (the bottom part of the triangle; modest effect).

(Fig. 4C). By contrast, malaria results in very high mortality rates in the youngest children, but with no distinct peak in the elderly (L pattern). A similar L pattern was observed for tetanus and measles before vaccination became widely available. These two patterns of age-dependent incidence also apply to other infectious diseases, such as invasive pneumococcal disease and herpes simplex encephalitis (U pattern), and diphtheria and poliomyelitis (L pattern) (data not shown). There is thus overwhelming epidemiological evidence for an age-dependent incidence distribution of mortality due to infectious diseases worldwide and throughout history. The actual incidence of any infectious disease depends partly on the available preventive/curative measures available in the population considered. However, for almost all infectious diseases, young children are by far the most susceptible, with adolescents and young adults the least susceptible. In older subjects, susceptibility may remain low or may progressively increase with age. These clear epidemiologic patterns of incidence for severe infectious diseases suggest that different mechanisms of susceptibility operate in children and in people who have reached reproductive age. A distinct set of mechanisms may come to the fore in those of advanced age who are biologically senescent.

Mathematical evidence: modeling of the underlying genetic architecture

One of the most striking observations from Figure 4 is the sharp decrease of childhood mortality with increasing age. Age-dependent epigenetic or somatic modifications may have an effect. However, we nonetheless used a simulation-based procedure to test the potential impact of two main modes of inheritance of germline genetic predisposition (monogenic and polygenic) to death from infection. We considered an endemic disease with an overall prevalence of 1% and compared two equivalent groups of newborns displaying autosomal recessive (AR) or polygenic predisposition. The simulated data were generated as follows: (1) we generated genotypes at 20 loci for 1,000,000 newborns; (2) the first 10 loci were assigned an AR impact, and any individual carrying two risk alleles at any locus was considered to be disease prone (to fit with the overall prevalence, the frequency of the predisposing allele at each locus was fixed at ~0.022 so that the 10 loci accounted for an overall frequency of homozygous susceptible subjects in the population of 0.005); (3) the remaining 10 loci were considered to contribute to polygenic susceptibility, five acting dominantly and five recessively, and for each of these polygenic loci, the allelic frequency was calculated

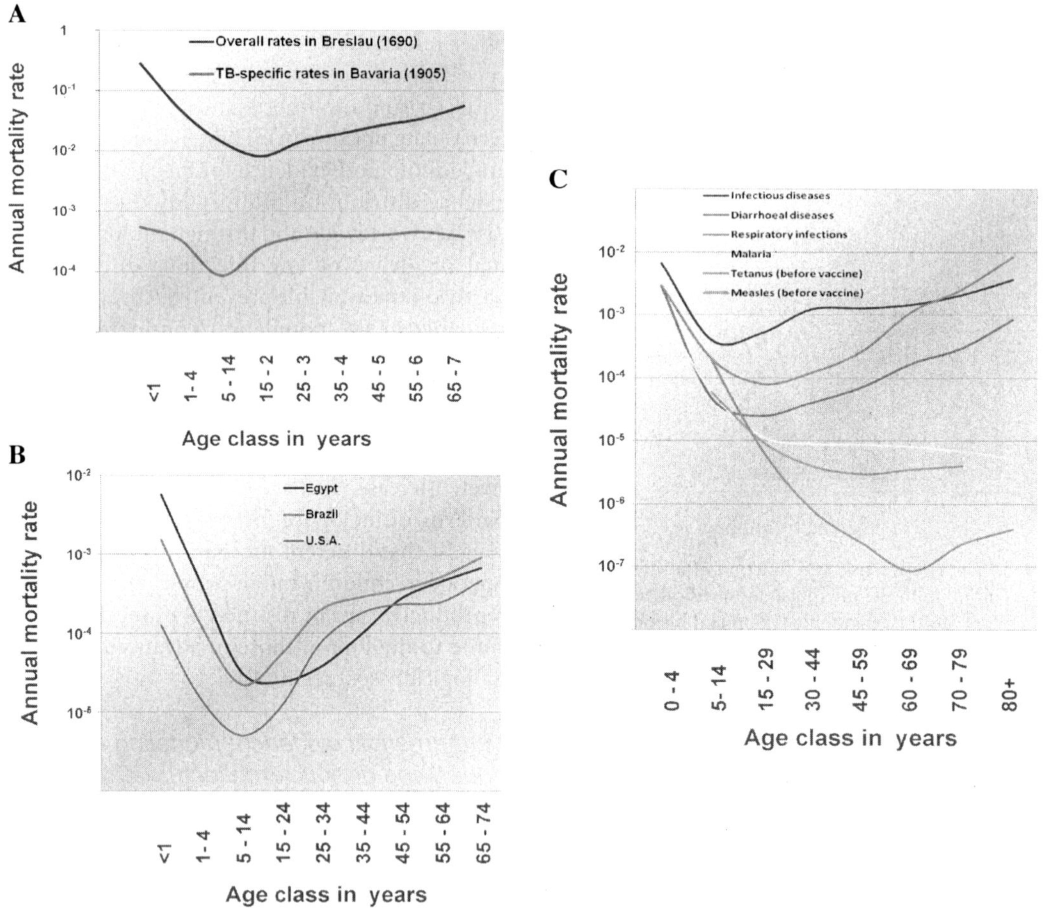

Figure 4. Annual mortality rates (log scale) as a function of age, before (A) and after (B and C) the development of hygiene, vaccination, and antiinfectious drugs. (A) Overall rates observed in Breslau (blue line) by the English astronomer Edmund Halley, who conducted the first reliable life table survey in 1690 (adapted from the book by John Cairns[22]), and tuberculosis-specific rates for the inhabitants of Bavaria, Germany, in 1905 (purple line) (adapted from the paper by Stead[114]). (B) Specific mortality rates due to infectious and parasitic diseases in 2000, observed in the United States (green line), Brazil (red line), and Egypt (blue line), as reported by the WHO (www.who.int/healthinfo/morttables/en/index.html). (C) Specific worldwide mortality rates reported in 2004, by the WHO, for the three main infection-related causes of death, infectious and parasitic diseases (dark blue line), diarrheal diseases (red line), and respiratory infections (green line). In addition to these three groups of diseases, rates of mortality due to malaria (yellow line) (2004 WHO), tetanus before vaccination (orange line) (United States 1900–1969[115]), and measles before vaccination (light blue line) (Brazil 1979, www.who.int/healthinfo/morttables/en/index.html) are shown. We excluded HIV/AIDS because the modes of transmission are age-dependent, and tuberculosis, which is now strongly influenced by vaccination/treatment and HIV infection.

so as to obtain a proportion of susceptible individuals of 20%; (4) in each, an individual was considered to have a probability of 0.5 of being infected with the microbe; (5) following infection, individuals displaying AR susceptibility would be expected to develop disease in either 100% or 50% of cases (i.e., the penetrance at the individual level was set at 1 or 0.5), whereas other individuals would develop disease with a variable probability, depending on the number of polygenic "susceptibility loci" (ranging from 0.001 for one susceptible allele to 0.85 for 10 susceptibility alleles); (6) individuals developing disease (cases) were removed from the population; and (7) we repeated steps 4–6 50 times (i.e., until surviving individuals reach the age of 50 years), and steps 1–7 were repeated 100 times. The mean distribution of cases displaying AR and polygenic predisposition, as a function of age, is shown in Figure 5. When comparing the observed (Fig. 4) and simulated data (Fig. 5), we observed that single-gene

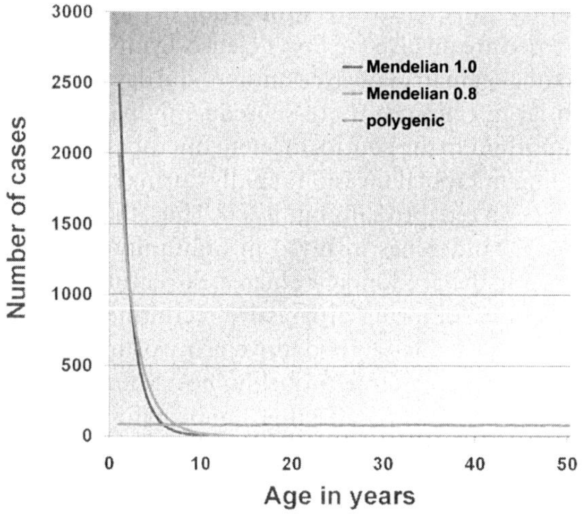

Figure 5. Estimated frequency of Mendelian/single-gene and polygenic cases of a fictitious infectious disease as a function of age. We fixed the prevalence of the disease to 1%, with a 50/50 ratio of Mendelian/single-gene and polygenic cases, a population size of one million, and for the Mendelian cases, a penetrance at the individual level of either 1 or 0.5 (see text for further details on the simulation parameters).

susceptibility, unlike polygenic predisposition, could account for the dynamics of the curve in childhood, consistent with our model (Fig. 1). The single-gene susceptibility hypothesis also closely matches the observed mortality curved for malaria, tetanus, and measles (L pattern, Fig. 4C). However, this simple model cannot account for the changes in the curve during adulthood for other infectious diseases (U pattern), highlighting a more complex determinism of gene-environment interactions in such cases. This process may involve polygenic predisposition, age-dependent somatic and epigenetic factors potentially leading to a progressive impairment of immunity, and a cumulative effect of microbial challenge and ecological variations (see Text Box).

Genetic evidence: pediatric infectious diseases due to single-gene variants

The human molecular genetic basis of susceptibility to most infectious diseases remains elusive in most patients. However, for a small but steadily increasing number of patients and infections, the available data are consistent with our hypothesis. In particular, there are well-documented examples of single-gene variants conferring a predisposition to severe pediatric infectious diseases.[9] This is the case for some of the least virulent pathogens, such as the fungus *Pneumocystis jiroveci*, which causes disease in children with multiple infections and rare, fully penetrant Mendelian PIDs.[13–16] Single-gene predisposition to more common pathogens in otherwise resistant children has also been documented, as illustrated by invasive pneumococcal disease (IPD) in patients with IRAK-4 and MyD88 deficiency,[30–32] and herpes simplex virus encephalitis (HSE) in patients with UNC-93B, TRAF3, and TLR3 deficiency.[33–36] HSE is a particularly informative disease, as the known *UNC93B1* and *TLR3* alleles display full penetrance at the cellular level (with impaired response to TLR3 agonists) but incomplete penetrance at the clinical level (with only two of seven TLR3-deficient individuals having developed HSE upon infection with HSV-1). Children with genetically determined IPD or HSE display a selective impairment of immunity to primary infection.[37] Even predisposition to very common infectious diseases of childhood (and adulthood) can be Mendelian, as illustrated by susceptibility to norovirus diarrhea in children with autosomal-dominant FUT2 proficiency (resistance to norovirus is conferred by AR FUT2 deficiency),[38–40] and susceptibility to *Plasmodium vivax* infection in children with autosomal-dominant DARC proficiency (resistance to *P. vivax* is conferred by AR DARC deficiency).[41–43] Susceptibility to HIV infection in adults also displays Mendelian inheritance, with resistance conferred by AR CCR5 deficiency.[44–46] The nature of genetic predisposition to most other infections in adult patients remains unknown although two major loci *TST1* and *TST2* were recently shown to be involved in the control of tuberculosis infection.[47] With the notable exception of malaria, which is extensively discussed in another section later, perhaps the best understood adult infectious disease is leprosy, which was found to be under the control of several genes, including *LTA* and *PARK2/PACRG*.[48–50] For the leprosy-associated variants of these two genes, odds ratios showed the effect to be much stronger in patients with early onset leprosy than in patients whose disease began later (Ref. 49 and unpublished data). By contrast, although much less is known about the genetic basis of tuberculosis, at least a few examples of single-gene predisposition have been identified in children, and more complex predisposition seems to occur in adults.[51–53] For example, a number of genes (e.g., *HLADQ*, *NRAMP1*, *IL12RB1*) have been reported to play a role in adult tuberculosis.[54,55]

However, it has never been determined whether these genes act independently and additively on the same phenotype in the same patient, or even in the same population. Several genes have repeatedly been reported to have a potential influence on the onset of a given infectious disease, but there is currently no experimental proof of polygenic predisposition per se in individual human beings.[9] The genetic basis of progression to AIDS in HIV-infected patients provides another beautiful example of complex determinism, with various common alleles, including those encoding HLA, KIR, and chemokines, making an important contribution.[56–58] Finally, recent studies have highlighted the contribution of common alleles, including alleles of *IL28B* (encoding IFN-λ3) in particular, to the control of infection with hepatitis C virus either under treatment[59] or spontaneously.[60–62] Thus, the few molecular genetic studies carried out to date have generated results consistent with single-gene predisposition in children and more complex predisposition in adults.

Immunological evidence: immunity to primary infection requires fewer genes than immunity to secondary or latent infections

The age-dependent U or L incidence distribution of death from infectious diseases mirrors the age-dependent interaction with microbes. Infectious diseases of childhood follow primary infection, whereas infectious diseases in adults more commonly result from a secondary infection or from the reactivation of a latent infection.[63,64] This pattern has been observed for many diseases, whether bacterial (tuberculosis, IPD), viral (HSE), or parasitic (malaria) in nature, almost regardless of their global incidence. This is relevant to our discussion because the number of genes involved in immunity to infection increases daily after microbial invasion, culminating in the recruitment of a large number of somatically diverse, antigen-specific T and B lymphocytes. In the first stage of infection, the nonhematopoietic cells of the body (e.g., epithelial cells at the host barrier, cells from inner tissues) rely on their own set of expressed genes, only some of which are involved in host defense (a process often referred to as intrinsic immunity). Myeloid cells expressing numerous genes contributing to host defense are rapidly recruited and play a key role in the course of primary infection (innate immunity). They precede the production of antigen-specific lymphoid cells, in which an even greater proportion of the genes expressed are involved in host defense. Lymphoid cells do not begin to play a role until several days after the initial infection, but are subsequently particularly important in the control of latent microbes and curtailing microbial invasion rapidly during secondary infection (adaptive immunity).[65] Thus, the number of cells and genes involved in immunity to infection and their redundancy increase over time during infection, following progressive recruitment of the intrinsic, innate, and adaptive arms of immunity. The host is therefore probably most vulnerable to monogenic defects affecting nonhematopoietic or myeloid cells during the initial stages of microbial invasion. Fewer genes are expressed during primary infection in childhood, and redundancy levels are lower. Single-gene defects are therefore more likely to manifest early in life. The broader range of cells and genes involved at later stages of primary infection, and the even greater range involved in latent and secondary infections, ensures more robust coverage and a greater degree of molecular and cellular redundancy. The greater genetic and functional redundancy in immunity to latent and secondary infections developing gradually with age would favor a more complex predisposition in adults. The infectious phenotype of children with profound developmental or functional defects of adaptive immunity is severe, presents early in life, and invariably deteriorates. By contrast, although initially equally severe, single-gene lesions in nonhematopoietic or myeloid cells may be progressively compensated by adaptive immunity. Indeed, the clinical consequences of single-gene disorders of intrinsic and innate immunity that affect the TLR3-UNC-93B pathway[33–35] or the MyD88-IRAK-4 pathway[30–32] have been shown to improve with age.[37] The occurrence of pathogen-specific T and B cells probably compensates for the poor innate immunity in such patients.[66,67] Current understanding of the differences in immunity between primary and latent/secondary infection is thus consistent with our model.

Clinical evidence: most single-gene traits first appear in childhood, whereas corresponding diseases in adults reflect more complex inheritance

Our hypothesis is also consistent with progress in clinical genetics, whether in pediatrics or internal medicine, since the work of Archibald Garrod.[68]

Indeed, the pattern of genetic inheritance of most human diseases depends almost entirely on age at onset. Inherited diseases of childhood are typically monogenic, and most of the 8,000 or so known monogenic traits have an onset during childhood.[69,70] There is a large overlap between pediatrics, clinical genetics, and Mendelian genetics. These diseases are generally defined clinically and then classified according to their genetic causes. Some monogenic diseases may be relatively common, with an incidence similar to that of common infectious diseases. For example, muscular dystrophies and congenital deafness result from a collection of single-gene disorders (>20 genes for muscular dystrophies and >100 genes for congenital deafness), with an overall incidence at birth of between 10^{-4} and 10^{-3} (see Refs. 71, 72), which is strikingly similar to the incidence of severe malaria in young children. Pediatric inborn diseases that do not result from single-gene defects often result from chromosomal aberrations, which may involve several genes but typically result from a single, large genetic lesion. By contrast, predisposition to disease in adults is more complex, with a less pronounced genetic component. Moreover, most known single-gene traits conferring predisposition to an adult-onset disease typically affect young adults.[73] This general pattern applies to virtually all fields of medicine, including inborn errors of metabolism, the clinical expression of which is highly dependent on nutrition, a critical environmental variable.[74] For example, phenylketonuria is clinically severe only if excessive amounts of phenylalanine are ingested. Moreover, the five major metabolic disorders that can strike at all ages—diabetes, obesity, hypertension, osteoporosis, and atherosclerosis—all show a dual pattern of inheritance. These disorders are usually caused by a single-gene defect in early childhood, whereas their mode of inheritance seems to become more complex with increasing age. For example, neonatal diabetes, and even type I diabetes occurring before the age of six months, may be monogenic,[75,76] whereas in older children, diabetes is usually due to a major HLA gene effect,[77] Similarly, type II diabetes in young adults is often monogenic, whereas late-onset type II diabetes is apparently more complex.[78] Similarly, single-gene causes of obesity,[79–81] osteoporosis,[82] hypertension,[83] hypercholesterolemia,[84] and atherosclerosis[85] have been identified in children, whereas the corresponding syndromes in adults appear to display a more complex pattern of inheritance. However, there is still some uncertainty about whether this pattern results from genetic heterogeneity or truly polygenic inheritance (at the individual level), or both. The "polygenic" model, ascribing an essential role to "common variants," is increasingly being called into question for various diseases in adults, and rare variants may also be involved.[23–28] Inborn errors conferring predisposition to the development of metabolic illnesses in the context of a specific nutritional context display an age-dependent inheritance pattern. Single-gene traits have a major impact in children, during their initial exposure to necessary but deleterious nutrients. The similarity between the pathogenesis of inborn errors of metabolism and that of inborn errors of immunity is striking, in terms of the respective contributions of predisposing genetic traits and environmental triggers and, by inference, possibly also their genetic architecture.

Evolutionary evidence: unequal selective pressures on genes underlying single-gene and complex disease risks

The hypothesis of a bimodal pattern of human genetics of infectious diseases finds further support in the observed evolutionary fate of alleles underlying single-gene and complex traits.[24,87,88] As morbid single-gene variants are not transmitted from individuals dying during childhood, accounting for their relatively low frequency in the general population (e.g., <1%), the corresponding genes should display low levels of variation and, therefore, signatures of purifying selection—a selective regime eliminating almost all newly occurring amino-acid variations. Indeed, at the entire genome scale, most single genes associated with disease (for both infectious and noninfectious diseases) have been shown to be under purifying selection, particularly in cases in which the disease-causing mutations are dominant.[88–90] In the context of infectious diseases, for example, single-gene TLR3 deficiency confers a predisposition to HSE in childhood,[34] and TLR3 is under strong purifying selection.[91] Conversely, the single-gene resistance alleles revealed at the start of new, epidemics of emerging, life-threatening diseases may be positively selected, even if resistance is AR, as shown by four classical examples of Mendelian resistance (DARC deficiency and *P.*

vivax infection, P antigen and parvovirus, CCR5 and HIV, and FUT2 and norovirus) (reviewed in Refs. 9, 92) Another interesting example is provided by the resistance to *Neisseria gonorrhoeae in vitro* of acidic sphingomyelinase-deficient cells from patients with Niemann-Pick disease.[93] In principle, Mendelian resistance may operate at all ages. Only rarely does such resistance result in a population being almost completely protected against an infectious agent, as reported in West Africa for the *DARC* mutation, which prevents *P. vivax* infection. Mostly, susceptibility to infection remains common in the population, with additional single-gene susceptibility genes conferring a predisposition to more severe, lethal forms of the disease, as discussed below for *P. falciparum* malaria. Complex susceptibility alleles generally have a lower penetrance and are therefore able to reach higher frequencies in the population (i.e., >5–10%) than single-gene susceptibility alleles.[24,87,94] As the alleles contributing to late-onset diseases will be transmitted from generation to generation, the corresponding genes should be under less pervasive purifying selection. Indeed, a global relaxation of selection constraints has been observed among genes contributing to complex disease risk (i.e., bearing weakly deleterious mutations) with respect to single-gene disease genes (i.e., bearing strongly deleterious mutations), and the selection coefficient of mutations associated with complex disease risk is lower than that for single-gene mutations.[90] Thus, the weaker evolutionary conservation of complex disease genes indicates that this class of genes may include loci targeted by purifying selection to various extents (e.g., genes under weak or no purifying selection) and loci subject to positive selection.[90] A higher frequency of targets of positive selection has been documented for complex disease genes than for single-gene disease genes.[90,95] In addition, genome-wide scans for selection have revealed that genes playing a role in immunity-related processes and host-pathogen interactions are among the most common targets of recent positive selection.[96–98] Thus, overall, evolutionary and population genetic studies are generally consistent with a dichotomy in the mode of evolution of purely single-gene and complex disease genes. The evolutionary footprints of natural selection in genes involved to various extents in both single-gene and complex diseases are likely to be much more complex. Nevertheless, in general, deleterious mutations in host defense genes under purifying selection are likely to predispose the individual to severe infectious diseases in childhood. Our model of single-gene inborn errors of immunity contributing to severe pediatric infectious diseases is thus consistent with several lines of evidence. This model, in turn, raises several questions.

Questions

How about malaria?

Malaria deserves special consideration, as it was the first infectious disease to be studied from the human genetics standpoint and remains the most thoroughly investigated, and these studies have paved the way to further evaluations of the contribution of various erythrocyte disorders.[20,99–101] In regions in which *P. falciparum* is highly endemic, host genetic factors are estimated to account for ~25% of the risk of severe malaria.[102,103] Heterozygosity for various recessive erythrocyte disorders confers protection against life-threatening forms of disease. For example, patients with HbS are heterozygous at the *HBB* locus and have a risk of *P. falciparum* malaria only one-tenth that of individuals lacking this allele.[20,99,100,104,105] However, while this and other red cell variants (e.g., HbC, HbE) have been strongly positively selected by malaria in various geographic regions worldwide,[20] most people living in regions of endemic malaria do not carry any of these variants (e.g., the HbS allele is maintained at a frequency of ~10% in several areas). Moreover, only a very small fraction of infected children (whether or not they carry erythrocyte disorders) are susceptible to severe malaria. The high malaria mortality rates in young children (~0.001 before the age of 4 years), followed by the rapid decrease in malaria mortality in older children (Fig. 4C), thus remain unexplained. Initial GWA studies of severe malaria in Gambia[103] and Ghana (R. Horstman, personal communication, 2010) identified only the *HBB* gene as a major signal of association, despite the conclusions of epidemiological studies of African children estimating that the HbS allele makes a minor contribution (<10%) to the total impact of host genetics on the risk of severe malaria.[102] Further genome-wide association (GWA) studies on larger samples, based on arrays with broader SNP coverage to account for the known low level of linkage disequilibrium (LD) displayed by African populations, will probably identify additional genes with more

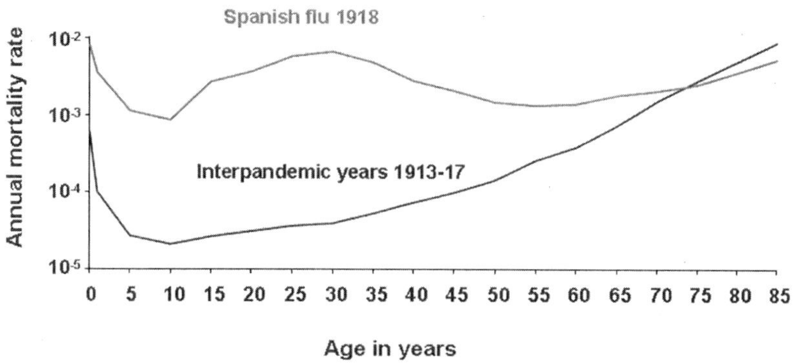

Figure 6. Specific death rates due to influenza per 100,000 persons in each age group in the United States, 1913–1918. Influenza-specific death rates are plotted for the interpandemic years 1913–1917 (dashed line) and for the pandemic year 1918 (solid line).[116]

modest effects. However, all the common variants identified by these GWA studies are likely to account for only a small proportion of the overall genetic variation, as reported for other diseases and described as "missing heritability."[26,27] The many non-mutually exclusive explanations for the observed missing heritability include the contribution of rare variants.[28] For example, we recently showed, by a simple calculation, that a small proportion (5%) of Mendelian forms (dominant risk allele with a penetrance of 20%) in Crohn's disease may largely account for missing heritability.[106] It is therefore entirely possible that severe malaria in young children results from a collection of rare single-gene variants accounting for the pattern of malaria mortality rates in children (Fig. 4C). The observed mortality curve for malaria is well fitted by our simulated curve generated under a simple Mendelian hypothesis (Fig. 5). As an illustration of the definition and role of single-gene variants (Fig. 2), a morbid variant with $q_V = 0.002$ and RR $= 10$, and therefore a clinical penetrance of 0.01 (assuming an overall prevalence K of 0.001 for severe malaria) would account for about 2% of all cases of severe malaria. A few dozen variants exerting a similar effect could therefore potentially account for the human genetic contribution to malaria.

Does age matter per se?

In this model of single-gene variations accounting for severe diseases in the course of primary infection, late-onset cases in low-exposure settings may also result from single-gene variations and merely reflect late infection. Moreover, assuming that single-gene disease genes are generally subject to purifying selection, we would predict that the observation of single-gene cases (regardless of their incidence) would not be correlated with age per se in a naive population recently exposed to a new and potentially lethal microbe. Thus, individuals carrying the deleterious genotype would be expected to be affected at the time of infection, regardless of their age. The example of the exceptionally severe "Spanish" influenza pandemic of 1918–1919, which killed between 2% and 20% of infected individuals and caused ~50 million deaths worldwide, provides a good opportunity to address this issue. An unusual feature of the 1918 flu pandemic was that it mostly killed young adults, with 99% of pandemic influenza deaths occurring in people under the age of 65, and more than half in young adults between the ages of 20 and 40 years.[107] Figure 6 contrasts influenza-specific mortality rates in each age group between the interpandemic years 1913–1917 and the pandemic year, 1918. Consistent with the predictions of our model, the classical U-curve has been replaced by a flatter one. A similarly high mortality rate is observed across all age groups. As a consequence, the sharp decrease until young adulthood observed during the interpandemic years (a pattern similarly observed in Fig. 4) is much less pronounced in 1918, with a ratio of mortality between young adults and children of ~10 in 1918, rather than the ratio of ~100 observed during the interpandemic years (Fig. 6). A similar pattern was observed in terms of the age distribution of the 318 patients (280 of whom died) diagnosed with Ebola hemorrhagic fever during the 1976 outbreak in Zaire.[108] These observations suggest that age at primary infection and disease onset, rather than the age per se, is the determinant of the incidence distribution of mortality due to infectious diseases in human populations and, possibly, of the corresponding genetic architecture. Emerging, virulent pathogens strike across a broad range of ages. The corresponding cases may reflect single-gene case predisposition, and the ensuing selective pressure against single-gene case susceptibility alleles accounts for both the U/L incidence distribution pattern seen and our model of a dual genetic architecture proposed on the basis of data for more ancient microbes. This process would be much slower and less efficient if susceptibility to emerging pathogens were polygenic.

Perspectives

After many generations of exposure to a particular microbe, a large fraction of the corresponding single-gene susceptibility alleles may still account for much of the disease incidence. For example, purifying selection inefficiently purges the genome of AR alleles, as most alleles are carried by heterozygous individuals who are not vulnerable to the infection and therefore not subject to evolutionary counterselection. Moreover, autosomal-dominant lesions may occur *de novo* in each generation, possibly balancing out the loss of mutations from deceased individuals who inherited the susceptibility allele from a parent. The inheritance of a susceptibility allele from a parent is, in turn, made possible by the incomplete clinical penetrance of most autosomal-dominant traits, particularly those conferring a predisposition to infectious diseases (Fig. 3). Examples of autosomal-dominant predisposition to specific infectious agents include IFNGR1, STAT1, TRAF3, and TLR3 deficiencies.[34,36,109,110] The hypothesis that life-threatening infectious diseases in most children result from single-gene variations may at first glance appear provocative but actually merely represents the extension to the field of infectious diseases of a paradigm well-established in human medicine. There are two probable reasons for which a single-gene theory of pediatric infectious diseases was not proposed earlier, at odds with most other fields in medicine. First, the germ theory of disease holds such sway that there is still some reluctance to think of infectious diseases in terms of human genetics. This has impeded the development of models of genetic architecture. Second, the popularity of the "common disease-common variant" hypothesis and the fact that infectious diseases are the most common pediatric illnesses have resulted in resistance to thinking about pediatric infections as monogenic diseases. Rare pediatric illnesses are thought to be Mendelian, whereas common illnesses are thought to be infectious. However, many individual infectious diseases in children are less common than some well-known genetic diseases. Severe infectious diseases are globally common because there are many of them and some are common. Moreover, the incomplete clinical penetrance of single-gene variations can largely account for the lack of a clear Mendelian pattern of infection for most infections. Our model unifies pediatrics, recasting both rare and common infectious diseases as the result of single-gene variations, like most other pediatric illnesses. The model of the dual architecture of human genetics of infectious diseases, with single-gene traits predominating before puberty and more complex predisposition developing with increasing age, is theoretically plausible. Equally important, the proposed model is now experimentally testable at a large scale (i.e., at the population level), following the recent development of new-generation deep resequencing techniques and the possibility of sequencing the "whole exome" and, soon, the "whole genome" from patients[111,112] as illustrated by the recent discovery of STIM1 deficiency in a single child with HHV8-driven Kaposi sarcoma.[113] This model provides a conceptual framework for experimental exploration of the genetic basis of human infectious diseases, a field with considerable biological and clinical implications. We predict that the investigation of death from infection in childhood will reveal some of the most interesting impacts of single-gene variations on human phenotypes.

Acknowledgments

We thank the members of our laboratories for helpful discussions, critical reading, and experimental testing of the model proposed. Special thanks go to Danielle Malo, Sylvia Vidal, and Marcel Behr for critical reading; Jerome Flatot for the simulation study; and Michel Garenne (Unité d'Epidémiologie des Maladies Emergentes, Institut Pasteur, Paris, France) for the data on Spanish flu. We also thank the INSERM, ANR, Institut Pasteur, CNRS, FRM, The Rockefeller University, HHMI, Dana Foundation, March of Dimes, The Bill and Melinda Gates Foundation, St. Giles Foundation, Jeffrey Modell Foundation and Talecris Biotherapeutics, The Sloan

Foundation, CIHR, FRSQ, and L'Ordre de Malte for support. Funding source: A.A. and L.A. are supported in part by Assistance-Publique-Hôpitaux de Paris.

Conflicts of interest

The authors declare no conflicts of interest.

References

1. Pasteur, L. 1922–1939. *Oeuvres Complètes de Louis Pasteur, Réunies par Pasteur Vallery-Radot*. Masson et Cie. Paris.
2. Nicolle, C. 1933. Les infections inapparentes. *Scientia* **33:** 181–271.
3. Nicolle, C. 1937. *Destin des Maladies Infectieuses*. Alcan. Paris.
4. Garnham, P. 1977. Charles Nicolle and inapparent infections. *Am. J. Trop. Med. Hyg.* **26:** 1101–1104.
5. Casanova, J.L. & L. Abel. 2004. The human model: a genetic dissection of immunity to infection in natural conditions. *Nat. Rev. Immunol.* **4:** 55–66.
6. Marmot, M. 2005. Social determinants of health inequalities. *Lancet* **365:** 1099–1104.
7. Alcais, A., L. Abel & J.L. Casanova. 2009. Human genetics of infectious diseases. In *Human Genetics: Problems and Approaches*. F. Vogel, *et al.*, Eds.: 403–415. Springer-Verlag, Heidelberg, Dordrecht, London, & New York.
8. Casanova, J.L. & L. Abel. 2005. Inborn errors of immunity to infection: the rule rather than the exception. *J. Exp. Med.* **202:** 197–201.
9. Alcais, A., L. Abel & J.L. Casanova. 2009. Human genetics of infectious diseases: between proof of principle and paradigm. *J. Clin. Invest.* **119:** 2506–2514.
10. Vidal, S.M. *et al.* 2008. Forward genetic dissection of immunity to infection in the mouse. *Annu. Rev. Immunol.* **26:** 81–132.
11. Quintana-Murci, L. *et al.* 2007. Immunology in natura: clinical, epidemiological and evolutionary genetics of infectious diseases. *Nat. Immunol.* **8:** 1165–1171.
12. Casanova, J.L. & L. Abel. 2007. Human genetics of infectious diseases: a unified theory. *EMBO. J.* **26:** 915–922.
13. Fischer, A. 2007. Human primary immunodeficiency diseases. *Immunity* **27:** 835–845.
14. Casanova, J.L. & L. Abel. 2007. Primary immunodeficiencies: a field in its infancy. *Science* **317:** 617–619.
15. Notarangelo, L.D. *et al.* 2009. Primary immunodeficiencies: 2009 update. *J. Allergy Clin. Immunol.* **124:** 1161–1178.
16. Ochs, H., C.I.E. Smith, J. Puck. 2007. *Primary Immunodeficiencies: A Molecular and Genetic Approach*. Oxford University Press. New York.
17. Kwiatkowski, D. 2000. Science, medicine, and the future: susceptibility to infection. *BMJ* **321:** 1061–1065.
18. Hill, A.V. 2001. The genomics and genetics of human infectious disease susceptibility. *Annu. Rev. Genomics Hum. Genet.* **2:** 373–400.
19. Hill, A.V.S. 2006. Aspects of genetic susceptibility to human infectious diseases. *Annu. Rev. Genet.* **40:** 469–486.
20. Kwiatkowski, D.P. 2005. How malaria has affected the human genome and what human genetics can teach us about malaria. *Am. J. Hum. Genet.* **77:** 171–192.
21. Reich, D.E. & E.S. Lander. 2001. On the allelic spectrum of human disease. *Trends Genet.* **17:** 502–510.
22. Cairns, J. 1997. *Matters of Life and Death*. Princeton University Press. Princeton, NJ.
23. Pritchard, J.K. 2001. Are rare variants responsible for susceptibility to complex diseases? *Am. J. Hum. Genet.* **69:** 124–137.
24. Pritchard, J.K. & N.J. Cox. 2002. The allelic architecture of human disease genes: common disease-common variant. . .or not? *Hum. Mol. Genet.* **11:** 2417–2423.
25. Wright, A. *et al.* 2003. A polygenic basis for late-onset disease. *Trends. Genet.* **19:** 97–106.
26. Maher, B. 2008. The case of the missing heritability. *Nature* **456:** 18–21.
27. Manolio, T.A. *et al.* 2009. Finding the missing heritability of complex diseases. *Nature* **461:** 747–753.
28. Goldstein, D.B. 2009. Common genetic variation and human traits. *N. Engl. J. Med.* **360:** 1696–1698.
29. Dubos, R.J. & J. Dubos. 1952. *The White Plague; Tuberculosis, Man, and Society*. Little Brown. Boston.
30. Picard, C. *et al.* 2003. Pyogenic bacterial infections in humans with IRAK-4 deficiency. *Science* **299:** 2076–2079.
31. von Bernuth, H. *et al.* (2008. Pyogenic bacterial infections in humans with MyD88 deficiency. *Science* **321:** 691–696.
32. Ku, C.L. *et al.* 2007. Selective predisposition to bacterial infections in IRAK-4 deficient children: IRAK-4 dependent TLRs are otherwise redundant in protective immunity. *J. Exp. Med.* **10:** 2407–2422.
33. Casrouge, A. *et al.* 2006. Herpes simplex virus encephalitis in human UNC-93B deficiency. *Science* **314:** 308–312.
34. Zhang, S.Y. *et al.* 2007. TLR3 deficiency in patients with herpes simplex encephalitis. *Science* **317:** 1522–1527.
35. Zhang, S.Y. *et al.* 2008. Inborn errors of interferon (IFN)-mediated immunity in humans: insights into the respective roles of IFN-alpha/beta, IFN-gamma, and IFN-lambda in host defense. *Immunol. Rev.* **226:** 29–40.
36. Perez de Diego, R. *et al.* 2010. Human TRAF3 adaptor molecule deficiency leads to impaired Toll-like receptor 3 response and susceptibility to herpes simplex encephalitis. *Immunity* **33:** 400–411.
37. Bousfiha, A. *et al.* 2010. Primary immunodeficiencies of protective immunity to primary infections. *Clin. Immunol.* **135:** 204–209.
38. Lindesmith, L. *et al.* 2003. Human susceptibility and resistance to Norwalk virus infection. *Nat. Med.* **9:** 548–553.
39. Le Pendu, J. *et al.* 2006. Mendelian resistance to human norovirus infections. *Semin. Immunol.* **18:** 375–386.
40. Kindberg, E. *et al.* 2007. Host genetic resistance to symptomatic norovirus (GGII.4) infections in Denmark. *J. Clin. Microbiol.* **45:** 720–722.
41. Miller, L.H. *et al.* 1976. The resistance factor to Plasmodium vivax in blacks. The Duffy-blood-group genotype, FyFy. *N. Engl. J. Med.* **295:** 302–304.
42. Miller, L.H. *et al.* 1975. Erythrocyte receptors for (Plasmodium knowlesi) malaria: Duffy blood group determinants. *Science* **189:** 561–563.
43. Tournamille, C. *et al.* 1995. Disruption of a GATA motif in the Duffy gene promoter abolishes erythroid gene

expression in Duffy-negative individuals. *Nat. Genet.* **10:** 224–228.
44. Dean, M. *et al.* 1996. Genetic restriction of HIV-1 infection and progression to AIDS by a deletion allele of the CKR5 structural gene. *Science* **273:** 1856–1862.
45. Samson, M. *et al.* 1996. Resistance to HIV-1 infection in Caucasian individuals bearing mutant alleles of the CCR5 chemokine receptor gene. *Nature* **382:** 722–725.
46. Liu, R. *et al.* 1996. Homozygous defects in HIV-1 coreceptor accounts for resistance of some multiply-exposed individuals to HIV-1 infection. *Cell* **86:** 367–377.
47. Cobat, A. *et al.* 2009. Two loci control tuberculin skin test reactivity in an area hyperendemic for tuberculosis. *J. Exp. Med.* **206:** 2583–2591.
48. Mira, M.T. *et al.* 2004. Susceptibility to leprosy is associated with PARK2 and PACRG. *Nature* **427:** 636–6340.
49. Alcais, A. *et al.* 2007. Stepwise replication identifies a low-producing lymphotoxin-alpha allele as a major risk factor for early-onset leprosy. *Nat. Genet.* **39:** 517–522.
50. Zhang, F.R. *et al.* 2009. Genomewide association study of leprosy. *N. Engl. J. Med.* **361:** 2609–2618.
51. Özbek, N. *et al.* 2005. Interleukin-12 receptor beta 1 chain deficiency in a child with disseminated tuberculosis. *Clin. Infect. Dis.* **40:** e55–e58.
52. Alcais, A. *et al.* 2005. Tuberculosis in children and adults: two distinct genetic diseases. *J. Exp. Med.* **202:** 1617–1621.
53. Baghdadi, J.E. *et al.* 2006. An autosomal dominant major gene confers predisposition to pulmonary tuberculosis in adults. *J. Exp. Med.* **203:** 1679–1684.
54. Schurr, E. 2007. Is susceptibility to tuberculosis acquired or inherited? *J. Intern. Med.* **261:** 106–111.
55. Schurr, E. & I. Kramnik. 2008. Genetic control of host susceptibility to tuberculosis. In *Handbook of Tuberculosis, Immunology and Cell Biology*. S.H. Kaufmann & W. Britton, Eds.: 305–346. Wiley-VCH, Weinheim.
56. Hunt, P.W. & M. Carrington. 2008. Host genetic determinants of HIV pathogenesis: an immunologic perspective. *Curr. Opin. HIV AIDS* **3:** 342–348.
57. Goldstein, D.B. 2008. Genomics and biology come together to fight HIV. *PLoS Biol.* **6:** e76.
58. Fellay, J. *et al.* 2009. Common genetic variation and the control of HIV-1 in humans. *PLoS Genet.* **5:** e1000791.
59. Ge, D. *et al.* 2009. Genetic variation in IL28B predicts hepatitis C treatment-induced viral clearance. *Nature* **461:** 399–401.
60. Thomas, D.L. *et al.* 2009. Genetic variation in IL28B and spontaneous clearance of hepatitis C virus. *Nature* **461:** 798–801.
61. Suppiah, V. *et al.* 2009. IL28B is associated with response to chronic hepatitis C interferon-alpha and ribavirin therapy. *Nat. Genet.* **41:** 1100–1104.
62. Tanaka, Y. *et al.* 2009. Genome-wide association of IL28B with response to pegylated interferon-alpha and ribavirin therapy for chronic hepatitis C. *Nat. Genet.* **41:** 1105–1109.
63. Feigin, R.D. & J.D. Cherry. 1998. *Textbook of Pediatric Infectious Diseases*. W.B. Saunders Company. Philadelphia.
64. Mandell, G., J. Bennett & R. Dolin. 2004. *Principles and Practice of Infectious Diseases*. Elsevier Health Sciences, New York.
65. Paul, W.E. 2008. *Fundamental Immunology*. Lippincott Williams & Wilkins. Philadelphia.
66. Gavin, A.L. *et al.* 2006. Adjuvant-enhanced antibody responses in the absence of toll-like receptor signaling. *Science* **314:** 1936–1938.
67. Meyer-Bahlburg, A., S. Khim & D.J. Rawlings. 2007. B cell intrinsic TLR signals amplify but are not required for humoral immunity. *J. Exp. Med.* **204:** 3095–3101.
68. Garrod, A.E. 1923. *Inborn Errors of Metabolism*. Henry Frowden and Hodder & Stoughton. London.
69. McKusick, V.A. 1998. *Mendelian Inheritance in Man. Catalogs of Human Genes and Genetic Disorders*. Johns Hopkins University Press. Baltimore, MD.
70. Antonarakis, S.E. & J.S. Beckmann. 2006. Mendelian disorders deserve more attention. *Nat. Rev. Genet.* **7:** 277–282.
71. Emery, A.E. 2002. The muscular dystrophies. *Lancet* **359:** 687–695.
72. Nance, W.E. 2003. The genetics of deafness. *Ment. Retard. Dev. Disabil. Res. Rev.* **9:** 109–119.
73. Speicher, M., S.E. Antonarakis & A.G. Motulsky. 2009. *Vogel and Motulsky's Human Genetics*. Springer. Berlin.
74. Scriver, C. *et al.* 2001. *The Metabolic and Molecular Bases of Inherited Disease*. McGraw-Hill. New York.
75. Slingerland, A.S. 2006. Monogenic diabetes in children and young adults: challenges for researcher, clinician and patient. *Rev. Endocr. Metab. Disord.* **7:** 171–185.
76. Vaxillaire, M. *et al.* 2009. Breakthroughs in monogenic diabetes genetics: from pediatric forms to young adulthood diabetes. *Pediatr. Endocrinol. Rev.* **6:** 405–417.
77. Onengut-Gumuscu, S. & P. Concannon. 2006. Recent advances in the immunogenetics of human type 1 diabetes. *Curr. Opin. Immunol.* **18:** 634–638.
78. McIntyre, E.A. & M. Walker. 2002. Genetics of type 2 diabetes and insulin resistance: knowledge from human studies. *Clin. Endocrinol. (Oxf.)* **57:** 303–311.
79. Farooqi, I.S. & S. O'Rahilly. 2005. Monogenic obesity in humans. *Annu. Rev. Med.* **56:** 443–458.
80. Mutch, D.M. & K. Clement. 2006. Unraveling the genetics of human obesity. *PLoS Genet.* **2:** e188.
81. Friedman, J.M. 2004. Modern science versus the stigma of obesity. *Nat. Med.* **10:** 563–569.
82. Ralston, S.H. & B. de Crombrugghe. 2006. Genetic regulation of bone mass and susceptibility to osteoporosis. *Genes Dev.* **20:** 2492–2506.
83. Lifton, R.P. 2004. Genetic dissection of human blood pressure variation: common pathways from rare phenotypes. *Harvey Lect.* **100:** 71–101.
84. Hobbs, H.H. *et al.* 2002. Genetic defenses against hypercholesterolemia. *Cold Spring Harb. Symp. Quant. Biol.* **67:** 499–505.
85. Lusis, A.J., R. Mar & P. Pajukanta. 2004. Genetics of atherosclerosis. *Annu. Rev. Genomics Hum. Genet.* **5:** 189–218.
86. Zwick, M.E., D.J. Cutler & A. Chakravarti. 2000. Patterns of genetic variation in Mendelian and complex traits. *Annu. Rev. Genomics Hum. Genet.* **1:** 387–407.
87. Di Rienzo, A. 2006. Population genetics models of common diseases. *Curr. Opin. Genet. Dev.* **16:** 630–636.

88. Bustamante, C.D. *et al.* 2005. Natural selection on protein-coding genes in the human genome. *Nature* **437:** 1153–1157.
89. Barreiro, L.B. *et al.* 2008. Natural selection has driven population differentiation in modern humans. *Nat. Genet.* **40:** 340–345.
90. Blekhman, R. *et al.* 2008. Natural selection on genes that underlie human disease susceptibility. *Curr. Biol.* **18:** 883–8839.
91. Barreiro, L.B. *et al.* 2009. Evolutionary dynamics of human toll-like receptors and their different contributions to host defense. *PLoS Genet.* **5:** e1000562.
92. Picard, C., J.L. Casanova & L. Abel. 2006. Mendelian traits that confer predisposition or resistance to specific infections in humans. *Curr. Opin. Immunol.* **18:** 383–390.
93. Grassme, H. *et al.* 1997. Acidic sphingomyelinase mediates entry of N. gonorrhoeae into nonphagocytic cells. *Cell* **91:** 605–615.
94. Kryukov, G.V., L.A. Pennacchio & S.R. Sunyaev. 2007. Most rare missense alleles are deleterious in humans: implications for complex disease and association studies. *Am. J. Hum. Genet.* **80:** 727–739.
95. Sabeti, P.C. *et al.* 2006. Positive natural selection in the human lineage. *Science* **312:** 1614–1620.
96. Williamson, S.H. *et al.* 2007. Localizing recent adaptive evolution in the human genome. *PLoS Genet.* **3:** e90.
97. Wang, E.T. *et al.* 2006. Global landscape of recent inferred Darwinian selection for Homo sapiens. *Proc. Natl. Acad. Sci. USA* **103:** 135–140.
98. Barreiro, L.B. & L. Quintana-Murci. From evolutionary genetics to human immunology: how selection shapes host defence genes. *Nat. Rev. Genet.* **11:** 17–30.
99. Allison, A.C. 2009. Genetic control of resistance to human malaria. *Curr. Opin. Immunol.* **21:** 499–505.
100. Min-Oo, G. & P. Gros. 2005. Erythrocyte variants and the nature of their malaria protective effect. *Cell Microbiol.* **7:** 753–763.
101. Louicharoen, C. *et al.* 2009. Positively selected G6PD-Mahidol mutation reduces Plasmodium vivax density in Southeast Asians. *Science* **326:** 1546–1549.
102. Mackinnon, M.J. *et al.* 2005. Heritability of malaria in Africa. *PLoS Med.* **2:** e340.
103. Jallow, M. *et al.* 2009. Genome-wide and fine-resolution association analysis of malaria in West Africa. *Nat Genet.* **41:** 657–665.
104. Allison, A.C. 2002. The discovery of resistance to malaria of sickle-cell heterozygotes. *Biochem. Mol. Biol. Educ.* **30:** 279–287.
105. Bongfen, S.E. *et al.* 2009. Genetic and genomic analyses of host-pathogen interactions in malaria. *Trends Parasitol.* **25:** 417–422.
106. Casanova, J.L. & L. Abel. 2009. Revisiting Crohn's disease as a primary immunodeficiency of macrophages. *J. Exp. Med.* **206:** 1839–1843.
107. Taubenberger, J.K. 2006. The origin and virulence of the 1918 "Spanish" influenza virus. *Proc. Am. Philos. Soc.* **150:** 86–112.
108. World Health Organization. 1978. Ebola haemorrhagic fever in Zaire, 1976. *Bull. World Health Organ.* **56:** 271–293.
109. Dupuis, S. *et al.* 2001. Impairment of mycobacterial but not viral immunity by a germline human STAT1 mutation. *Science* **293:** 300–303.
110. Jouanguy, E. *et al.* 1999. A human IFNGR1 small deletion hotspot associated with dominant susceptibility to mycobacterial infection. *Nature Genet.* **21:** 370–378.
111. Ng, S.B. *et al.* 2010. Exome sequencing identifies the cause of a Mendelian disorder. *Nat. Genet.* **42:** 30–35.
112. Ng, S.B. *et al.* 2009. Targeted capture and massively parallel sequencing of 12 human exomes. *Nature* **461:** 272–276.
113. Byun, M. *et al.* 2010. Whole-exome sequencing-based discovery of STIM1 deficiency in a child with fatal classic Kaposi sarcoma. *J. Exp. Med.* **207:** 2307–2312.
114. Stead, W.W. 1992. Genetics and resistance to tuberculosis. *Ann. Int. Med.* **116:** 937–941.
115. Fraser, D.W. 1972. Tetanus in the United States, 1900–1969. Analysis by cohorts. *Am. J. Epidemiol.* **96:** 306–312.
116. U.S. Department of Health, Education, & Welfare, National Office of Vital Statistics. 1956. Death rates by age, race, and sex, United States, 1900–1953. *Vital Statistics-Special Reports* **43:** 1–493.

Ann. N.Y. Acad. Sci. ISSN 0077-8923

ANNALS OF THE NEW YORK ACADEMY OF SCIENCES
Issue: *The Year in Human and Medical Genetics*

Disruption of long-distance highly conserved noncoding elements in neurocristopathies

Jeanne Amiel, Sabina Benko, Christopher T. Gordon, and Stanislas Lyonnet
Department of Genetics, University Paris Descartes and INSERM U-781, Necker-Enfants Malades APHP, Paris, France

Address for correspondence: Stanislas Lyonnet, Department of Genetics, University Paris Descartes and INSERM U-781, Hôpital Necker-Enfants Malades, 149, rue de Sèvres, 75743 Paris cedex 15, France. stanislas.lyonnet@inserm.fr

One of the key discoveries of vertebrate genome sequencing projects has been the identification of highly conserved noncoding elements (CNEs). Some characteristics of CNEs include their high frequency in mammalian genomes, their potential regulatory role in gene expression, and their enrichment in gene deserts nearby master developmental genes. The abnormal development of neural crest cells (NCCs) leads to a broad spectrum of congenital malformation(s), termed neurocristopathies, and/or tumor predisposition. Here we review recent findings that disruptions of CNEs, within or at long distance from the coding sequences of key genes involved in NCC development, result in neurocristopathies via the alteration of tissue- or stage-specific long-distance regulation of gene expression. While most studies on human genetic disorders have focused on protein-coding sequences, these examples suggest that investigation of genomic alterations of CNEs will provide a broader understanding of the molecular etiology of both rare and common human congenital malformations.

Keywords: malformation; genetics; neural crest; evolution; regulation

Introduction

One of the key discoveries of vertebrate genome sequencing projects has been the identification of highly conserved noncoding elements (CNEs). CNEs are predicted to occur at high frequency (3% when comparing human and mouse genomes, according to the ENCODE pilot project[1]). In particular, CNEs are enriched in gene deserts near key developmental genes.[2] A proportion of CNEs are potentially regulator elements (e.g., enhancers or silencers) influencing spatiotemporal gene expression during development.[3] While the maximum distance over which long-range regulatory elements can act is unknown, enhancers can be located over 1 Mb away from a gene.[4]

Historically, mutations that contribute to human malformation have been sought mostly in the coding regions of the genome (exons), whereas functional genomic elements independent of coding sequences and untranslated regions remain poorly investigated. However, a number of human developmental anomalies resulting from disruption of long-range regulatory CNEs have been reported, suggesting a potentially important role in normal growth, development, and disease.[5,6]

As a proof of principle, a variation in a CNE located in an intron of the *RET* proto-oncogene was recently identified as a frequent hypomorphic allele in Hirschsprung disease,[7] and deletions and a point mutation have been discovered in regulatory long-distance CNEs surrounding the *SOX9* gene in patients with Pierre Robin sequence (PRS), an endophenotype of a much more complex and lethal syndrome (campomelic dysplasia).[8] Both disorders feature among neurocristopathies and lead to the postulate that other congenital defects of neural crest-derived cells could be ascribed to mutations in long-distance regulatory CNEs. In this review, we summarize these findings and elaborate on their possible significance for other Mendelian disorders.

Defining conserved noncoding elements

The comparison of multiple genome sequences, now available across mammals and other species,

doi: 10.1111/j.1749-6632.2010.05878.x

revealed that in very distant species, both the number of genes and their coding regions were rather similar. Thus, great differences between distant species cannot be attributed to variations within coding sequences only. Conversely, the obvious difference between the simplest and the most complex organisms is the amount of noncoding sequence: the more complex the species, the greater the proportion of noncoding DNA, reaching about 98% in the human genome.[9] However, thus far, while the molecular bases of a great number of genetic disorders remain unknown, noncoding DNA has not attracted so much attention as a possible source of genome alteration in Mendelian disorders. Obviously, the limitations are not only technical, but also conceptual, since it is difficult to search for mutations in vast DNA domains without clues that might indicate their function. This is why conservation of sequences across species became the most obvious way to postulate that noncoding DNA might be functional, based on the assumption that evolutionary constraint for selection impacts on a DNA sequence, preventing genetic drift. Cross-species sequence comparisons would thereby potentially highlight noncoding DNA regions of putative biological importance (reviewed in Ref. 10). Those highly conserved noncoding elements (also reviewed in another chapter of the present volume) might be of great interest in genetic medicine, because, while their conservation might indicate function, their mutation might in turn be relevant to diseases in humans.

Screening highly conserved noncoding sequences for mutations is partly a blind approach, bypassing our ignorance of the function of those DNA segments, with the simple assumption that DNA variation located within them could impair their unknown biological properties. It goes without saying that, as for Mendelian disorders with high penetrance of mutant alleles, the usual criteria for significance of a DNA variant should also apply, such as (i) familial segregation according to the presumed mode of inheritance, (ii) *de novo* occurrence in a patient, and (iii) absence in controls from databases featuring single nucleotide polymorphisms. Further criteria for validation of a DNA variant occurring in a CNE would then involve both computational and biological investigations aimed at elucidating the function of the conserved sequence.

Comparative genome analysis, using mostly human and mouse genomes, demonstrated that evolution shapes the conservation of about 5% of the human genome, of which about 2% would contain protein-coding regions (exons and UTR sequences) and noncoding RNAs, whereas 3% of our genome would consist of conserved noncoding sequences.[11,12] Today, no particular nomenclature or anthology of those CNEs is defined, given that they can only be distinguished on a 2D rationale: length of the conserved sequences and depth of the conservation, that is, the evolutionary distance between the compared species. It is important to note that these CNEs do not define a family of sequences featuring particular motifs of primary DNA sequence. However, a family of ultraconserved elements has been defined according to criteria of 100% homology between human, mouse, and rat, over a distance greater than 200 bp;[13] for some of these elements, an enhancer (or possibly silencer) activity was demonstrated in reporter assays in embryonic mice.[3] Subsequently, a larger compendium of mammalian ultraconserved elements was identified.[14] The stringent definitions in these studies do not preclude the relevance of a more limited depth of conservation. A generally admitted, although disputable, threshold for qualification as a CNE would be sequences >100 bp with >70% conservation between human and mouse. In the case of regions of noncoding DNA with primate-specific functions, the disruption of which may result in behavioral or cognitive phenotypes in humans, the utility of cross-species comparisons as a means of highlighting functional elements may be limited, given the reduced evolutionary distance among primate genomes. The depth of conservation and the species included in the comparison may need tailoring for each genomic interval and phenotype studied in clinical genetics.

The functions proposed for CNEs fall into several categories. First, they may be regulatory elements involved in binding transcription factors, acting in *cis* with respect to the transcribed sequences. In this category, CNEs could activate or inhibit tissue-specific transcription, insulate regulatory regions, or participate within locus control regions involved in the concerted transcriptional regulation of a group of genes.[5] Such *cis*-regulatory CNEs could be localized at a very long distance to the coding sequence of their target gene, whose expression they regulate.

This raises the great difficulty of determining which gene is the target of a regulatory element. In the second category, CNEs may act as chromatin organizers. Indeed, about 8% of highly conserved CNEs (as described by Woolfe[2]) were found to bind SUZ12 (a core component of polycomb repressive complexes) in human ES cells, highlighting the role of CNEs in chromatin modification.[15] Third, CNEs could function as genomic regulatory blocks (GRB) dispersed throughout the genome in nongenic regions, gene-dense regions, or blocks of grouped genes. Finally, CNEs may function as regulatory elements acting in trans.[16]

Other hypotheses regarding the function of CNEs could be considered on theoretical grounds, including participation in the transcription or regulation of transcription of noncoding long interspersed transcripts, influence on the timing and origin of DNA replication, involvement in a higher order genome organization of chromosomes as far as their respective nuclear localization is concerned, and participation in the epigenetic phenomenon known as transvection, which results from an interaction between homologous alleles (reviewed in Ref. 17).

Impact on malformation of the neural crest

Neural crest cells

Neural crest cells (NCCs) form in the human embryo during the third to fifth weeks of pregnancy, within the neural folds that delineate the neural plate from the ectoderm. During the fusion of the neural folds, which ultimately yields a tube that will become the central nervous system, NCCs detach and become mesenchymal. They migrate throughout the body, integrating into nearly every organ.

NCC derivatives include the neurons and support cells of the entire peripheral nervous system (sensory and autonomic), adrenergic, and other endocrine cells, and all pigment cells except those arising from the retina (reviewed by Ref. 18). In the head, in addition to the cell types mentioned above, NCCs differentiate into connective and structural tissues such as dermis,[19] bones, and cartilage of most of the skull,[20] and muscle tendons.[21] They also infiltrate and are essential for the function of glandular and vascular elements such as the thymus, the thyroid and parathyroid glands, the conotruncal region of the heart, and the entire branchial vascular sector,[21–23] giving rise to connective, adipose, and smooth muscle cells.

The ultimate choice in phenotype made at a given site of differentiation is the result of a combination of extrinsic factors in the embryonic microenvironment and cell-intrinsic properties that modify the responsiveness of NCCs to these external influences. Both their migration pathways and fate are imposed on NCCs by surrounding tissues as they leave the neural primordium; these are less dependent on intrinsic properties regionally distributed along the neuraxis than had initially been presumed (reviewed in Ref. 18). For instance, truncal NCC transplanted at the vagal level colonize the gut and differentiate into enteric ganglia in which neurons synthesize acetylcholine rather than catecholamines, as they would have done normally. On the other hand, the cephalic NCC population contributing to the branchial arch-derived facial skeleton has some intrinsic positional information and commitment,[24] while the rostral endoderm imparts signals to the adjacent NCCs and thereby patterns craniofacial cartilages;[25,26] these studies highlight the importance of cross-talk between NCCs and their environment.

Definition of a neurocristopathy

Abnormal migration, differentiation, division, or survival of NCCs leads to highly diverse clinical and pathological features. Referring to their proposed common embryological origin, Bolande first introduced the concept of "neurocristopathy" in order to highlight the potential for shared pathogenetic mechanisms in the fields of both congenital malformations and tumor predisposition.[27]

Classification as a neurocristopathy has evolved from a "purist" definition to a larger view, including disorders involving tissues composed of both NCC and non-NCC-derived cell types, such as DiGeorge syndrome.[28] Additionally, endocrine, melanocytic and PNS tumors arising from NCC cell types have been described; these can be isolated or associated with anomalies of other derivatives of the neural crest (i.e., Multiple endocrine neoplasia type 2 A and B) or non-NCC cell types (i.e., Congenital central hypoventilation syndrome).

CNE variation within a gene: Hirschsprung disease

Hirschsprung disease (HSCR), or aganglionic megacolon, is a congenital malformation characterized by the absence of enteric ganglia, which are NCC derived, along a variable length of the intestine. HSCR has served as a paradigm for dissectible diseases with complex inheritance since the 1980s. Indeed, it is frequent (1/5,000 live births), is diagnosed mostly in the neonatal period by objective histological criteria, and heritability is high (200).[29] Expressivity is an objective criterion that depends on the length of the aganglionic segment and is classified into short-segment (SS), long-segment (LS), and total colonic aganglionosis (TCA) forms. Among sporadic cases (80% of the cases), SS-HSCR is the most frequent (80%), and the sex bias is high (5.5/1) in favor of females. Conversely, familial cases are less frequent (20%), with both vertical and horizontal transmission, and the more severe the disease (LS and TCA), the more even the sex ratio. Therefore, the mode of inheritance is complex and sex dependant with a recurrence risk in sibs of 4% but with great variability (from 1 to 33%) when considering the gender, segment length, and familiarity of the index case. Indeed, according to Carter's paradox, the less frequent the presentation (i.e., a female with LS-HSCR), the higher the recurrence risk in sibs (reviewed in Ref. 29). Epidemiological studies favored an autosomal dominant, sex-dependant mode of inheritance in LS-HSCR, while autosomal dominant and autosomal recessive modes of inheritance were equally likely in SS-HSCR.[30]

The major disease-causing gene is *RET*, which encodes a tyrosine kinase receptor.[31] HSCR is caused by *RET* loss-of-function, due to large deletions, indels, frameshift, nonsense, and missense mutations scattered along the coding sequence of *RET*, with no mutational hot spot. A low-coding sequence mutation detection rate became rapidly obvious, because *RET* mutations were only identified in 50% of familial cases and 20% of sporadic cases.[32] Two sets of experiments were then performed: linkage analysis in vertical familial cases with at least three generations and transmission disequilibrium test (TDT) in familial cases with recurrence in sibs. Modifier loci were mapped at 9q31, 3p21, and 19q12.[33,34] Importantly, both experiments also showed that the impact of a genetic event at the *RET* locus was underestimated. Indeed, the vast majority of vertical and horizontal familial cases were compatible with linkage to *RET*, and almost all affected sibs shared at least one identical allele by descent at *RET*.

During the same period, overtransmission of a silent polymorphism encoded by *RET* exon 2 in sporadic cases had been observed by several groups (rs1800858, c.135G > A; p.A45A, reviewed in Ref. 29). Interestingly, the frequency of the mutant allele varies with the incidence of the disease in various ethnic groups; it is rare in Africa (about 1%), frequent in Europe (24%), and reaches 45% in Asia. The A allele falls within a 30 kb haplotype lying between the *RET* promoter and IVS5. The first intron of the *RET* gene is 23 kb and remarkably poor in regions conserved when comparing human and chick, with a unique conserved sequence of 250 bp at 9.7 kb downstream of the transcription start site, denoted MCS+9.7[7] (Fig. 1). A SNP (rs2435357) lies in the conserved sequence and the maximum transmission disequilibrium is found for the mutant T allele in HSCR patients.[7] In transgenic reporter mice, the sequence recapitulates the pattern of *RET* expression and can therefore be regarded as a target for binding tissue-specific protein(s).[35]

Recently, it has been demonstrated that MCS+9.7 binds SOX10,[36] a key transcription factor in NCC development that leads to Waardenburg syndrome (WS) type IV and II when mutated (reviewed in Ref. 37). Moreover, MCS+9.7 is the most powerful enhancer of reporter gene expression among all conserved sequences at the *RET* locus tested in luciferase reporter assays in a neuroblastoma (NB) cell line.[35]

Interestingly, two polymorphic SNPs (rs2435357 and rs2506004) in almost complete disequilibrium are found on the enhancer element. The dissection of the respective roles of each SNP is a difficult task (see Section 4). However, one or the combination of both mutant alleles is a low-penetrant predisposing allele to HSCR (hypomorphic allele) and the most frequent molecular event at the *RET* locus in cases presenting the common form of the disease, that is, males with nonfamilial SS-HSCR.[7,36]

HSCR is observed in combination with other congenital malformation(s) in 20% of neonates. Some are poorly defined associations and some may occur by chance, but others are well-characterized syndromes with Mendelian inheritance and with a growing number of known disease-causing genes.[29] Interestingly, the penetrance of the HSCR trait is extremely variable, ranging from 3–5% (Down syndrome) to over 80% (WS IV with *SOX10* mutation). These observations point to a putative role

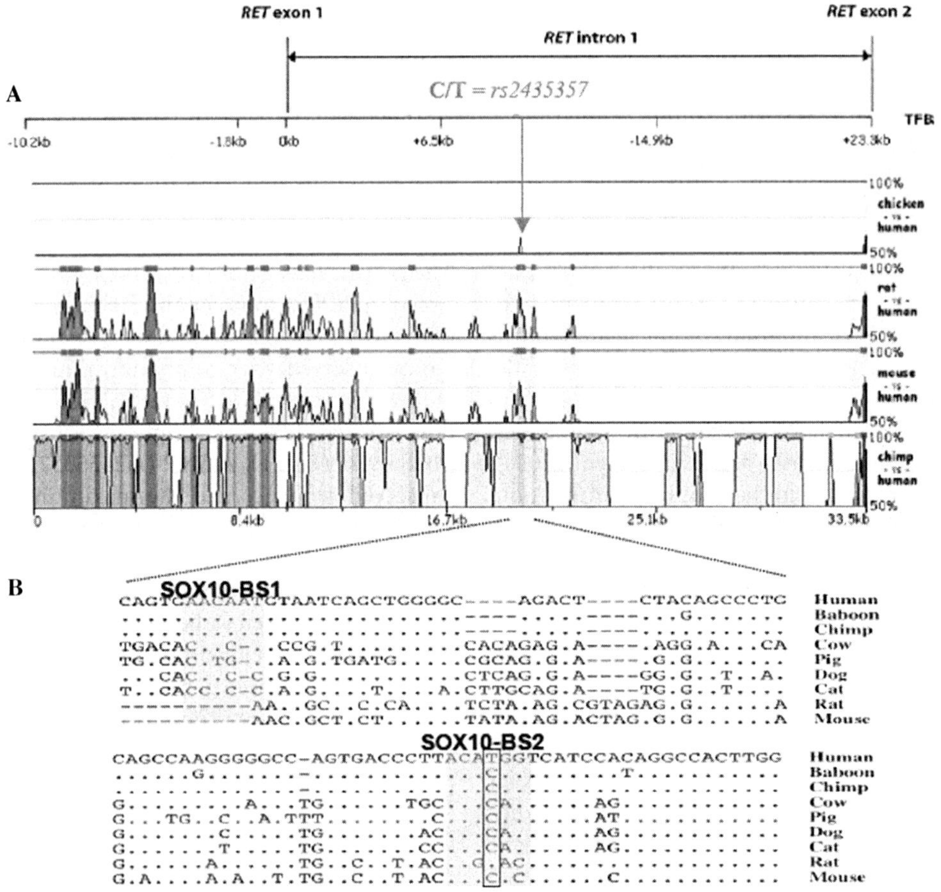

Figure 1. Overview of the sequence conservation within the first intron of *RET* (A), and the position of an HSCR-associated variant within the intronic CNE, MCS+9.7 (B). In A, graphs depicting conservation between human and other vertebrates were obtained from the ECR browser (http://ecrbrowser.dcode.org/). The position of MCS+9.7 is indicated by a red arrow. In B, a portion of the MCS+9.7 sequence is aligned with the corresponding region in other mammals, and predicted SOX10 binding sites (SOX10-BS1 and SOX10-BS2) identified by Emison *et al.*[36] are highlighted in yellow. The HSCR-associated C/T variant (boxed) falls within SOX10-BS2. Reproduced from Emison *et al.*[36]

of modifier loci for the HSCR traits for each syndrome. An obvious candidate is *RET*, which can be easily tested by genotyping the hypomorphic allele on IVS1 in all predisposing syndromes for which the gene (or chromosome) is known and a large enough series of affected patients with or without HSCR can be gathered. Thus far, such analysis had been carried out for Down, congenital central hypoventilation, Bardet-Biedl, Mowat-Wilson, and Waardenburg IV syndromes. The latter two were classified as RET-independent as the frequency of the hypomorphic allele was similar in the groups with and without HSCR for the same syndrome while the three others were found RET-dependent due to a significantly greater frequency of the hypomorphic allele in the group with HSCR when compared to the group without.[38,39] Interestingly, Bardet-Biedl syndrome being *RET*-dependent for the HSCR trait points to the role of cilia and the SHH signaling pathway for the normal development of the enteric nervous system.[40] Finally, these findings illustrate how the same gene can be both a major and a modifier gene depending on the genome.

A 4/1 biased sex ratio in isolated HSCR is suggestive of an X-linked modifier gene. However, neither linkage analyses nor TDT and case-control studies have pointed to the X chromosome. The only X-linked syndrome predisposing to HSCR is

MASA, which is due to mutation in the *L1CAM* gene. No *L1CAM* gene mutation was found in isolated HSCR.[41] Along the same lines, we genotyped seven SNPs with an informativity above 20% scattered along the *L1CAM* gene and found similar frequencies in 90 males with HSCR compared to controls (personal unpublished data). Interestingly, the upregulation of L1cam by Sox10 has been shown *in vitro* and loss of *L1cam* increases the incidence of aganglionosis of $Sox10^{+/lacZ}$ male mice.[42] A role for *RET* itself has been proposed based on the observation that among sib pairs the shared allele was significantly more often maternal.[7,34] However, the International HSCR Consortium data showed that 87% of the *RET* alleles in HSCR patients encode a loss-of-function protein when pooling the hypomorphic *RET* allele and coding sequence mutation alleles and therefore favoring an autosomal recessive mode of inheritance.[36] Interestingly, a biased sex ratio is also observed in *RET*-independent autosomal syndromes predisposing to HSCR.[39] This argues for genetic factors unlinked to *RET*. A possible explanation to the missing X-linked locus in HSCR could be that predisposing allele(s) have a frequency sufficient for homozygote mothers to be masking the locus. Indeed, a predisposing allele with a frequency of 70% would result in 50% of the females being homozygote and 70% of the males being hemizygotes. Interestingly, the penetrance of a *RET* CDS mutation is indeed of 70% in males and 50% in females. Altogether, the sex bias observed in HSCR remains largely unexplained at present.

Pierre Robin sequence

Neurocristopathies affecting the cardio-cephalic pole often leave their most visible mark on the face, since most facial tissues (bone, cartilage, teeth, vascular walls, and dermis) are direct NCC derivatives.

The Pierre Robin anomaly is regarded as a sequence since it combines three features that might result one from another as a cascade: micrognathia (mandibular hypoplasia) or retrognathia, leading to glossoptosis, itself preventing midline fusion of the palatal shelves and leading to a posterior U-shaped cleft palate (reviewed in Ref. 43). Pierre Robin sequence patients also feature functional anomalies mostly corresponding to hindbrain dysfunction: central apneas, sucking and swallowing defects, esophageal reflux, and cardiac rhythm dysregulation.[44] There is still an open debate on the embryological bases of PRS, that is, whether it can be regarded as a craniofacial neurocristopathy resulting from an abnormal development of the first branchial arch, or whether originating in an abnormal patterning of the hindbrain.

Based on a large series of 110 consecutive cases (1987–1996), it was apparent that PRS patients could present with other congenital malformation(s) in 52% of cases or could have isolated PRS (48%).[45] Among those cases with associated anomalies, several syndromes are recognized including a number of chondrodysplasias, branchial arch anomalies, neuromuscular disorders, or embryofoetopathies. Interestingly, in the chondrodysplasia group, the most frequently observed syndromes are Stickler and related syndromes (ascribed to anomalies of the genes encoding collagens 2A1, 11A1, and 11A2) and campomelic dysplasia, ascribed to coding sequence mutation of the *SOX9* gene on chromosome 17q24 (MIM114290).

When isolated, PRS cases are mostly sporadic. However, 13% are familial with a mode of inheritance that is most likely autosomal dominant with incomplete penetrance and variable expression.[45] Based on such familial cases, a PRS locus was mapped by a traditional linkage approach to chromosome 17q24.2-q25.1 in a large four-generation PRS family and later confirmed in smaller families.[8] Interestingly enough, that mapping region centered on the *SOX9* locus in the middle of a vast gene-desert, encompassing about 2.5 Mb of DNA (Fig. 2). However, no coding sequence mutation of the *SOX9* gene (nor of the flanking genes) could be detected in PRS cases linked to 17q24. Several observations of reciprocal translocations, for which one of the breakpoints in each case maps to 17q24, have been reported in families segregating PRS as an isolated trait.[8,46] Remarkably, all translocations clustered in a small DNA region, 1.2Mb upstream from the *SOX9* coding sequence. These data suggested that a genomic alteration responsible for the PRS phenotype might not impair coding sequence, but, rather, noncoding DNA. Comparative genomic alignments had previously indicated that the *SOX9* gene-desert is packed with CNEs conserved between humans and other mammals and, in some cases, between mammals and fish.[47] Comparative genomic hybridization and sequencing of candidate elements resulted in the discovery of several deletions and a point mutation involving or located within CNEs in

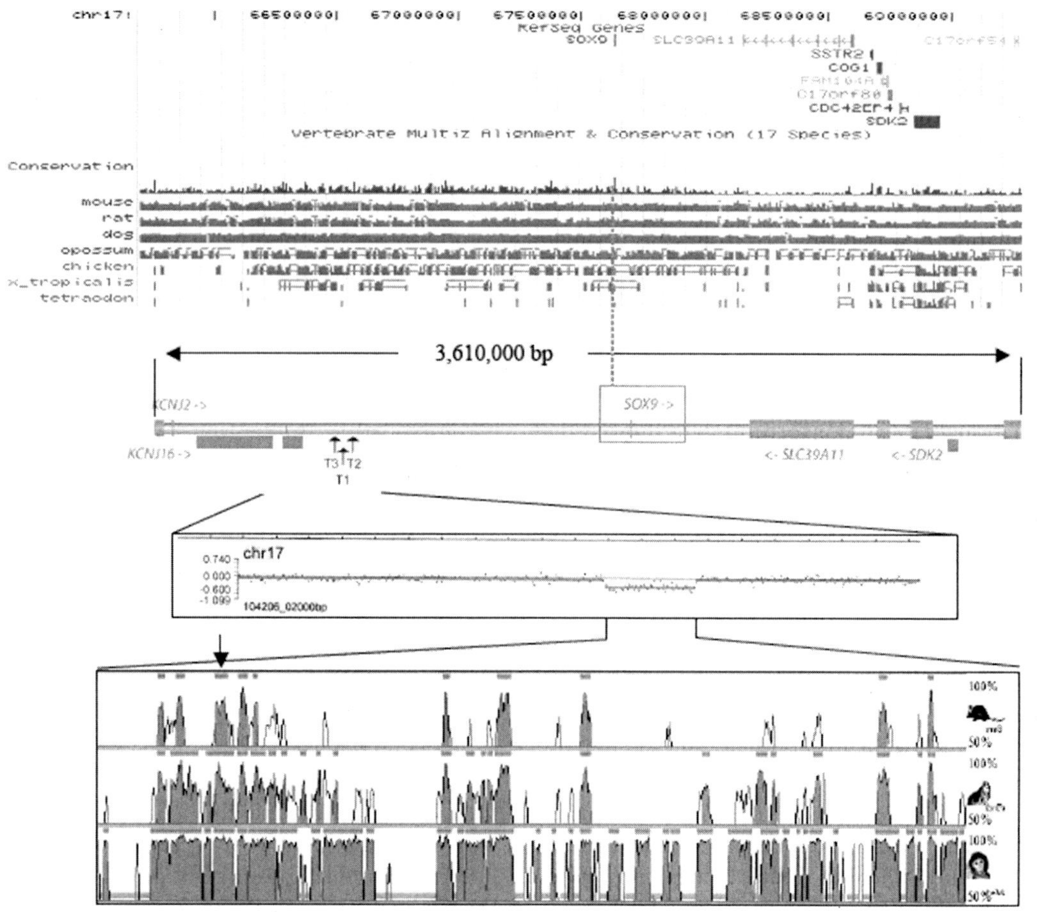

Figure 2. Distribution of CNEs within the gene desert surrounding *SOX9* and noncoding lesions identified in isolated PRS patients. The Multiz alignment was obtained from the UCSC genome browser (http://genome.ucsc.edu/). Beneath the alignment, the position of translocation breakpoints identified in Benko *et al.*[8] and Jakobsen *et al.*[46] are indicated by T1–3, and deletions identified in Benko *et al.*[8] are indicated with red boxes. In the lower half of the figure, the CGH data for one of the familial deletions is depicted, as is the distribution of conserved elements within the deleted region. A black arrow indicates the position of a point mutation identified within a CNE in a family with isolated PRS. Reproduced from Benko *et al.*[8]

sporadic and familial PRS cases mapping to 17q24 (Fig. 2), with either inherited or *de novo* occurrence of those DNA variations.[8] The wild-type versions of some of these CNEs, located at a long distance from the *SOX9* coding sequence, were able to drive reporter gene expression in the branchial arches of transgenic mice, suggesting they can indeed function as enhancers. The CNE sequence variation identified in a PRS family affected the binding of a transcription factor important for craniofacial development, MSX1. While the genetic data involving alteration of long-distance CNEs in PRS are overwhelming, the difficult question to address is the identity of the gene whose expression is impaired by such noncoding genomic alterations located deep in the middle of a gene-desert. Several lines of evidence strongly point toward *SOX9*, although definite proof may require further 3-C experiments or the generation of a mouse model whereby a targeted CNE deletion or mutation segregates with a PRS-related phenotype. There are several other arguments that support *SOX9* as the target of the enhancer(s) disrupted in the genomic lesions reported in.[8] The *SOX9* gene is conserved in all gnathostomes, and *SOX9* expression is conserved in the branchial arches of several animal models (and in the human embryo). *Sox9* is involved in early NCC development and in mandibular chondrogenesis,[48–50] while cleft palate occurs in mice heterozygous null for *Sox9* and in mice with a conditional knockout of *Sox9* in NCCs.[48,49] The *SOX9* coding sequence is mutated in campomelic dysplasia, a frequently lethal syndrome combining bone dysplasia, sex reversal, and PRS in more than 80% of cases.[51,52] SOX9 functions as a transcription factor, activating the expression of the *COL2A1* gene, whose mutations result in a group of syndromes featuring PRS as an endophenotype.[53,54] Finally, the conformation of chromatin across a broad domain encompassing the *SOX9* locus, as shown by interphase FISH experiments,[8] is consistent with long-range interactions between the PRS locus and *SOX9*, and ChIP analysis in mouse cell-lines derived from the mandibular arch demonstrated the association of the PRS locus with markers consistent with a regulatory function of the locus.[8]

It was thus proposed that the long-distance genomic alteration in PRS patients may represent a site- and stage-specific loss of transcription, resulting in a restricted reduction of *SOX9* expression in human cranial NCCs and/or mandibular arch (reviewed in Ref. 55). This fits well with the tissue-specific regulatory nature of the CNEs within the *SOX9* gene-desert. Finally, these findings may be more broadly applicable, whereby the disruption of highly conserved sequences might explain the effect of genomic alteration in *cis* at long distances up or downstream of disease gene loci.

Other examples

A number of diseases or malformations have been ascribed to genomic alterations, located in noncoding DNA, sometimes at a very long distance from gene coding sequences. These conditions have been reviewed elsewhere,[6] including in the present volume, and the most relevant examples are listed in Table 1. It is remarkable that a number of genomic alterations involving CNEs are duplications; this could be an emerging theme in the molecular investigation of congenital malformations. Indeed, duplications may lead to a tissue-specific modification in gene dosage because of an increase in the number of enhancers or silencers lying in the duplicated segment. Duplications could also result in a general disruption of the chromatin conformation.

Position effects

Correct spatiotemporal expression of developmental genes requires the concerted action of a vast number of regulatory elements, located in *cis*, upstream or downstream of target gene coding sequences. These elements could be schematically involved in two processes: (a) direct control of gene expression via the binding of transcription factors and thereby recognized as silencers, enhancers, or promoter regions; or (b) higher order regulation of genome structure, such as chromatin organization.

Indeed, a number of possible mechanisms have been proposed to explain position effects of chromosomal rearrangements at considerable distance from the coding sequence of a known locus, including separation of the gene from an enhancer or promoter, juxtaposition with a heterologous enhancer in a novel chromosomal environment, removal of a long-range insulator, competition by a heterologous gene for an enhancer that would normally interact with the disease gene, and position-effect variegation resulting from spread of heterochromatic DNA.[6,56]

Implication for gene identification and mechanisms of disease

It is likely that the long-range control of gene expression is a concordant feature of a limited number of genes involved in multiple tissues at several stages of development. These genes, many of which may be transcription factors, are predicted to feature a highly complex regulation of their expression pattern, both spatially and temporally.

Implication for Mendelian disorders: a model

Given that the location of CNEs at a distance from the gene coding sequence challenges the definition of a gene, the domain of DNA to investigate in Mendelian disorders might be much broader than traditionally investigated and difficult to limit, possibly extending beyond neighboring gene(s). Since master developmental genes tend to lie in gene-deserts enriched in CNEs,[2] it is tempting to hypothesize that these are candidate regions for high-throughput DNA screening in congenital malformations in humans. Importantly, the PRS model may suggest that the alteration of long-distance noncoding sequences might result in endophenotypes, each of which would depend on the loss of tissue-specific expression that is driven by the cognate CNE, and the sum of which would recapitulate the complete syndrome phenotype ascribed to loss-of-function coding sequence

Table 1. A selection of human disorders ascribed to disruption of long-distance CNEs

Gene locus	Phenotype	CNE mutation	CNE function	Distance to CDS (kb)	Reference
SHH	Preaxial polydactyly	Point mutation/ duplication	Limb-specific enhancer (ZRS)	1,000	4
SHH	Holoprosencephaly	Point mutation	Brain-specific enhancer	460	69
RET	Hirschsprung	Point mutation	Intragenic enhancer	9.7	7
IRF6	Cleft lip	Point mutation	Enhancer	14	67
BMP2	Brachydactyly A2	Duplication	Enhancer	110	70
SOX9	Pierre Robin	Deletion/point mutation	Enhancer	up to 1,450	8
SOX9	Brachydactyly-anonychia	Duplication	?	~1,200	71
FOXL2	BPES	Deletion	Enhancer	280	72, 73
POU3F4	Deafness	Deletion	Enhancer	900	74
PAX6	Aniridia	Translocation	Enhancer	~150	75
SHOX	Leri-Weill dyschondrosteosis	Deletion	Enhancer	~250	58

Note: ZRS, zone of polarizing activity regulatory sequence; BPES, blepharophimosis ptosis epicanthus inversus syndrome. A question mark indicates uncertainty regarding the identity of the regulatory element disrupted.

mutations (Fig. 3). Such a model could be tested for major developmental genes for which a syndrome is ascribed to coding sequence mutations. The disruption of CNEs in human disorders could occur by many processes, including mutation, deletion/duplication and translocation, and should encourage fewer "coding-centric" studies.

Experimental attempts to map and characterize functional regulatory elements

A number of techniques are available to identify and map active *cis* regulatory elements, such as DNase hypersensitive mapping and ChIP experiments for the detection of protein-DNA interaction and histone modification. Besides those approaches, the regulatory potential of conserved sequences can be tested by *lacZ* reporter assays in transgenic animals such as zebrafish or mouse.[2,3] In addition to these traditional reporter models, transient transfection of enhancer-reporter constructs in chick embryos has enabled rapid screening of putative enhancers within a genomic region of interest, such as those at the *SOX2* and *SHOX* loci.[57,58] Indeed, the ease of electroporation of neural crest in the chick makes this cell type particularly suitable for enhancer analysis, as demonstrated recently for a novel *SOX10* enhancer.[59] Targeted disruption or mutation of the CNE of interest can be performed in various animal models although redundancy of regulatory elements/protein binding sites may dilute phenotypes. 3C and derived techniques can also be appropriate given that relevant tissue/cell lines are available.

Genes influenced by noncoding genomic alterations

The models for normal long-range regulation by *cis* regulatory elements, modulating the expression of gene coding sequences, are numerous. Several proposals have been made, including a looping model and a facilitated tracking model.[5] Today, these models are difficult to address and distinguish experimentally, and could require innovative technologies. Indeed, the question of which gene (or genes) is influenced by the disruption of noncoding sequences remains challenging. In other words, while the proof of the importance of the genomic disruption in noncoding regions of the genome might be overwhelming (segregation of the lesion with affected individuals, *de novo* occurrence, absence from controls or CNV or SNP databases, sequence conservation across evolution, *in vitro* and *in vivo* demonstration of reporter expression consistent with regulatory element function), the identity of the gene or genes with which those CNEs interact might often remain elusive, especially as the regulatory elements can often be located more than 1 Mb upstream or

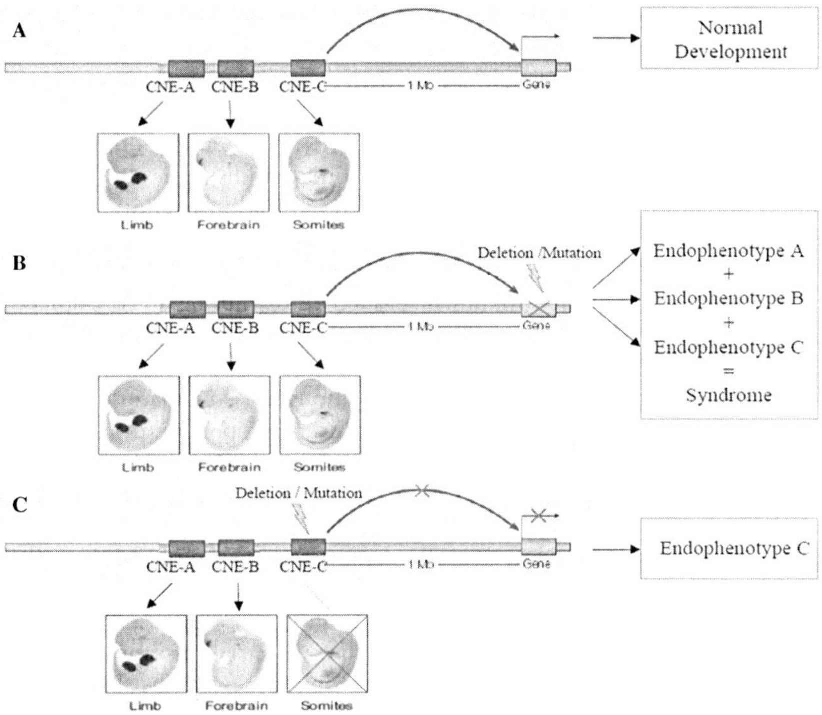

Figure 3. A model predicting the existence of endophenotypes of human syndromes caused by mutations in CNEs. Expression of key developmental genes requires the activity of tissue- and stage-specific CNEs located at a long distance from the transcription start site (A). Three examples of tissue-specific enhancers driving *lacZ* reporter expression in transgenic mice are illustrated. Mutation within the coding sequence of the gene causes a syndrome composed of several tissue-specific endophenotypes (B). A lesion disrupting a single CNE may result in an endophenotype caused by loss of expression of the gene in an isolated tissue (C). Images of transgenic mouse embryos are reproduced from Pennacchio *et al.*[3]

downstream of the gene coding sequences. In addition to the types of evidence listed above, other arguments may add weight to the likelihood that a candidate regulatory element targets a specific gene. The information to be considered could include the expression pattern and function of the candidate target gene, with respect to the disease phenotype; the identity of the transcription factors interacting with the mutated/deleted noncoding sequence, their relevance to the developmental events that are predicted to be disturbed in the disorder, and their hierarchical relationship to the candidate target gene; the general organization of the genomic regulatory blocs with respect to the gene coding sequence; and the general chromatin context.

Perhaps the best method for determining whether a particular gene is the relevant target of an enhancer is via targeted disruption of that enhancer in mice. Subsequently, expression analysis of the genes surrounding the deletion would potentially indicate the relevant in vivo target. Essential functions for several tissue-specific enhancers have been demonstrated by targeted deletion, including those at the *Emx2* and *Shh* loci.[60,61] In a recent example, a knockout of a noncoding interval, known to contain SNPs associated with coronary heart disease in humans, resulted in defects in vascular cells and affected cardiac expression of genes neighboring the interval; although discrete enhancers were not identified, this study provided strong evidence for the *cis*-regulatory influence of the noncoding region.[62] On the other hand, targeted deletion of predicted enhancers can result in mice with no obvious abnormalities.[63] This may be an indication of enhancer redundancy, or may be due to limitations of phenotypic screening. In contrast to reverse genetics techniques in mice, the identification of genomic alterations in noncoding DNA in humans with cognitive and behavioral phenotypes may be the most sensitive means of identifying elements that regulate gene expression in the mammalian brain.

Regarding the investigation of genomic interactions, a number of high-throughput techniques have been devised to investigate chromatin structure in detail over large genomic domains. In particular, ChIP experiments with microarrays as well as the 3C and derived techniques[64] are aimed at detecting specific patterns of modification that might allow a more global view of the organization of *cis*-regulatory elements across a particular region.

It is clear, however, that several other unknown phenomena might play an important role, which could be more difficult to investigate. Such mechanisms might involve the cross-talk between chromatin structure and noncoding RNA transcribed over large regions of noncoding DNA, the 3D nuclear architecture of chromosomes and gene loci as well as the spatial organization of the chromatin fibers in the interphase nucleus and the location of a gene relative to other chromosome territories.

Conservation of function without conservation of primary DNA sequence

It has been recently shown that some DNA regions might contain regulatory functions, without having a clear degree of conservation across multiple genome species. In particular, using transgenic reporter assays in zebrafish, it was shown that DNA sequences within noncoding regions, and with no conservation of the primary DNA sequence, could be functional.[65,66]

However, conservation of sequences, across the most divergent genomes possible, remains a good standard for deduction of function. A careful analysis of such sequences involves the length of the DNA in question, the depth of conservation and the proportion of consistent nucleotides across species. A number of bio-informatics tools are available to align such sequences. The question of the nomenclature of evolutionary conserved elements is still open.

Implication for common malformations and complex disorders

As suggested by the PRS mode, the alteration of long-distance noncoding sequences might result in a limited endophenotype of a full-blown syndrome phenotype ascribed to loss-of-function coding sequence mutations in a given gene. Importantly, it may open new routes to the molecular dissection of common malformations such as cleft lip and palate, neural tube defects, and congenital cardiac defects. In order to test this hypothesis, large series of patients presenting a single common malformation need to be investigated for alterations in a number of CNEs at master developmental gene loci, keeping in mind that disruption of *cis*-regulatory elements may even lead to phenotypes not observed within the full-blown phenotype caused by CDS mutation. This hypothesis already has a few examples, such as a SNP disrupting an AP2alpha binding site in a *cis* enhancer element of the *IRF6* gene and resulting in a higher risk for cleft lip.[67] Similarly, sequence variations in *cis*-regulator elements may underlie a significant proportion of disease-associated QTLs. One of the best known examples is SNP variants in *cis* regulatory elements of the lactase promoter gene allowing persistent high-level expression of lactase. A more recent example involves allele-specific chromatin remodeling at a locus associated with asthma and autoimmune disease.[68] The vast clinical consequences of variations in spatiotemporal regulation of coding genes are just beginning to emerge and should become as frequent as, or even more frequent than, coding sequence mutations.

Acknowledgments

The authors are supported by the Institut National de la Santé et de la Recherche Médicale, the Association pour la Recherche sur le Cancer, the FRM, and the ANR. We especially thank Drs. David Fitzpatrick, Michel Vekemans, Sophie Thomas, Heather Etchevers, Tania Attié-Bitach, Peter Farlie, and Arnold Munnich for their insight and assistance.

Conflicts of interest

The authors declare no conflicts of interest.

References

1. Birney, E., J.A. Stamatoyannopoulos, A. Dutta, *et al.* 2007. Identification and analysis of functional elements in 1% of the human genome by the ENCODE pilot project. *Nature* **447:** 799–816.
2. Woolfe, A., M. Goodson, D.K. Goode, *et al.* 2005. Highly conserved non-coding sequences are associated with vertebrate development. *PLoS Biol.* **3:** e7.
3. Pennacchio, L.A., N. Ahituv, A.M. Moses, *et al.* 2006. In vivo enhancer analysis of human conserved non-coding sequences. *Nature* **444:** 499–502.
4. Lettice, L.A., S.J. Heaney, L.A. Purdie, *et al.* 2003. A long-range Shh enhancer regulates expression in the developing

limb and fin and is associated with preaxial polydactyly. *Hum. Mol. Genet.* **12:** 1725–1735.
5. Kleinjan, D.A. & V. van Heyningen. 2005. Long-range control of gene expression: emerging mechanisms and disruption in disease. *Am. J. Hum. Genet.* **76:** 8–32.
6. Kleinjan, D.J. & P. Coutinho. 2009. Cis-ruption mechanisms: disruption of cis-regulatory control as a cause of human genetic disease. *Brief Funct. Genomic. Proteomic.* **8:** 317–332.
7. Emison, E.S., A.S. McCallion, C.S. Kashuk, *et al.* 2005. A common sex-dependent mutation in a RET enhancer underlies Hirschsprung disease risk. *Nature* **434:** 857–863.
8. Benko, S., J.A. Fantes, J. Amiel, *et al.* 2009. Highly conserved non-coding elements on either side of SOX9 associated with Pierre Robin sequence. *Nat. Genet.* **41:** 359–364.
9. Taft, R.J., M. Pheasant & J.S. Mattick. 2007. The relationship between non-protein-coding DNA and eukaryotic complexity. *Bioessays* **29:** 288–299.
10. Boffelli, D., M.A. Nobrega & E.M. Rubin. 2004. Comparative genomics at the vertebrate extremes. *Nat. Rev. Genet.* **5:** 456–465.
11. Mattick, J.S. & I.V. Makunin. 2006. Non-coding RNA. *Hum. Mol. Genet.* **15:** R17–R29.
12. Pheasant, M. & J.S. Mattick. 2007. Raising the estimate of functional human sequences. *Genome Res.* **17:** 1245–1253.
13. Bejerano, G., M. Pheasant, I. Makunin, *et al.* 2004. Ultraconserved elements in the human genome. *Science* **304:** 1321–1325.
14. Stephen, S., M. Pheasant, I.V. Makunin & J.S. Mattick. 2008. Large-scale appearance of ultraconserved elements in tetrapod genomes and slowdown of the molecular clock. *Mol. Biol. Evol.* **25:** 402–408.
15. Lee, T.I., R.G. Jenner, L.A. Boyer, *et al.* 2006. Control of developmental regulators by Polycomb in human embryonic stem cells. *Cell* **125:** 301–313.
16. Lomvardas, S., G. Barnea, D.J. Pisapia, *et al.* 2006. Interchromosomal interactions and olfactory receptor choice. *Cell* **126:** 403–413.
17. Duncan, I.W. 2002. Transvection effects in Drosophila. *Annu. Rev. Genet.* **36:** 521–556.
18. Le Douarin, N. & C. Kalcheim. 1999. *The Neural Crest.* Cambridge University Press. Cambridge.
19. Johnston, M.C. 1966. A radioautographic study of the migration and fate of cranial neural crest cells in the chick embryo. *Anat. Rec.* **156:** 143–155.
20. Couly, G.F., P.M. Coltey & N.M. Le Douarin. 1993. The triple origin of skull in higher vertebrates: a study in quail-chick chimeras. *Development* **117:** 409–429.
21. Bockman, D.E. & M.L. Kirby. 1984. Dependence of thymus development on derivatives of the neural crest. *Science* **223:** 498–500.
22. Etchevers, H.C., C. Vincent, N.M. Le Douarin & G.F. Couly. 2001. The cephalic neural crest provides pericytes and smooth muscle cells to all blood vessels of the face and forebrain. *Development* **128:** 1059–1068.
23. Bockman, D.E., M.E. Redmond, K. Waldo, *et al.* 1987. Effect of neural crest ablation on development of the heart and arch arteries in the chick. *Am. J. Anat.* **180:** 332–341.
24. Schneider, R.A. & J.A. Helms. 2003. The cellular and molecular origins of beak morphology. *Science* **299:** 565–568.
25. Couly, G., S. Creuzet, S. Bennaceur, *et al.* 2002. Interactions between Hox-negative cephalic neural crest cells and the foregut endoderm in patterning the facial skeleton in the vertebrate head. *Development* **129:** 1061–1073.
26. Ruhin, B., S. Creuzet, C. Vincent, *et al.* 2003. Patterning of the hyoid cartilage depends upon signals arising from the ventral foregut endoderm. *Dev. Dyn.* **228:** 239–246.
27. Bolande, R.P. 1974. The neurocristopathies; a unifying concept of disease arising in neural crest maldevelopment. *Hum. Pathol.* **5:** 409–420.
28. Bolande, R.P. 1997. Neurocristopathy: its growth and development in 20 years. *Pediatr. Pathol. Lab. Med.* **17:** 1–25.
29. Amiel, J., E. Sproat-Emison, M. Garcia-Barcelo, *et al.* 2008. Hirschsprung disease, associated syndromes and genetics: a review. *J. Med. Genet.* **45:** 1–14.
30. Badner, J.A., W.K. Sieber, K.L. Garver & A. Chakravarti. 1990. A genetic study of Hirschsprung disease. *Am. J. Hum. Genet.* **46:** 568–580.
31. Edery, P., S. Lyonnet, L.M. Mulligan, *et al.* 1994. Mutations of the RET proto-oncogene in Hirschsprung's disease [see comments]. *Nature* **367:** 378–380.
32. Attie, T., A. Pelet, P. Edery, *et al.* 1995. Diversity of RET proto-oncogene mutations in familial and sporadic Hirschsprung disease. *Hum. Mol. Genet.* **4:** 1381–1386.
33. Bolk, S., A. Pelet, R.M. Hofstra, *et al.* 2000. A human model for multigenic inheritance: phenotypic expression in hirschsprung disease requires both the RET gene and a new 9q31 locus. *Proc. Natl. Acad. Sci. USA* **97:** 268–273.
34. Gabriel, S.B., R. Salomon, A. Pelet, *et al.* 2002. Segregation at three loci explains familial and population risk in Hirschsprung disease. *Nat. Genet.* **31:** 89–93.
35. Grice, E.A., E.S. Rochelle, E.D. Green, *et al.* 2005. Evaluation of the RET regulatory landscape reveals the biological relevance of a HSCR-implicated enhancer. *Hum. Mol. Genet.* **14:** 3837–3845.
36. Emison, E.S., M. Garcia-Barcelo, E.A. Grice, *et al.* 2010. Differential contributions of rare and common, coding and noncoding Ret mutations to multifactorial Hirschsprung disease liability. *Am. J. Hum. Genet.* **87:** 60–74.
37. Pingault, V., D. Ente, F. Dastot-Le Moal, *et al.* 2010. Review and update of mutations causing Waardenburg syndrome. *Hum. Mutat.* **31:** 391–406.
38. de Pontual, L., A. Pelet, D. Trochet, *et al.* 2006. Mutations of the RET gene in isolated and syndromic Hirschsprung's disease in human disclose major and modifier alleles at a single locus. *J. Med. Genet.* **43:** 419–423.
39. de Pontual, L., A. Pelet, M. Clement-Ziza, *et al.* 2007. Epistatic interactions with a common hypomorphic RET allele in syndromic Hirschsprung disease. *Hum. Mutat.* **28:** 790–796.
40. de Pontual, L., N.A. Zaghloul, S. Thomas, *et al.* 2009. Epistasis between RET and BBS mutations modulates enteric innervation and causes syndromic Hirschsprung disease. *Proc. Natl. Acad. Sci. USA* **106:** 13921–13926.
41. Hofstra, R.M., P. Elfferich, J. Osinga, *et al.* 2002. Hirschsprung disease and L1CAM: is the disturbed sex ratio caused by L1CAM mutations? *J. Med. Genet.* **39:** E11.
42. Wallace, A.S., C. Schmidt, M. Schachner, *et al.* 2010. L1cam acts as a modifier gene during enteric nervous system development. *Neurobiol. Dis.* **40:** 622–633.

43. Cohen, M.M., Jr. 1999. Robin sequences and complexes: causal heterogeneity and pathogenetic/phenotypic variability. *Am J. Med. Genet.* **84:** 311–315.
44. Abadie, V., M.P. Morisseau-Durand, C. Beyler, *et al.* 2002. Brainstem dysfunction: a possible neuroembryological pathogenesis of isolated Pierre Robin sequence. *Eur. J. Pediatr.* **161:** 275–280.
45. Holder-Espinasse, M., V. Abadie, V. Cormier-Daire, *et al.* 2001. Pierre Robin sequence: a series of 117 consecutive cases. *J. Pediatr.* **139:** 588–590.
46. Jakobsen, L.P., R. Ullmann, S.B. Christensen, *et al.* 2007. Pierre Robin sequence may be caused by dysregulation of SOX9 and KCNJ2. *J. Med. Genet.* **44:** 381–386.
47. Bagheri-Fam, S., C. Ferraz, J. Demaille, *et al.* 2001. Comparative genomics of the SOX9 region in human and Fugu rubripes: conservation of short regulatory sequence elements within large intergenic regions. *Genomics* **78:** 73–82.
48. Bi, W., W. Huang, D.J. Whitworth, *et al.* 2001. Haploinsufficiency of Sox9 results in defective cartilage primordia and premature skeletal mineralization. *Proc. Natl. Acad. Sci. USA* **98:** 6698–6703.
49. Mori-Akiyama, Y., H. Akiyama, D.H. Rowitch & B. de Crombrugghe. 2003. Sox9 is required for determination of the chondrogenic cell lineage in the cranial neural crest. *Proc. Natl. Acad. Sci. USA* **100:** 9360–9365.
50. Kist, R., H. Schrewe, R. Balling & G. Scherer. 2002. Conditional inactivation of Sox9: a mouse model for campomelic dysplasia. *Genesis* **32:** 121–123.
51. Foster, J.W., M.A. Dominguez-Steglich, S. Guioli, *et al.* 1994. Campomelic dysplasia and autosomal sex reversal caused by mutations in an SRY-related gene. *Nature* **372:** 525–530.
52. Wagner, T., J. Wirth, J. Meyer, *et al.* 1994. Autosomal sex reversal and campomelic dysplasia are caused by mutations in and around the SRY-related gene. *Cell* **79:** 1111–1120.
53. Bell, D.M., K.K. Leung, S.C. Wheatley, *et al.* 1997. SOX9 directly regulates the type-II collagen gene. *Nat. Genet.* **16:** 174–178.
54. Lefebvre, V., W. Huang, V.R. Harley, *et al.* 1997. SOX9 is a potent activator of the chondrocyte-specific enhancer of the pro alpha1(II) collagen gene. *Mol. Cell Biol.* **17:** 2336–2346.
55. Gordon, C.T., T.Y. Tan, S. Benko, *et al.* 2009. Long-range regulation at the SOX9 locus in development and disease. *J. Med. Genet.* **46:** 649–656.
56. Kleinjan, D.J. & V. van Heyningen. 1998. Position effect in human genetic disease. *Hum. Mol. Genet.* **7:** 1611–1618.
57. Uchikawa, M., Y. Ishida, T. Takemoto, *et al.* 2003. Functional analysis of chicken Sox2 enhancers highlights an array of diverse regulatory elements that are conserved in mammals. *Dev Cell* **4:** 509–519.
58. Sabherwal, N., F. Bangs, R. Roth, *et al.* 2007. Long-range conserved non-coding SHOX sequences regulate expression in developing chicken limb and are associated with short stature phenotypes in human patients. *Hum. Mol. Genet.* **16:** 210–222.
59. Betancur, P., M. Bronner-Fraser & T. Sauka-Spengler. 2010. Genomic code for Sox10 activation reveals a key regulatory enhancer for cranial neural crest. *Proc. Natl. Acad. Sci. USA* **107:** 3570–3575.
60. Suda, Y., K. Kokura, J. Kimura, *et al.* 2010. The same enhancer regulates the earliest Emx2 expression in caudal forebrain primordium, subsequent expression in dorsal telencephalon and later expression in the cortical ventricular zone. *Development* **137:** 2939–2949.
61. Sagai, T., T. Amano, M. Tamura, *et al.* 2009. A cluster of three long-range enhancers directs regional Shh expression in the epithelial linings. *Development* **136:** 1665–1674.
62. Visel, A., Y. Zhu, D. May, *et al.* 2010. Targeted deletion of the 9p21 non-coding coronary artery disease risk interval in mice. *Nature* **464:** 409–412.
63. Ahituv, N., Y. Zhu, A. Visel, *et al.* 2007. Deletion of ultraconserved elements yields viable mice. *PLoS Biol.* **5:** e234.
64. Dostie, J., Y. Zhan & J. Dekker. 2007. Chromosome conformation capture carbon copy technology. *Curr. Protoc. Mol. Biol.* Chapter 21, Unit 21 14.
65. McGaughey, D.M., Z.E. Stine, J.L. Huynh, *et al.* 2009. Asymmetrical distribution of non-conserved regulatory sequences at PHOX2B is reflected at the ENCODE loci and illuminates a possible genome-wide trend. *BMC Genomics* **10:** 8–21.
66. Fisher, S., E.A. Grice, R.M. Vinton, *et al.* 2006. Conservation of RET regulatory function from human to zebrafish without sequence similarity. *Science* **312:** 276–279.
67. Rahimov, F., M.L. Marazita, A. Visel, *et al.* 2008. Disruption of an AP-2alpha binding site in an IRF6 enhancer is associated with cleft lip. *Nat. Genet.* **40:** 1341–1347.
68. Verlaan, D.J., S. Berlivet, G.M. Hunninghake, *et al.* 2009. Allele-specific chromatin remodeling in the ZPBP2/GSDMB/ORMDL3 locus associated with the risk of asthma and autoimmune disease. *Am. J. Hum. Genet.* **85:** 377–393.
69. Jeong, Y., F.C. Leskow, K. El-Jaick, *et al.* 2008. Regulation of a remote Shh forebrain enhancer by the Six3 homeoprotein. *Nat. Genet.* **40:** 1348–1353.
70. Dathe, K., K.W. Kjaer, A. Brehm, *et al.* 2009. Duplications involving a conserved regulatory element downstream of BMP2 are associated with brachydactyly type A2. *Am. J. Hum. Genet.* **84:** 483–492.
71. Kurth, I., E. Klopocki, S. Stricker, *et al.* 2009. Duplications of noncoding elements 5′ of SOX9 are associated with brachydactyly-anonychia. *Nat. Genet.* **41:** 862–863.
72. Beysen, D., J. Raes, B.P. Leroy, *et al.* 2005. Deletions involving long-range conserved nongenic sequences upstream and downstream of FOXL2 as a novel disease-causing mechanism in blepharophimosis syndrome. *Am. J. Hum. Genet.* **77:** 205–218.
73. D'Haene, B., C. Attanasio, D. Beysen, *et al.* 2009. Disease-causing 7.4 kb cis-regulatory deletion disrupting conserved non-coding sequences and their interaction with the FOXL2 promotor: implications for mutation screening. *PLoS Genet.* **5:** e1000522.
74. de Kok, Y.J., E.R. Vossenaar, C.W. Cremers, *et al.* 1996. Identification of a hot spot for microdeletions in patients with X-linked deafness type 3 (DFN3) 900 kb proximal to the DFN3 gene POU3F4. *Hum. Mol. Genet.* **5:** 1229–1235.
75. Kleinjan, D.A., A. Seawright, A. Schedl, *et al.* 2001. Aniridia-associated translocations, DNase hypersensitivity, sequence comparison and transgenic analysis redefine the functional domain of PAX6. *Hum. Mol. Genet.* **10:** 2049–2059.

Ann. N.Y. Acad. Sci. ISSN 0077-8923

ANNALS OF THE NEW YORK ACADEMY OF SCIENCES
Issue: *The Year in Human and Medical Genetics*

Modifier genes in Mendelian disorders: the example of hemoglobin disorders

Vijay G. Sankaran,[1,2] Guillaume Lettre,[3,4] Stuart H. Orkin,[1,2] and Joel N. Hirschhorn[5,6]

[1]Division of Hematology/Oncology, Children's Hospital Boston, Boston, Massachusetts. [2]Department of Pediatric Oncology, Dana-Farber Cancer Institute, Boston, Massachusetts. [3]Montreal Heart Institute, Montréal, Québec, Canada. [4]Département de Médecine, Université de Montréal, Succursale Centre-ville, Montréal, Québec, Canada. [5]Divisions of Genetics and Endocrinology and Program in Genomics, Children's Hospital Boston, Boston, Massachusetts. [6]Program in Medical and Population Genetics, Broad Institute, Cambridge, Massachusetts

Addresses for correspondence: Joel N. Hirschhorn, M.D., Ph.D., Divisions of Genetics and Endocrinology and Program in Genomics, Children's Hospital Boston, Boston, MA 02115; and Program in Medical and Population Genetics, Broad Institute, Cambridge, MA 02142. joelh@broadinstitute.org
Stuart H. Orkin, M.D., Division of Hematology/Oncology, Children's Hospital Boston, MA 02115; and Department of Pediatric Oncology, Dana-Farber Cancer Institute, Boston, MA 02115. stuart_orkin@dfci.harvard.edu

The disorders of hemoglobin, including sickle cell disease (SCD) and β-thalassemia, are the most common "Mendelian" genetic diseases in the world. Numerous studies have demonstrated the complexity in making genotype–phenotype correlations in both SCD and β-thalassemia. Indeed, patients with exactly the same set of pathogenic globin mutations can have dramatically variable clinical courses. We discuss natural history studies that have attempted to delineate the factors responsible for the variability among the numerous clinical complications noted in these diseases. We then discuss, in depth, two well characterized ameliorating factors in the β-hemoglobin disorders, concomitant α-thalassemia, and elevated levels of fetal hemoglobin (HbF). We use the study of HbF regulation to illustrate how important insights into the genetic modifiers in Mendelian diseases can be achieved through the study of such factors. We finally go on to discuss future avenues of research that may allow us to gain further insight into the poorly understood clinical heterogeneity of this fascinating set of common genetic diseases.

Keywords: sickle cell disease; β-thalassemia; fetal hemoglobin; association studies

Overview

The most common set of monogenic diseases in the world are the inherited hemoglobin disorders. These disorders are thought to have reached such a high prevalence due in large part to the carrier protection afforded against malaria.[1] Approximately 7% of the world's population is a carrier for at least one of these diseases.[2] It is estimated that each year over 500,000 babies with severe forms of these disorders are born.[3] Unfortunately, many children born with these diseases do not receive adequate treatment and therefore death occurs in the first few years of life. The burden of these diseases is undoubtedly going to increase in the coming years, as many developing countries are making an epidemiologic transition as a consequence of the reduction in infant mortality from nutritional and infectious causes.

There are two major categories of the inherited hemoglobin disorders: the structural hemoglobin variants and the thalassemia syndromes. Generally, carriers or heterozygotes for these variants are asymptomatic, whereas those who inherit two copies of these mutations are often affected by this spectrum of diseases. The structural hemoglobin variants result primarily from amino acid substitutions in one of the two polypeptide chains, α- or β-globin, that compose the hemoglobin (Hb) molecule. These mutations can be innocuous or alternatively can structurally destabilize or alter functional properties of the assembled Hb molecule. On a worldwide basis, only three structural variants are widespread. All three of these structural

doi: 10.1111/j.1749-6632.2010.05821.x

variants are due to missense mutations in the β-globin polypeptide and include HbS, HbC, and HbE.[2] The homozygous state of HbS results in sickle cell anemia, whereas the compound heterozygous state for HbS and HbC results in HbSC disease, which shares many features with sickle cell anemia, though it is generally milder.[4,5] HbS may also be found in the compound heterozygous state with β-thalassemia mutations causing HbS-β-thalassemia that also has extensive clinical overlap with sickle cell anemia. Collectively, these disorders are grouped together with the term *sickle cell disease* (SCD). The etiology of the complications in SCD is primarily attributable to the propensity of this sickle Hb to polymerize in the deoxygenated state, which results in the red blood cells becoming deformed into sickle shapes.[4] These misshapen cells then occlude small blood vessels, which leads to attacks of pain in bones from ischemia; sequestration of blood in the spleen, liver, or lungs; and thrombosis of cerebral vessels, causing strokes.[5] In addition to these acute consequences of sickle Hb polymerization, long-term changes can also result. These alterations include red cell membrane damage, chronic hemolysis, altered interactions between sickle red blood cells and vascular endothelial cells, and inflammatory responses from such interactions.[6]

The thalassemia syndromes are attributable to defective production of the globin polypeptides. This can either be due to impaired production of the α- or β-globin chains, which causes α- or β-thalassemia, respectively. In the case of the α-thalassemia syndromes, unpaired β-globin chains can assemble into an unstable Hb molecule known as HbH, which results in hemolytic anemia. There is a great deal of heterogeneity in this disease, which has been discussed in other review articles.[2,7] In the case of β-thalassemia, unbalanced globin chain synthesis from a reduced production of β-globin chains results in the presence of highly unstable α-globin polypeptides. These polypeptides precipitate in precursors of mature red blood cells, resulting in toxicity, death, and ineffective maturation into red blood cells.[8,9] This causes anemia because of ineffective production of red blood cells (known as ineffective erythropoiesis) and results in an expansion of the erythroid precursor populations within the bone marrow, spleen, as well as in other sites within the body. At the molecular level, it is important to recognize that the mutations causing β-thalassemia are extremely heterogeneous, with over 200 different mutations described in the literature.[2,9] Generally, a few mutations predominate in a particular region of the world. The missense mutation in HbE is structurally innocuous, but it does create a novel splice site resulting in defective production of this Hb.[10] When HbE is found in the compound heterozygous state with a β-thalassemia mutation, a clinically heterogeneous disease known as HbE-β-thalassemia can occur.[1,11] This disease ranges in severity from being essentially asymptomatic, with the presence of a moderate anemia, all the way to a transfusion-dependent form. HbE-β-thalassemia is considered to be part of the spectrum of the β-thalassemia syndromes.

Although many children with the various β-thalassemia syndromes cannot survive without regular blood transfusions, a major source of morbidity and mortality in patients who either are transfused regularly or can survive without transfusions is due to iron overload. This is attributable to both the iron burden resulting from regular transfusions, as well as from the increased physiologic tendency of these patients to absorb iron.[12,13] Indeed, many patients with transfusion-independent forms of thalassemia (thalassemia intermedia) still succumb to the consequences of increased iron overload from deregulated iron absorption.

Natural history

Both SCD and β-thalassemia were among the first single-gene disorders whose molecular etiology was unraveled by using the tools of molecular biology.[1] However, although the field of molecular medicine began to burgeon soon after the elucidation of these initial "molecular" diseases, many years passed before natural history studies began to reveal the actual extent of phenotypic variability in these diseases. As a result of this work, it became evident that these diseases were far from the seemingly simple Mendelian disorders that many workers in the field had come to believe; among patients with known causal mutations, the clinical course can be extremely variable.

The understanding of the natural history of SCD stems from a rather limited number of studies. The majority of children with SCD in natural history studies in the developing world have died early.[14,15] Older studies suggested that the majority of these deaths were attributable to increased susceptibility to infectious diseases among these children,[15,16] likely secondary to the functional loss of the spleen in this disease. In the developed world, a series of important studies showed that this increased susceptibility was attributable to specific bacterial infections among children with SCD,[2,5,17] leading to arguably the most important intervention to date in patients with SCD, the prophylactic use of penicillin to prevent bacteremia.[18,19] However, it was not known whether the same causes of early death were present in SCD patients in the developing world, until the recent ground-breaking work from Williams *et al.*[20] showing that the spectrum of bacteremia in these children with SCD in Africa mimics what has been observed in other settings.[21]

As a result of the above early interventions, coupled with newborn screening in some countries, the life expectancy for patients with SCD has increased considerably.[2] However, these patients continue to face a number of complications of this disease. Cohort studies in countries including the United States and Jamaica examined the causes and risk factors for various SCD-related complications, including vaso-occlusive crises, acute chest syndrome, strokes, death, and others.[5,22–25] An important common theme of all of these studies is the extent of diversity among patients for these various complications. For example, although a subset of SCD patients may present with frequent vaso-occlusive crises, many patients rarely or almost never suffer from this complication.[23] It is also apparent that other populations with SCD may have a great diversity of clinical manifestations.[26,27] Moreover, most studies have been performed in the developed world and few studies exist to describe the clinical manifestations of SCD in the developing world, where the majority of patients with this disease live.[21]

A similar heterogeneity in the natural history of disease is present for the β-thalassemia syndromes.[2,28] In the majority of cases, patients with homozygous or compound heterozygous β-thalassemia mutations require regular transfusions to maintain Hb levels compatible with survival. However, it is apparent that there are occasional patients, even with mutations normally considered to create transfusion-dependent forms of thalassemia, who have a much milder clinical course and who may not require regular transfusions.[29] The precise reasons for this heterogeneity are not well understood in many instances,[30] although certain modifiers, such as concomitant α-thalassemia mutations or increased fetal hemoglobin (HbF) production (see below), play a role in this phenotypic variability.[31,32] When children with β-thalassemia are given adequate transfusions along with appropriate iron chelation, development can occur in a nearly normal fashion and survival into adulthood is possible.[2] Nonetheless, an important degree of heterogeneity is observed even among patients who receive regular treatment. For example, β-thalassemia patients with a similar degree of apparent iron burden (as assessed by either serum ferritin or liver iron measurements) may have extremely heterogeneous manifestations of iron overload.[33,34] For example, some patients with a similar degree of iron burden may develop endocrine complications such as diabetes or impaired growth, whereas others manifest primarily with cardiac or other complications of iron overload.

This situation is further complicated in the case of β-thalassemia syndromes that are characterized as being intermediate in severity. The most notable and common group of these disorders are the HbE-β-thalassemia syndromes. As discussed above, HbE-β-thalassemia can range in severity, from being asymptomatic to being a transfusion-dependent disease. Until recently, the full extent of this diversity was not well appreciated.[11,35] It is apparent that many patients who were once thought to require transfusions for survival can actually do extremely well in the absence of such treatment.[11] However, the study of this clinical heterogeneity is complicated by the important observation that even a single patient can have a dramatically variable clinical course over his or her lifetime.[1,11] Such observations stress the important role that nongenetic factors, such as environmental exposures, may have upon the clinical course of Hb diseases, and also highlight an important complexity that must be considered when attempting to develop scoring systems for severity in these diseases.[1,35]

Hb disease modifiers

As is apparent from our discussion on the natural history of these diseases, we currently have a rather limited grasp on the etiology of heterogeneity in these disorders, even among patients who acquire the exact same set of Mendelian mutations. However, there are a few cases in which clearly identified modifiers of clinical severity in the β-hemoglobin disorders are known. These well-characterized modifiers primarily act directly to alter the known pathophysiology of SCD or β-thalassemia. In the case of SCD, factors that reduce HbS deoxygenation and/or polymerization appear to be important and primary modifiers of the disease. In the case of β-thalassemia, any factor that reduces the imbalance between α- and β-globin polypeptide chain production is an important modifier of these diseases.

α-Thalassemia

A major well characterized modifier of severity in both SCD and β-thalassemia is the concomitant presence of α-thalassemia. Higgs *et al.*[36] and Embury *et al.*[37] first described this important clinical effect in SCD, where a milder clinical course was observed with a greater reduction in α-globin production. In SCD, the reduction in α-globin polypeptides results in a reduced concentration of HbS within a red blood cell, which then reduces the likelihood that HbS will become deoxygenated and undergo polymerization.[38] As a result, there is a reduced severity of certain SCD-related complications, particularly those associated with hemolysis, although compound heterozygous patients can still be severely affected by many of the complications of SCD. In the context of β-thalassemia, a reduced concentration of α-globin chains reduces the globin chain imbalance and therefore allows for more effective erythropoiesis.[1,39,40] Although this interaction among α- and β-thalassemia has been appreciated for many years, the critical nature of globin chain balance has been highlighted by the finding that the presence of α-globin duplications can interact with heterozygous β-thalassemia to cause a syndrome resembling homozygous β-thalassemia.[41,42]

Fetal hemoglobin

In 1948, Janet Watson[43] hypothesized that infants with SCD were protected from the clinical manifestations of their disease secondary to the presence of a unique form of Hb found in the fetus and in early infancy. Similar hypotheses were put forth to explain the delayed onset of symptoms in patients with β-thalassemia.[29] Furthermore, observations in certain populations with Hb disorders, such as SCD patients from Saudi Arabia[27,44,45] and India,[26] indicated that higher steady-state levels of HbF tend to result in less severe disease. The conclusions drawn from these disparate observations were subsequently confirmed in larger epidemiological studies, which demonstrated the quantitative partial ameliorating effect of HbF levels in SCD[22–24,46] and the β-thalassemia syndromes.[1,8,9,11,28]

With the identification and characterization of different Hb isoforms, it became evident that a unique β-like globin subunit, γ-globin, is expressed in much of fetal life and declines soon after birth.[1] Together with two α-globin chains, two γ-globin polypeptides combine with a heme moiety to form HbF. γ-Globin polypeptides compensate for reduced production of β-globin chains in β-thalassemia, thus ameliorating the symptoms of this disease. In SCD, HbF inhibits polymerization of HbS and therefore serves as a powerful antisickling agent.

In the following sections, we use the regulation of HbF production to illustrate the benefit of modern human genetics of complex traits in understanding clinical heterogeneity in "simple" monogenic diseases. Specifically, although HbF levels themselves have a strong genetic influence, it has been possible to apply modern methods to study the genetic determinants of HbF levels, and thereby identify modifiers of disease severity.

Genetic regulation of HbF

Although the level of HbF falls promptly after birth (to reach <1% in adults), the eventual level maintained in adults varies among individuals. This interindividual variation in HbF levels is highly heritable ($h^2 \sim 0.60$–0.89),[47,48] suggesting that genetic investigation could identify factors that control HbF production. In fact, early DNA mapping and resequencing at the β-globin gene cluster had identified deletions and rare point mutations that prevent globin gene switching and are characterized by very high HbF levels (10–40%) throughout adulthood. This Mendelian form of high γ-globin gene expression and HbF production, termed *hereditary persistence of fetal hemoglobin* (HPFH), is rare and cannot explain the variation in HbF levels observed in the general population. However, HPFH provided the first evidence that HbF production is genetically controlled, and suggested that common DNA polymorphisms at the β-globin locus could act in *cis* to modulate HbF levels in normal individuals. This prediction was supported by identification of the *Xmn*I polymorphism (rs7482144) in the proximal promoter of HBG2 ($^{G}\gamma$-globin).[45,49,50] The T allele at rs7482144 is associated with increased HbF levels, and is frequent in individuals of Northern European (HapMap CEU T allele frequency: 28%) and west African (HapMap YRI T allele frequency: 5%) ancestry. The *Xmn*I-rs7482144 polymorphism explains 2–10% of the variation in HbF levels, depending on the frequency of the T allele in the different populations considered.[51–53] Despite 25 years of research, it is still unclear whether the *Xmn*I-rs7482144 itself, or a marker in linkage disequilibrium (LD) with it controls HbF levels at the β-globin locus. Analysis of the five classic sickle cell haplotypes also suggested the presence of additional DNA sequence variants at the β-globin locus that influence HbF production, whereas both the Senegal and Arab–Indian haplotypes carry the *Xmn*I-rs7482144 T allele; SCD patients with the Arab–Indian haplotype have much higher HbF levels.[27,44,45] Extensive DNA resequencing in Arab–Indian SCD populations could therefore yield additional DNA polymorphisms associated with HbF levels.

Linkage studies

Despite its strong effect on HbF levels, the *Xmn*I-rs7482144 polymorphism is not sufficient to classify individuals in the low or high ends of the HbF level distribution; some individuals have the *Xmn*I-rs7482144 T allele and low HbF levels, and others have high HbF levels but the C allele at *Xmn*I-rs7482144, which is otherwise associated with decreased HbF. This incomplete correlation suggested that HbF is a complex genetic trait, influenced by multiple DNA sequence variants of modest effect, as well as environmental factors. Specifically, because estimates of narrow-sense heritability for HbF are substantial ($h^2 = 0.60$–0.89),[47,48] and are only partially explained by the *Xmn*I-rs7482144 polymorphism, there are likely other *cis* and *trans* genetic factors that regulate HbF production.

Linkage studies analyze genetic markers across the human genome in families to identify chromosomal regions linked to phenotypic variation within the families. Several linkage scans were carried out to find loci linked to HbF, leading to the identification of HbF linkage peaks on chromosomes 6q23,[54] 8q,[55,56] and Xp22.2.[57] Linkage signals on chromosomes 6q23 and 8q have since then been replicated in additional families; replication data are still lacking for the Xp22.2 linkage peak. Because linkage analysis is better suited to find alleles of moderate or strong effect on complex traits and diseases,[58,59] these results, if correct, suggested that variants might exist with moderate or strong effects on HbF levels. These results therefore supported the notion that HbF is a polygenic trait controlled by genetic factors in *trans* with the β-globin locus on chromosome 11.

Candidate–gene association studies

Linkage scans implicate large chromosomal regions that segregate together within families, which are typically on the order of several megabases of DNA, and usually include many genes. Higher resolution mapping can be achieved by association studies, which search for correlations (LD) between specific alleles and traits or diseases of interest; these correlations usually span much shorter genomic regions. Thein *et al.*[52] used association studies in healthy Northern European twins with F cell measurements available (F cells represent the fraction of erythrocytes that contain HbF, and are highly correlated with HbF levels) to fine-map the 6q23 HbF linkage peak to a small intergenic interval (~80 kb) between HBS1L and MYB. This association has been replicated in healthy Sardinians,[32] Jamaicans,[60] and in African-American and Brazilian patients with SCD,[53] establishing the HBS1L–MYB intergenic region as the second bona fide HbF-associated locus (in addition to the *XmnI* polymorphism at the β-globin locus itself). Interestingly, genetic variants within the HBS1L–MYB region have pleiotropic effects and have been shown to associate not only with HbF levels, but also with many other hematologic traits,[61–63] suggesting a general role for DNA polymorphisms in the HBS1L–MYB intergenic region in hematopoiesis.

HBS1L and MYB encode, respectively, a Guanosine-5′-triphosphate (GTP)-binding protein and a transcriptional regulator of hematopoiesis. It is currently unclear whether HBS1L, MYB, both, or any of these gene products directly regulate HbF production. HbF-associated variants in the HBS1L–MYB intergenic region were suggested to control the expression of HBS1L in erythroid progenitor cells and lymphoblasts.[52,64] However, MYB is a key regulator of erythropoiesis, and its overexpression is associated with inhibition of γ-globin gene expression.[65] Recent functional work in erythroid precursor cells using DNase I sensitivity assays and chromatin immunoprecipitation suggested the presence of multiple distal regulatory in the HBS1L–MYB intergenic region.[66] Importantly, additional functional studies will be needed to link how HbF-associated single nucleotide polymorphisms (SNPs) in the HBS1L–MYB region influence the activity of these transcriptional regulatory sequences, and how this relates to HBS1L and MYB activity, as well as HbF production.

Genome-wide association studies

The extensive repertoire of common human DNA sequence variants, such as the HapMap project,[67,68] and low-cost genotyping arrays allow researchers to move away from candidate–gene-based association studies and perform genome-wide association (GWA) experiments in an unbiased manner.[69] Genome-wide association GWA studies interrogate millions of common SNPs for association to human complex traits. The GWA approach has proven extremely useful to find robust genetic associations, highlighting promising and often unanticipated underlying biology.[70] For HbF variation, two GWA studies were initially carried out: one in 179 healthy Northern Europeans selected from the tails of the F cell distribution (lower 5th and upper 95th percentile) among 5,184 individuals with F cell measures available,[51] and one in 4,305 healthy Sardinians.[32] Both studies replicated the previously known associations between HbF levels and the β-globin locus on chromosome 11 and the HBS1L–MYB intergenic region on chromosome 6. They also identified and replicated a new HbF locus on chromosome 2: SNPs in intron 2 of the zinc-finger protein encoded by BCL11A were strongly associated with HbF levels and explained ~15% of the phenotypic variation in HbF levels.[32,51] This genetic association to HbF levels, originally described in populations of European ancestry, has since then been replicated in African-American and Brazilian SCD patients,[53] and in Chinese and Thai patients with β-thalassemia or the HbE trait.[71,72] The identification of BCL11A is a clear example of the advantage of GWA as an unbiased discovery tool in human genetics: BCL11A would not be an *a priori* strong candidate gene for modulating HbF levels, and the BCL11A locus did not emerge from any of the HbF linkage scans.

Following up on the observation that genetic variants in the BCL11A gene were significantly associated with HbF levels, we demonstrated that the gene product encoded by BCL11A serves as a direct transcriptional regulator of the fetal to adult Hb switch in humans.[73] In particular, we provided evidence that the variants associated with high HbF levels were associated with lower levels of BCL11A expression. Additionally, BCL11A protein expression varies at different developmental stages, such that stages when the γ-globin genes were robustly expressed had low or absent expression of the full-length forms of this protein. Additionally, short variant forms were noted to be present in these earlier stages of ontogeny. BCL11A proteins interact in erythroid cells with specific transcriptional regulators of erythropoiesis, including GATA-1 and FOG-1. Additionally, BCL11A interacts with a repressor complex, known as the NuRD complex, which may contribute to its repressive activity at the γ-globin genes. Knockdown of BCL11A expression (and hence, protein) in primary human erythroid precursor cells leads to dramatic elevations in the expression of the γ-globin genes without resulting in a major perturbation of erythropoiesis. Moreover, knockout of BCL11A in mice prevents appropriate shutoff of endogenous embryonic β-like globin gene expression, and also failure to properly silence human γ-globin transgenes.[74] In further mechanistic studies, we showed that BCL11A acts at a distance from the γ-globin gene promoters within the β-globin locus and cooperates with other factors such as GATA-1 and the transcription factor SOX6 to repress γ-globin expression.[75]

Together, these recent findings establish BCL11A as the first genetically and biochemically validated regulator of the fetal to adult globin switch in humans. Further work should lead to an improved understanding of the mechanisms by which this transcription factor functions to silence the γ-globin genes. Our studies in the mouse suggest that BCL11A mediates the evolutionarily divergent globin gene switches in other mammals, strongly supporting the notion that this factor is a critical mediator of these processes.[74] These findings also show how genetic studies of disease modifiers can identify new potential targets for therapy because downregulating or inhibiting BCL11A *in vivo* could be a desirable therapeutic approach.

Recent work has extended these findings and revealed further informative connections to the human genetics of HbF regulation. By studying a family from Malta where elevations in HbF levels were noted in certain family members (ranging from 3–19%), a new locus linked to the HbF elevation was found on chromosome 19.[76] By using mapping and sequencing the genes in the region, a heterozygous nonsense mutation in the KLF1 gene was found in affected family members. By gene expression profiling and chromatin analysis, the investigators found that haploinsufficiency of KLF1 appears to affect HbF expression directly and also appears to act indirectly through the downregulation of BCL11A.[76] These findings were also supported by a companion study that utilized mouse models to show that downregulation of KLF1 affected BCL11A expression and globin gene expression.[77]

HbF-associated SNPs and hemoglobinopathy severity

HbF is a strong modifier of disease severity in β-thalassemia and SCD patients. Because the SNPs identified at the BCL11A, HBS1L–MYB, and β-globin explain a significant fraction of the interindividual variation in HbF levels (20–50%), genotypes at these SNPs could become useful predictive tests. Uda *et al.*[32] genotyped the BCL11A rs11886868 SNPs in β-thalassemia patients, and showed that the C allele, associated with increased HbF levels, was found significantly more often in transfusion-independent patients (β-thalassemia intermediate) than in patients that depend on blood transfusions for survival (β-thalassemia major). Modulation of β-thalassemia severity by HbF-associated SNPs has now been extended to the β-globin and HBS1L–MYB loci.[31,72] One limitation in these studies of β-thalassemia severity is that extreme clinical differences are used, and clinically important subtle differences among patients are ignored for the sake of simplicity. Although these studies are promising, further studies will be necessary to dissect to what extent these genetic variants explain the diversity of clinical manifestations noted in these diseases.

In SCD, one of the best measures of overall morbidity is the pain crisis rate.[23] We tested the effect of SNPs at the three known HbF loci on pain crisis rate in African-American patients from the Cooperative Study of Sickle Cell Disease (CSSCD); in aggregate, these SNPs were associated with pain crisis rate, even after considering HbF measurements.[53] This suggests that HbF-associated SNP genotypes provide additional predictive information, beyond steady-state HbF levels. Validation of these preliminary results in independent cohorts are warranted, as they might lead to insight in the identification and follow-up of hemoglobinopathy patients more at risk of severe complications. This would be particularly useful in young patients (<5 years old) in whom HbF levels have not yet reached a steady-state.

Genetics and genomics: beyond HbF

Most of the current efforts to identify modifiers of the major hemoglobinopathies have been focused around the regulation and reactivation of HbF production. There are other important modifiers of severity for β-thalassemia and SCD, including red blood cell counts, which are genetically controlled; identifying loci that control these modifiers could highlight new therapeutic approaches. Moreover, β-thalassemia and SCD are extremely heterogeneous diseases, with some patients presenting severe complications and others not.[78] The identification of genetic variation that protects some hemoglobinopathy patients from most severe complications should improve risk stratification and treatment strategies. Fortunately, large GWA studies in well-phenotyped cohorts, such as the CSSCD, are ongoing and should yield new exciting insights into the clinical heterogeneity of these diseases. As we move to the next step, hemoglobinopathy modifier genetics should also benefit from second-generation DNA resequencing technologies, allowing investigators to consider not only common genetic variation, but also all forms (common, rare, and structural) of genetic variation, and their contribution to disease heterogeneity.

It is expected that the major limitation in our search to understand how genetic modifiers influence severity in β-thalassemia and SCD patients will come more from a lack of appropriately large and well-phenotyped cohorts rather than genetic or technical challenges. As such, and given the healthcare burden that hemoglobinopathies represent worldwide,[2,29] priorities need to be set to create new large prospective patient cohorts to allow continued genetic research. A major challenge that remains will be to define the range of clinical phenotypes in these diseases in large enough samples to provide adequate power, and to better understand the full extent of variability that exists in these disorders. This will be a critically needed step before further and more comprehensive studies on the genetic basis for variation in these model Mendelian disorders can be more fully appreciated. Clearly, the future appears to be quite fruitful and will require an integrated approach by clinicians caring for patients with these disorders, epidemiologists studying the global burden of these diseases, geneticist interested in understanding this variability, clinical researchers designing studies to better understand the nature of this variation, and basic scientists who will study the mechanisms that underlie the poorly understood variability of the Hb disorders.

Conflicts of interest

The authors declare no conflicts of interest.

References

1. Weatherall, D.J. 2001. Phenotype-genotype relationships in monogenic disease: lessons from the thalassaemias. *Nat. Rev. Genet.* **2:** 245–255.
2. Weatherall, D., O. Akinyanju, S. Fucharoen, *et al.* 2006. Inherited disorders of hemoglobin. In *Disease Control Priorities in Developing Countries* 2nd ed.: 663–680. Oxford University Press. New York.
3. Weatherall, D.J. 2010. The inherited diseases of hemoglobin are an emerging global health burden. *Blood* **115:** 4331–4336.
4. Bunn, H.F. 1997. Pathogenesis and treatment of sickle cell disease. *N. Engl. J. Med.* **337:** 762–769.
5. Serjeant, G.R. 1997. Sickle-cell disease. *Lancet* **350:** 725–730.
6. Stuart, M.J. & R.L. Nagel. 2004. Sickle-cell disease. *Lancet* **364:** 1343–1360.
7. Fucharoen, S. & V. Viprakasit. 2009. Hb H disease: clinical course and disease modifiers. *Hematol. Am. Soc. Hematol. Educ. Prog.* 26–34.
8. Olivieri, N.F. 1999. The beta-thalassemias. *N. Engl. J. Med.* **341:** 99–109.
9. Rund, D. & E. Rachmilewitz. 2005. Beta-thalassemia. *N. Engl. J. Med.* **353:** 1135–1146.
10. Orkin, S.H., H.H. Kazazian Jr. S.E. Antonarakis, *et al.* 1982. Abnormal RNA processing due to the exon mutation of beta E-globin gene. *Nature* **300:** 768–769.
11. Premawardhena, A., C.A. Fisher, N.F. Olivieri, *et al.* 2005. Haemoglobin E beta-thalassaemia in Sri Lanka. *Lancet* **366:** 1467–1470.
12. Tanno, T., N.V. Bhanu, P.A. Oneal, *et al.* 2007. High levels of GDF15 in thalassemia suppress expression of the iron regulatory protein hepcidin. *Nat. Med.* **13:** 1096–1101.
13. Nathan, D.G. 2005. Thalassemia: the continued challenge. *Ann. N.Y. Acad. Sci.* **1054:** 1–10.
14. Ibidapo, M.O. & O.O. Akinyanju. 2000. Acute sickle cell syndromes in Nigerian adults. *Clin. Lab. Haematol.* **22:** 151–155.
15. Fleming, A.F., J. Storey, L. Molineaux, *et al.* 1979. Abnormal haemoglobins in the Sudan savanna of Nigeria. I. Prevalence of haemoglobins and relationships between sickle cell trait, malaria and survival. *Ann. Trop. Med. Parasitol.* **73:** 161–172.
16. Molineaux, L., A.F. Fleming, R. Cornille-Brogger, *et al.* 1979. Abnormal haemoglobins in the Sudan savanna of Nigeria. III. Malaria, immunoglobulins, and antimalarial antibodies in sickle cell disease. *Ann. Trop. Med. Parasitol.* **73:** 301–310.
17. Powars, D., G. Overturf, J. Weiss, *et al.* 1981. Pneumococcal septicemia in children with sickle cell anemia. Changing trend of survival. *JAMA* **245:** 1839–1842.
18. Gaston, M.H., J.I. Verter, G. Woods, *et al.* 1986. Prophylaxis with oral penicillin in children with sickle cell anemia. A randomized trial. *N. Engl. J. Med.* **314:** 1593–1599.
19. Falletta, J.M., G.M. Woods, J.I. Verter, *et al.* 1995. Discontinuing penicillin prophylaxis in children with sickle cell anemia. Prophylactic Penicillin Study II. *J. Pediatr.* **127:** 685–690.
20. Williams, T.N., S. Uyoga, A. Macharia, *et al.* 2009. Bacteraemia in Kenyan children with sickle-cell anaemia: a retrospective cohort and case-control study. *Lancet* **374:** 1364–1370.
21. Hankins, J. & R.E. Ware. 2009. Sickle-cell disease: an ounce of prevention, a pound of cure. *Lancet* **374:** 1308–1310.
22. Platt, O.S., D.J. Brambilla, W.F. Rosse, *et al.* 1994. Mortality in sickle cell disease. Life expectancy and risk factors for early death. *N. Engl. J. Med.* **330:** 1639–1644.
23. Platt, O.S., B.D. Thorington, D.J. Brambilla, *et al.* 1991. Pain in sickle cell disease. Rates and risk factors. *N. Engl. J. Med.* **325:** 11–16.
24. Castro, O., D.J. Brambilla, B. Thorington, *et al.* 1994. The acute chest syndrome in sickle cell disease: incidence and risk factors. The cooperative study of sickle cell disease. *Blood* **84:** 643–649.
25. Adams, R., V. McKie, F. Nichols, *et al.* 1992. The use of transcranial ultrasonography to predict stroke in sickle cell disease. *N. Engl. J. Med.* **326:** 605–610.
26. Kar, B.C., R.K. Satapathy, A.E. Kulozik, *et al.* 1986. Sickle cell disease in Orissa State, India. *Lancet* **2:** 1198–1201.
27. Perrine, R.P., M.J. Brown, J.B. Clegg, *et al.* 1972. Benign sickle-cell anaemia. *Lancet* **2:** 1163–1167.
28. Cao, A. & R. Galanello. 2010. Beta-thalassemia. *Genet. Med.* **12:** 61–76.
29. Weatherall, D.J. & J.B. Clegg. 2001. Inherited haemoglobin disorders: an increasing global health problem. *Bull. World Health Organ.* **79:** 704–712.
30. Taher, A.T., K.M. Musallam, A. El-Beshlawy, *et al.* 2010. Age-related complications in treatment-naive patients with thalassaemia intermedia. *Br. J. Haematol.* **150:** 486–489.
31. Galanello, R., S. Sanna, L. Perseu, *et al.* 2009. Amelioration of Sardinian beta0 thalassemia by genetic modifiers. *Blood* **114:** 3935–3937.
32. Uda, M., R. Galanello, S. Sanna, *et al.* 2008. Genome-wide association study shows BCL11A associated with persistent fetal hemoglobin and amelioration of the phenotype of beta-thalassemia. *Proc. Natl. Acad. Sci. USA* **105:** 1620–1625.
33. Taher, A., C. Hershko, M.D. Cappellini. 2009. Iron overload in thalassaemia intermedia: reassessment of iron chelation strategies. *Br. J. Haematol.* **147:** 634–640.

34. Viprakasit, V., C. Lee-Lee, Q.T. Chong, *et al.* 2009. Iron chelation therapy in the management of thalassemia: the Asian perspectives. *Int. J. Hematol.* **90:** 435–445.
35. Sripichai, O., W. Makarasara , T. Munkongdee, *et al.* 2008. A scoring system for the classification of beta-thalassemia/HbE disease severity. *Am. J. Hematol.* **83:** 482–484.
36. Higgs, D.R., B.E. Aldridge, J. Lamb, *et al.* 1982. The interaction of alpha-thalassemia and homozygous sickle-cell disease. *N. Engl. J. Med.* **306:** 1441–1446.
37. Embury, S.H., A.M. Dozy, J. Miller, *et al.* 1982. Concurrent sickle-cell anemia and alpha-thalassemia: effect on severity of anemia. *N. Engl. J. Med.* **306:** 270–274.
38. Embury, S.H., M.R. Clark, G. Monroy & N. Mohandas. 1984. Concurrent sickle cell anemia and alpha-thalassemia. Effect on pathological properties of sickle erythrocytes. *J. Clin. Invest.* **73:** 116–123.
39. Kan, Y.W. & D.G. Nathan. 1970. Mild thalassemia: the result of interactions of alpha and beta thalassemia genes. *J. Clin. Invest.* **49:** 635–642.
40. Higgs, D.R. 2004. Gene regulation in hematopoiesis: new lessons from thalassemia. *Hematol. Am. Soc. Hematol. Educ. Prog.* 1–13.
41. Premawardhena, A., C.A. Fisher, N.F. Olivieri, *et al.* 2005. A novel molecular basis for beta thalassemia intermedia poses new questions about its pathophysiology. *Blood* **106:** 3251–3255.
42. Faa, V., M. Masala, A. Cao, *et al.* 2010. Alpha globin gene duplications in beta thalassemia patients with intact beta globin gene. *Blood Cells Mol. Dis.* **44:** 156–158.
43. Watson J. 1948. The significance of the paucity of sickle cells in newborn Negro infants. *Am. J. Med. Sci.* **215:** 419–423.
44. Pembrey, M.E., W.G. Wood, D.J. Weatherall & R.P. Perrine. 1978. Fetal haemoglobin production and the sickle gene in the oases of Eastern Saudi Arabia. *Br. J. Haematol.* **40:** 415–429.
45. Miller, B.A., N. Olivieri, M. Salameh, *et al.* 1987. Molecular analysis of the high-hemoglobin-F phenotype in Saudi Arabian sickle cell anemia. *N. Engl. J. Med.* **316:** 244–250.
46. Higgs, D.R. & W.G. Wood. 2008. Genetic complexity in sickle cell disease. *Proc. Natl. Acad. Sci. USA* **105:** 11595–11596.
47. Garner, C., T. Tatu, J.E. Reittie, *et al.* 2000. Genetic influences on F cells and other hematologic variables: a twin heritability study. *Blood* **95:** 342–346.
48. Pilia, G., W.M. Chen, A. Scuteri, *et al.* 2006. Heritability of cardiovascular and personality traits in 6,148 Sardinians. *PLoS Genet.* **2:** e132.
49. Labie, D., O. Dunda-Belkhodja, F. Rouabhi, *et al.* 1985. The -158 site 5' to the G gamma gene and G gamma expression. *Blood* **66:** 1463–1465.
50. Labie, D., J. Pagnier, C. Lapoumeroulie, *et al.* 1985. Common haplotype dependency of high G gamma-globin gene expression and high HbF levels in beta-thalassemia and sickle cell anemia patients. *Proc. Natl. Acad. Sci. USA* **82:** 2111–2114.
51. Menzel, S., C. Garner, I. Gut, *et al.* 2007. A QTL influencing F cell production maps to a gene encoding a zinc-finger protein on chromosome 2p15. *Nat. Genet.* **39:** 1197–1199.
52. Thein, S.L., S. Menzel, X. Peng, *et al.* 2007. Intergenic variants of HBS1L-MYB are responsible for a major quantitative trait locus on chromosome 6q23 influencing fetal hemoglobin levels in adults. *Proc. Natl. Acad. Sci. USA* **104:** 11346–11351.
53. Lettre, G., V.G. Sankaran, M.A. Bezerra, *et al.* 2008. DNA polymorphisms at the BCL11A, HBS1L-MYB, and beta-globin loci associate with fetal hemoglobin levels and pain crises in sickle cell disease. *Proc. Natl. Acad. Sci. USA* **105:** 11869–11874.
54. Garner, C., J. Mitchell, T. Hatzis, *et al.* 1998. Haplotype mapping of a major quantitative-trait locus for fetal hemoglobin production, on chromosome 6q23. *Am. J. Hum. Genet.* 62: 1468–1474.
55. Garner, C., N. Silver, S. Best, *et al.* 2004. Quantitative trait locus on chromosome 8q influences the switch from fetal to adult hemoglobin. *Blood* **104:** 2184–2186.
56. Garner, C.P., T. Tatu, S. Best, *et al.* 2002. Evidence of genetic interaction between the beta-globin complex and chromosome 8q in the expression of fetal hemoglobin. *Am. J. Hum. Genet.* **70:** 793–799.
57. Dover, G.J., K.D. Smith, Y.C. Chang, *et al.* 1992. Fetal hemoglobin levels in sickle cell disease and normal individuals are partially controlled by an X-linked gene located at Xp22.2. *Blood* **80:** 816–824.
58. Risch, N.J. 2000. Searching for genetic determinants in the new millennium. *Nature* **405:** 847–856.
59. Risch, N. & K. Merikangas. 1996. The future of genetic studies of complex human diseases. *Science* **273:** 1516–1517.
60. Creary, L.E., P. Ulug, S. Menzel, *et al.* 2009. Genetic variation on chromosome 6 influences F cell levels in healthy individuals of African descent and HbF levels in sickle cell patients. *PLoS one.* **4:** e4218.
61. Menzel, S., J. Jiang, N. Silver, *et al.* 2007. The HBS1L-MYB intergenic region on chromosome 6q23.3 influences erythrocyte, platelet, and monocyte counts in humans. *Blood* **110:** 3624–3626.
62. Ganesh, S.K., N.A. Zakai, F.J. van Rooij, *et al.* 2009. Multiple loci influence erythrocyte phenotypes in the CHARGE Consortium. *Nat. Genet.* **41:** 1191–1198.
63. Soranzo, N., T.D. Spector, M. Mangino, *et al.* 2009. A genome-wide meta-analysis identifies 22 loci associated with eight hematological parameters in the HaemGen consortium. *Nat. Genet.* **41:** 1182–1190.
64. Dixon, A.L., L. Liang, M.F. Moffatt, *et al.* 2007. A genome-wide association study of global gene expression. *Nat. Genet.* **39:** 1202–1207.
65. Jiang, J., S. Best, S. Menzel, *et al.* 2006. cMYB is involved in the regulation of fetal hemoglobin production in adults. *Blood* **108:** 1077–1083.
66. Wahlberg, K., J. Jiang, H. Rooks, *et al.* 2009. The HBS1L-MYB intergenic interval associated with elevated HbF levels shows characteristics of a distal regulatory region in erythroid cells. *Blood* **114:** 1254–1262.
67. Altshuler, D., L.D. Brooks, A. Chakravarti, *et al.* 2005. A haplotype map of the human genome. *Nature* **437:** 1299–1320.

68. Frazer, K.A., D.G. Ballinger, D.R. Cox, *et al.* 2007. A second generation human haplotype map of over 3.1 million SNPs. *Nature* **449:** 851–861.
69. Hirschhorn, J.N. & M.J. Daly. 2005. Genome-wide association studies for common diseases and complex traits. *Nat. Rev. Genet.* **6:** 95–108.
70. Hindorff, L.A., P. Sethupathy, H.A. Junkins, *et al.* 2009. Potential etiologic and functional implications of genome-wide association loci for human diseases and traits. *Proc. Natl. Acad. Sci. USA* **106:** 9362–9367.
71. Sedgewick, A.E., N. Timofeev, P. Sebastiani, *et al.* 2008. BCL11A is a major HbF quantitative trait locus in three different populations with beta-hemoglobinopathies. *Blood Cells Mol. Dis.* **41:** 255–258.
72. Nuinoon, M., W. Makarasara, T. Mushiroda, *et al.* 2009. A genome-wide association identified the common genetic variants influence disease severity in beta(0)-thalassemia/hemoglobin E. *Hum. Genet.* **127:** 303–314.
73. Sankaran, V.G., T.F. Menne, J. Xu, *et al.* 2008. Human fetal hemoglobin expression is regulated by the developmental stage-specific repressor BCL11A. *Science* **322:** 1839–1842.
74. Sankaran, V.G., J. Xu, T. Ragoczy, *et al.* 2009. Developmental and species-divergent globin switching are driven by BCL11A. *Nature* **460:** 1093–1097.
75. Xu, J., V.G. Sankaran, M. Ni, *et al.* 2010. Transcriptional silencing of {gamma}-globin by BCL11A involves long-range interactions and cooperation with SOX6. *Genes Dev.* **24:** 783–798.
76. Borg, J., P. Papadopoulos, M. Georgitsi, *et al.* 2010. Haploinsufficiency for the erythroid transcription factor KLF1 causes hereditary persistence of fetal hemoglobin. *Nat. Genet.* **42:** 801–805.
77. Zhou, D., K. Liu, C.W. Sun, *et al.* 2010. KLF1 regulates BCL11A expression and gamma- to beta-globin gene switching. *Nat. Genet.* **42:** 742–744.
78. Serjeant, G.R., D.R. Higgs & I.R. Hambleton. 2007. Elderly survivors with homozygous sickle cell disease. *N. Engl. J. Med.* **356:** 642–643.

Ann. N.Y. Acad. Sci. ISSN 0077-8923

ANNALS OF THE NEW YORK ACADEMY OF SCIENCES
Issue: *The Year in Human and Medical Genetics*

Modifier genes in Mendelian disorders: the example of cystic fibrosis

Garry R. Cutting
McKusick-Nathans Institute of Genetic Medicine, Johns Hopkins University School of Medicine, Baltimore, Maryland

Address for correspondence: Garry R. Cutting, M.D., BRB 559, 733 North Broadway, Johns Hopkins Medical Institute, Baltimore, MD 21205. gcutting@jhmi.edu

In the past three decades, scientists have had immense success in identifying genes and their variants that contribute to an array of diseases. While the identification of such genetic variants has informed our knowledge of the etiologic bases of diseases, there continues to be a substantial gap in our understanding of the factors that modify disease severity. Monogenic diseases provide an opportunity to identify modifiers as they have uniform etiology, detailed phenotyping of affected individuals, and familial clustering. Cystic fibrosis (CF) is among the more common life-shortening recessive disorders that displays wide variability in clinical features and survival. Considerable progress has been made in elucidating the contribution of genetic and nongenetic factors to CF. Allelic variation in *CFTR*, the gene responsible for CF, correlates with some aspects of the disease. However, lung function, neonatal intestinal obstruction, diabetes, and anthropometry display strong genetic control independent of *CFTR*, and candidate gene studies have revealed genetic modifiers underlying these traits. The application of genome-wide techniques holds great promise for the identification of novel genetic variants responsible for the heritable features and complications of CF. Since the genetic modifiers are known to alter the course of disease, their protein products become immediate targets for therapeutic intervention.

Keywords: genome wide; candidate gene; heritability; variation

Overview

It is generally recognized that genetic, environmental, and stochastic factors contribute to phenotype variation. However, the relative effect of each component is difficult to assess, especially for common diseases where a myriad of environmental factors may play a role. Consider, for instance, the role of drugs (an environmental factor) in the treatment of diseases and the effect they can have in modifying outcome. Parsing patient populations based on multiple environmental exposures can lead to substantial drops in power to detect the responsible genetic variants. Despite the aforementioned challenges, progress has been made in dissecting genetic and nongenetic factors underlying disease variability for several of the more common Mendelian disorders. The "monogenic" diseases provide unique opportunities to dissect components as they each have a single etiology, relatively uniform treatments, and the contribution of the disease-causing gene is known to some degree.[1] In many cases, some portion of phenotype variability can be associated with the nature of the mutations in the disease-causing gene. However, disease variability in patients bearing the same combination of mutations emphasizes the role of genetic background and environment, which appears to be the rule rather than the exception.[2] Thus, the current challenge for many studying monogenic disorders is to assess the relative contribution of genetic factors distinct from the disease-causing gene and to identify those genes that modify outcome.[3]

Identification of such modifier genes increases our understanding of the elements that affect disease variability and thereby identifies new targets for therapy. Furthermore, it is possible that genes that modify single-gene disorders contribute to the development and/or progression of common diseases in the general population. These genes may be

doi: 10.1111/j.1749-6632.2010.05879.x

minimally penetrant in healthy individuals, whereas the altered homeostasis of patients with single-gene disorders may unmask the effect of such modifiers. Thus, the search for genetic modifiers of single-gene disorders could benefit individuals beyond those afflicted with monogenic conditions. Finally, by their very nature, single-gene disorders have one highly penetrant disease-causing gene that could serve as a starting point for modeling of gene–gene and gene–environment interactions.

Rationale for cystic fibrosis

Cystic fibrosis (CF), a single-gene-recessive disorder that affects 60,000 individuals worldwide, is an ideal model for the identification and characterization of factors that cause disease variation.[4,5] First, the unfortunate high prevalence of CF provides a large number of accessible patients to perform detailed phenotypic analyses necessary to identify modifier genes in humans. Second, the commonness of CF and its monogenic etiology enable family-based studies to dissect genetic and nongenetic influences upon disease variation. Third, the diagnosis of CF can be confirmed using an objective laboratory measure: the sweat chloride concentration assay. In the vast majority of CF patients, chloride levels in the sweat are raised well beyond the distribution of values observed in the normal and CF carrier population, thereby distinguishing the disease from other conditions with similar features.[6]

Patients with CF manifest disease in the lungs, pancreas, intestine, liver, male reproductive tract, and sweat gland.[6] CFTR, the dysfunctional protein in CF patients, conducts chloride across the apical membranes of polarized epithelia.[7] Loss of CFTR function affects the transport of chloride, sodium, and water across epithelial tissues, leading to inadequate hydration of mucous secretions of CF patients. Obstruction of luminal space follows, and recurrent cycles of inflammation and fibrosis ultimately destroys affected organs.[6,8] Obstruction of the exocrine pancreas causes intestinal malabsorption and an abnormal nutritional status in almost all CF patients. Obstructive lung disease is the cause of death in almost 90% of patients.[9]

As noted above, the large number of families with multiple affected offspring facilitated identification of the CF transmembrane conductance regulator (*CFTR*) gene, an early and remarkable success in positional cloning. The identification of *CFTR* provided immense insight into the molecular pathophysiology of CF. A network of research labs that joined to form the CF Genetic Analysis Consortium has identified over 1,800 mutations in *CFTR* (http://www.genet.sickkids.on.ca/cftr/app). One mutation, a deletion of three nucleotides causing the loss of a phenylalanine at codon 508 (p.Phe508del), accounts for approximately 70% of CF alleles in Caucasian patients.[10] CF patients homozygous for p.Phe508del constitute the most common *CFTR* genotype (about 50% of patients, designated as *F508del* homozygotes from here on), and, as such, have served as a reference population for genotype/phenotype correlations.[11] There is also a group of 15–20 less common mutations accounting for 15% of CF alleles in Caucasians.[10] A study of the effect of a number of less common mutations upon phenotype demonstrated that allelic variation in *CFTR* accounts for only a portion of phenotypic variability.[11] *CFTR* genotype is highly correlated with preservation of some function of the exocrine pancreas (termed "pancreatic sufficiency"). Furthermore, sweat chloride levels tend to be less elevated in patients with *CFTR* genotypes associated with preserved pancreatic function than in those with pancreatic insufficiency.[11,12] To date, minimal correlation has been found between *CFTR* genotype and severity of lung disease, the major cause of mortality for CF patients.[13,14] These observations prompted a search for the causes of disease variation independent of *CFTR* genotype.

Evidence for, and quantification of, genetic modifier role for CF traits

Before embarking upon a search for genetic modifiers, it is important to consider the importance of nongenetic contributions to a phenotype. In the case of CF, the impressive increase in mean years of survival from less than 1 year to the upper 30s currently is attributable almost exclusively to the treatments devised for these patients. From that perspective, environmentally mediated factors have a profound effect on disease variability. Specific treatments such as pancreatic enzyme supplementation, nutritional support, and vitamin supplementation have modified CF from a disease of malnutrition and early childhood demise to a disorder marked by respiratory compromise beginning in childhood and progressing into mid-adulthood.[6] Extension of longevity has also exposed dysfunction of the endocrine pancreas that manifests as diabetes.[15] On the other hand, treatment does not appear to have altered the prevalence of complications that occur in the newborn period (e.g., neonatal intestinal obstruction or meconium ileus) or early to mid-childhood (e.g., hepatic cirrhosis).[9]

Lung disease

A variety of measures have been devised to assess severity of lung disease in CF patients, and the most useful objective gauge is based on airflow. Pulmonary function testing is widely used, and methods for assessing ventilatory capacity and airflow rates are standardized. The forced expiratory volume in 1 sec (FEV_1) is an excellent measure of small and large airway obstruction, the major site of disease in CF lungs. Furthermore, FEV_1 can track progression of obstruction and is well correlated with survival.[16] This lung function measure is highly variable among CF patients with identical *CFTR* genotypes (e.g., *F508del* homozygotes).[17] With the exception of a few mutations that confer a milder pancreatic phenotype (e.g., p.Arg455Glu[18,19]), correlation between different *CFTR* genotypes associated with pancreatic insufficiency and lung function measures is minimal. The small numbers of patients with genotypes other than *F508del* homozygosity and the high degree of variability among patients with the same *CFTR* genotype may preclude observation of correlation.[11] Indeed, preliminary analysis of nearly 40,000 patients in the CFTR2 database reveals low correlation between *CFTR* mutations and FEV_1 (P. Sosnay, personal communication). Together, these studies indicate that factors other than *CFTR* genotype determine progression of airway obstruction in CF.

To parse genetic and nongenetic contributions to disease severity and complications, investigators have used family-based studies. Recurrence of complications in affected siblings at rates higher than in unrelated patients suggests genetic effect, although care must be taken to account for the effect of shared environment. A more powerful approach is to compare monozygous (MZ) and dizygous (DZ) twin pairs for concordance for qualitative traits and correlation for quantitative traits. Twin pairs raised together in the same household are controlled, to some degree, for environmental exposures. One then exploits the different rates of overall sharing of genetic variation between MZ pairs (100%) and DZ pairs (50%) to estimate genetic control. When MZ pairs demonstrate higher concordance (or correlation) for a clinical feature than DZ pairs, then genetic factors are assumed to be responsible.[20] The aforementioned classic twin study initially proposed by Galton and refined by Siemens has been used to estimate the proportion of phenotype variation attributable to genes (i.e., heritability or h^2) for numerous traits in humans.[21] This method has been adapted to estimate h^2 for variation in lung function and anthropometric measures and the complications of intestinal obstruction and diabetes in CF patients (see below).

The first twin-based assessment of gene modifier contribution to CF disease severity used a composite measure of lung function and body mass index (BMI). A higher correlation of this composite measure was observed in 29 MZ versus 12 DZ twins pairs, suggesting genetic control of this trait.[22] Analysis of lung function and weight for height as independent measures did not reveal significant differences between the MZ and DZ twin pairs. A subsequent comparison of 38 MZ pairs with six same-sex DZ pairs and 61 same-sex sibling pairs within 3 years of age drawn from the North American CF Twin and Sibling Study estimated heritability of lung function based on FEV_1 measurements to range from 0.54 to 1.0.[23] The heritability estimate was similar (0.56–0.86) when the analysis was restricted to patients with identical *CFTR* genotypes (*F508del* homozygotes). Variance analysis of 231 pairs of affected siblings generated a slightly higher estimate of heritability for the FEV_1 measures (0.68–1.0).[23] Together, these studies revealed that genetic modifiers play a substantial role in determining FEV_1, a key measure of lung function correlated with survival.

Factors in the environment also contribute heavily to variation in lung function in CF patients. As might be expected from a disorder heavily modified by treatment, socioeconomic factors such as access to care, parental income, and other factors that may compromise optimal treatment have been shown to have significant adverse effects on lung disease and survival.[24,25] Pollutants in the air including particulate matter and components of second-hand smoke can adversely affect lung function in CF.[26,27] To assess the relative contribution of genetic and

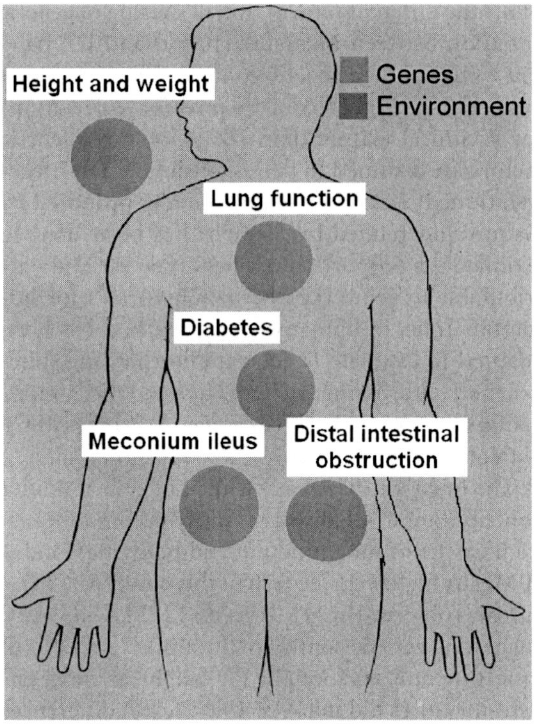

Figure 1. Relative contributions of non-*CFTR* genetic and environmental factors to major manifestations of cystic fibrosis.

environmental factors to variation in FEV_1 among CF patients, Collaco *et al.* evaluated 134 MZ twins and 272 DZ twins and siblings when living together and after moving apart. Differences in lung function between MZ twin pairs while living together in the same household provided an estimate of the contribution of unique environment and stochastic effects. The transition from the home environment to independent living was used to estimate the contribution of shared environment. Genetic contribution was assessed by comparing the similarity in lung function measures in MZ and DZ twin pairs when living together, and then when living apart. These methods revealed that genetic and nongenetic factors had almost equal contribution to variation in lung function (Fig. 1).[28] This estimate was unchanged when the analysis was restricted to twins and siblings bearing the same *CFTR* genotype (*F508del* homozygotes). Two thirds of the nongenetic effect could be attributed to unique environment and stochastic effects while shared environment was responsible for the remainder (Fig. 1). Analysis of 58 MZ twins and 568 DZ twins and siblings from the European CF Twin and Sibling Study generated similar estimates for the genetic and nongenetic contributions to lung function variance (~60% attributable to genetic factors in all subjects and ~50% in subjects who were *F508del* homozygotes).[29]

Exocrine pancreatic function

Pancreatic exocrine function is abnormal in almost all patients with CF. Most patients have severe dysfunction (pancreatic insufficiency) leading to steatorrhea and malabsorption that manifests in the first year of life.[6] As noted above, a fraction of CF patients (~10–20%) do not develop steatorrhea and are termed "pancreatic sufficient." *CFTR* genotype is known to be highly predictive of pancreatic exocrine status.[30] However, there are patients who appear to have delayed transition to pancreatic insufficiency suggesting a possible role for genetic modifiers. Serum immunoreactive trypsinogen (IRT) is a pancreatic enzyme precursor that is elevated in neonates with CF and, as such, serves as a marker for newborn screening for CF. In young CF patients that develop exocrine pancreatic insufficiency, IRT levels decline rapidly.[31] Heritability for variation in IRT levels was estimated in 23 sibling pairs with *CFTR* genotypes associated with pancreatic insufficiency using variance analysis. Evidence of genetic control was noted in patients at 2 months of age ($h^2 = 0.51$) and at 6 months of age (0.45).[32] Genetic effect could not be confirmed in older patients.

Diabetes

Due to the successful management of CF in childhood, more patients are surviving into adulthood. As an unfortunate result additional complications of the disease are becoming more prevalent. Dysfunction of the endocrine pancreas, generally thought to be a consequence of the exocrine pancreatic disease, leads to loss of beta cell function and decreased insulin secretion. Two percent of children with CF develop diabetes, while 19% of adolescents and over 40% of adults with CF develop this complication.[33] Diabetes in CF patients typically occurs in the absence of obesity and is associated with a significantly worse prognosis.[34] While CF-related diabetes (CFRD) has been viewed as a condition distinct from diabetes seen in the general population, there are a number of clinical and pathologic similarities with type 1 and type 2 diabetes.[35] Both type 1 and type 2 diabetes in the general population show evidence of strong genetic control. Numerous genes and DNA variants have been associated with each form of diabetes.

Twin study has used to assess genetic modifier contribution to CFRD. MZ twins displayed significantly higher rates of concordance than DZ twins and siblings of similar age and same sex generating an estimate of heritability approaching 1.0.[36] Pedigree analysis revealed that CFRD and type 2 diabetes may have similar genetic origins. The presence of type 2 diabetes in at least two first-degree relatives was associated with a significantly higher prevalence of diabetes in the CF proband (OR 3.1; $P = 0.0009$) compared to the absence of a similar family history of diabetes.[37] Thus, diabetes in CF patients is quite common, strongly determined by genetic modifiers (Fig. 1) and appears to share some genetic basis with type 2 diabetes in the general population.

Intestinal obstruction

Obstruction of the intestine by abnormal meconium in the neonatal period is a characteristic sign of CF. The condition, termed "meconium ileus" (MI), affects about 15% of newborns with CF.[6] While formally lethal, modern treatments consisting of enemas and/or surgery have reduced mortality to less than 10%.[38] *CFTR* genotype appears to make a sizable contribution to risk as patients bearing the missense mutation p.Gly551Asp are at considerably lower risk (6.4%) than p.Phe508del homozygotes (19.5%),[39] while patients carrying the nonsense mutation p.Gly542X may be at higher risk.[30] However, genetic modifiers make an important contribution as noted by two observations: (1) recurrence risk for MI in affected siblings (25%) is significantly higher than in unrelated CF patients (15%)[40] and (2) rates of intestinal obstruction in mouse models of CF differ by strain (i.e., genetic background).[41] Intriguingly, neonatal intestinal obstruction that is anatomically and temporally similar to humans occurs with a prevalence of 100% in a porcine model of CF[42] and 75% in a ferret model of CF.[43] Thus, it appears that humans may have genetic modifiers that protect from this complication.

Twins and sibling analysis has been used to estimate heritability of MI, and a similar phenotype observed in older CF patients termed "distal intestinal obstruction syndrome" (DIOS). Blackman and colleagues noted that concordance for MI in MZ twin pairs (82%) was significantly higher than in DZ and siblings of similar age and sex (22%; $P = 0.009$) suggesting that heritability for MI approaches 1.0[40] (Fig. 1). Nongenetic factors must play some role as concordance is not 100% in MZ twins. On the other hand, DIOS showed no differences in concordance rates among MZ and DZ twins and siblings indicating little, if any, genetic modifier effect.[40] The latter observation is notable as DIOS was formally termed MI equivalent due to anatomic and pathologic similarities to the neonatal condition.

The intestinal and pancreatic disease in CF leads to substantial disruptions in growth manifesting as short stature and low weight. Aggressive nutritional supplementation has improved both parameters, but chronic illness also affects patients' entry into puberty. Thus, the use of growth metrics such as BMI or percent of predicted weight for height (WfH%) that have been standardized in healthy individuals presents a challenge for assessing nutritional status in CF. Preliminary analysis of "average" BMI displays moderate heritability (~0.6).[44] Genetic control is also noted when BMI measures from select ages are used to avoid confounding by effects of the underlying disease process. Initial analysis of WfH% by the European CF Twin and Sibling Study demonstrated that intrapair discordance for this measure was similar between 29 MZ and 12 DZ twin pairs.[22] However, this issue was recently revisited, and a significant trend was noted when comparing intrapair differences in WfH% in 38 MZ, 24 DZ and 396 sibling pairs. Modeling of these three classes of CF siblings estimated that genetic factors account for ~80% of variation in WfH% in 466 related patients (~60% in *F508del* homozygotes).[29] Thus, anthropometric measures also appear to be under genetic control in CF patients. Further study is needed to assess the degree to which height and weight independently contribute to the heritability of these composite measures.

Other traits

The secretion of chloride via CFTR and other pathways across epithelial tissues has been studied in twin and sibling pairs. MZ twins were found to have higher rates of concordance (5 of 5 pairs) for the presence of cAMP-mediated chloride conductance in intestinal epithelia than DZ twins (9 of 15 pairs). Furthermore, alternative modes of chloride conductance were also more concordant in MZ versus DZ pairs.[45] In a companion study, MZ twins (six pairs) displayed greater concordance for baseline potential difference measurements across nasal epithelia than five sets of DZ twin pairs.[46] The latter generally measures the rate of sodium transport across the respiratory epithelia.[47] Concordance for other measurements that reflect chloride conductance (e.g., chloride-free solutions and isoproteronol response) across the nasal epithelia did not differ between MZ and DZ twins.[46] Abnormal ion transport abnormality in the sweat gland dysfunction is a characteristic finding of CF. Sweat chloride concentration shows a moderate degree of correlation with *CFTR* mutations grouped by pancreatic status (i.e., PI versus PS[48]). However, preliminary analysis of twins and siblings has not shown evidence of genetic effect beyond *CFTR* (Blackman, S. and G.R. Cutting, unpublished observations).

Liver disease is a common problem for CF patients due to obstruction of the common bile duct. A minor fraction (~3–5%) of patients develop severe hepatic cirrhosis and portal hypertension.[6] The occurrence of this complication in younger patients suggests that severe liver disease is not merely a complication of chronic liver damage but a discrete phenotype that may be genetically modified.[49] Sibling recurrence rates have not been extensively analyzed with the exception of discordance for liver disease in five sets of siblings.[50] Hepatic cirrhosis is too uncommon to derive heritability assessments using concordance analysis among the available twins with CF.

Identification of modifier genes

Intense study of the CFTR has uncovered numerous pathways and their components that have high biologic plausibility for modifying CF. Thus, the vast majority of modifier gene studies for CF traits to date have used the candidate gene approach. Testing of variants of known functional effect from carefully selected candidates is attractive for several reasons. The number of variants tested is generally small thereby avoiding severe penalties for multiple comparisons during statistical analysis. Detailed understanding of the candidate gene product and its variants provides mechanistic insight and facilitates experimental studies to evaluate modifier effects. In some cases, compounds that target the protein product of candidate genes may already be available for human trials. On the other hand, use of the candidate approach is fraught with bias due to our very incomplete understanding of CF biology. Since a significant association between a single gene variant and a CF trait can be obtained with very small differences in allele frequencies, it is essential that any association be replicated in a separate population (preferably multiple times). The latter approach is used extensively in genome-wide studies but is equally important in validating candidate gene associations.

In the case of CF, dozens of candidate genes have been evaluated, but replication has been reported for only a fraction of published associations between variants in these genes and the same CF trait. Of these, only a few variants have been tested in more than 1,000 CF patients. At least nine genes can be implicated as modifying some aspect of the CF phenotype based on replication in two independent studies in more than 500 patients in total (Table 1). Since a comprehensive review of all modifier gene studies would be beyond the scope of this review, details will be discussed only for the genes in Table 1. Several recent publications provide detailed lists of all of the genes that have been studied.[5,29,51–53]

Genome-wide approaches on chromosome 19 met with initial success in the localization of a region modifying risk for neonatal intestinal obstruction (MI).[54] Although subsequent studies were unable to replicate linkage to the MI locus (discussed in detail below), the success of genome-wide methods using both linkage and association for common diseases has rekindled enthusiasm for these approaches. The challenge of genome-wide methods for Mendelian diseases is to recruit sufficient patients to attain statistical power to detect variants of relatively small effect. Studies involving tens of thousands of patients as is becoming routine for common conditions are not practical for monogenic disorders, even diseases that are among the most frequent such as CF. However, compelling evidence of substantial genetic effect (i.e., heritability) for lung disease, anthropometry and intestinal obstruction, and diabetes justify the effort required for genome-wide studies. Furthermore, the magnitude of the modifier effect of variants is not known, therefore moderate numbers of CF patients may be sufficient to indentify significant loci. Finally, we do not know if variants conferring modifier effects are common or rare or some combination of both. Candidate gene studies support both concepts as the variants in

Table 1. **Genetic modifiers of different features of cystic fibrosis**[53]

Gene	Pulmonary Function (FEV_1)	*P. aeruginosa* Acquisition/ Colonization	Intestinal Obstruction	Diabetes	Liver Disease
ADIPOR2			Possible effect[87]		
EDNRA	Probable effect[79]				
IFRD1	Possible effect[74]				
IL8	Possible effect[75]				
MBL2	Probable effect[55,62–64,70,94]	Probable effect[60,70,94]			
MSRA			Probable effect[86]		
SERPINA1	No effect[57,95–98]	Likely no effect[97,99,100]			Possible effect[49]
TCF7L2				Probable effect[37]	
TGFB1	Probable effect[57,59,69,70,73,78]	No effect[70,72,73,78,101]			Likely no effect[49,101]

Probable effect: Association observed in ≥3 independent populations with ≥1000 participants in aggregate.
Possible effect: Association observed in ≥2 independent populations with ≥500 participants in aggregate.
Likely no effect: No association observed in ≥2 independent populations with ≥500 participants in aggregate.
No effect: No association observed in ≥3 more independent populations with ≥1000 participants in aggregate.
N.B.: Some studies include replication populations, which are treated as separate independent populations.

TGFb1 affecting lung function are common (~30–50% range) while the variant in *SERPINA1* that increases risk for hepatic cirrhosis is rare (1.2%).

Lung disease

One of first candidate modifier studies indicated that functional variants in *MBL2*, the gene that encodes mannose binding lectin, correlate with lung function measures (FEV_1 and FVC).[55] At least 12 studies have evaluated the effect of null variants in *MBL2* ("*O*" alleles), and 6 studies involving over 2,049 CF patients have reported that MBL-deficient genotypes (*OO* or *AO*) were associated with lower lung function measures (Table 1). Six studies (1,948 patients) did not detect association,[56–60] while one study of 109 Danish patients showed the reverse.[61] MBL was selected as a candidate due to its role in innate immunity and evidence that MBL deficiency results in predilection to bacterial and viral infection. As infection is a key aspect of CF lung disease, factors that alter resistance are reasonable biologic candidates for CF. Three studies involving 2003 CF patients demonstrated earlier age of infection with *Pseudomonas aeruginosa* (*Pa*) associated with MBL deficiency genotypes (Table 1). One study of 112 Swedish patients showed earlier acquisition with sufficient genotypes[56] while five studies (943 patients) did not detect association.[61–64] Lung disease severity (as measured by FEV_1) and infection status are correlated, and both are altered by the age of the patient and by CFTR genotype. Accounting for confounding among these related variables revealed that *MBL2* genotype was associated with infection status rather than the other variables.[60] Thus, it appears that deficiency in MBL predisposes to early infection with *Pa*, which, in turn, leads to more severe lung disease than observed in patients of the same age and CFTR genotype who are not MBL deficient.

Transforming growth factor beta 1 (*TGFb1*) has been investigated numerous times for potential modifier effect on CF. It is a compelling candidate as *TGFb1* genotypes have been shown to alter risk for other lung diseases such as asthma and chronic obstructive pulmonary disease.[65–67] Furthermore, TGFB1 plays a key role in processes central to CF lung pathophysiology, such as regulation of inflammation and tissue remodeling.[68] One of the largest CF genetic modifier studies to date, the Genetic Modifier Study (GMS) analyzed 808 *F508del* homozygotes drawn from the extreme of lung function (highest 30 percentile and lowest 30 percentile) and reported that alleles in the promoter (−509) and first exon (codon 10) of *TGFb1* are associated with worse lung function.[57] This finding was replicated in 498 patients with various CFTR genotypes and was independently confirmed when a haplotype composed of the opposite alleles at –509 and codon 10 were associated with improved lung function.[69] It also appears that gene and environmental interactions are important to consider in evaluating *TGFb1*, as variation in *CFTR* and in *MBL2* as well as exposure to second-hand smoke affect association with CF lung function.[69–71] Six studies encompassing over 2,500 CF patients report association between *TGFb1* and CF lung function (see Table 1) while one study composed of 118 patients did not,[72] and another involving 171 patients[73] found association with worse lung function and the opposite alleles than reported by Drumm *et al.*[57] and Bremer *et al.*[69] In aggregate, these studies indicate that alleles that increase *TGFb1* expression cause worse lung function in CF, although definitive experiments in patients remain to be performed.

Genome-wide methods have been used to identify loci and genes modifying CF lung disease. In a study of 160 mild and 160 severe lung disease *F508del* homozygous patients drawn from the GMS,[57] Karp and colleagues were able to identify 6 candidate loci that associated with lung disease.[74] These loci were prioritized using hierarchical clustering for features such as biologic plausibility and regional clustering of associated variants. Three SNPs in the highest ranking gene, the interferon-related developmental regulator 1 gene (*IFRD1*), showed replication in the entire GMS sample and demonstrated association using transmission-based methods in the family-based CF Twin and Sibling Study (TSS).[23] Cell- and mouse-based studies demonstrated that IFRD1 acts via transcriptional mechanisms to modulate neutrophil function in response to bacterial infection.

The concept that modification of CF lung disease may due to altered neutrophil response to infection is supported by evidence that variants in the interleukin-8 (*IL-8*) gene associate with lung function.[75] IL-8 is a mediator of neurophil chemotaxis and is variably and prominently elevated in the airway secretions of CF patients.[76,77] Intriguingly, the modifier effect of *IL-8* variation may be limited to males ($n = 608$). A second study of 329 patients failed to detect association between lung function and IL-8 variants, although analysis was not performed using patients separated by sex.[78] Taken together, the identification of *MBL2, TGFb1, IFRD1,* and *IL-8* (Table 1) suggests that genetic modifiers may alter lung disease severity by affecting host ability to tolerate infection.

Other mechanisms are likely to contribute to CF lung pathology, as illustrated by evidence that variants in the endothelin receptor type A (*EDNRA*) gene reproducibly associate with lung disease severity (Table 1). Association between a variant in the 3′ untranslated region of *EDNRA* was observed in 709 *F508del* homozygous patients in the GMS study and replicated in three separate samples of CF patients (769 combined).[79] Alleles of the *EDNRA* variant correlated with differences in RNA transcript level, suggesting a possible functional role.[79] Given that variation EDNRA has been implicated in vasoconstrictive diseases due to effects on smooth muscle function, it was speculated that this gene may modify CF lung disease by altering smooth muscle tone in the airways and/or vasculature.[79]

To maximize power to detect novel CF genetic modifiers using genome-wide approaches, the GMS[57] and TSS[23] have joined forces with a CF modifier study based in Toronto (Canadian Genetics Study; CGS)[70] to create the North American CF Gene Modifier Consortium. Three different study designs are incorporated in the consortium: extremes of phenotype (GMS), population based (CGS), and family based (TSS). The former two designs are well-suited to association methods as used in unrelated patients with common diseases. The family-based study can be used for both linkage and association. Using a complementary approach, seven loci meeting criteria for suggestive or significant association at the genome level identified by the two samples of unrelated patients (GMS and CGS) were tested for replication in the TSS. Two suggestive loci showed evidence of association in the TSS but did not meet correction for multiple testing. Sibling analysis revealed a prominent linkage peak on chromosome 20.[80–82]

Intestinal obstruction

Obstruction of the intestine in mice with a knockout of the CFTR gene provided an early opportunity to demonstrate the role of genetic modifiers. Strain-specific effects upon the rate of obstruction and survival of CF mice illustrate that genetic background played a significant role in the intestinal trait. By exploiting strain differences, Rozhmahel and colleagues were able to map a locus for obstruction to murine chromosome 7.[41] Chromosome 19 contains a region of synteny with the locus on mouse chromosome 7. Using markers from chromosome

19, Zielenski *et al.* demonstrated association with meconium ileus in 152 sibling pairs (7 concordant affected, 33 discordant, and 112 concordant unaffected).[54] Extensive analysis of the chromosome 19 region and the implicated genes failed to identify a causative variant. Two subsequent studies of MI performed in siblings with CF failed to find linkage to chromosome 19. Subsequent mapping studies have identified regions on several mouse chromosomes[83,84] as modifying the intestinal phenotype. Other genes that mediate chloride conductance in intestinal tissue have been evaluated in mice and support the conclusion that these genes could modify MI in humans.[85]

Recent studies using patients collected by the TSS and the CGS have uncovered regions of linkage and association that have been replicated. Blackman and colleagues performed a linkage study of 26 sibling pairs with CF who were both affected with MI. As expected, the CFTR gene was identified with high precision, demonstrating the effectiveness of linkage for genes of moderate to high effect size. Two regions of suggestive linkage (LOD >2.0) were identified on chromosome 8p23.1 and 11q25.[40] A subsequent association analysis of the most convincing linkage signal on 8p23.1 was performed using patients from the TSS, GMS, and CGS. Region-wide significance was obtained in the TSS for SNP and haplotype association that implicated the methionine sulfoxide reductase *(MSRA)* gene. Association between the haplotype (composed of three SNPs with a frequency of 15%) was replicated in a subset of patients from the GMS study and in the entire CGS sample.[86] *MSRA* reduces oxidized methionine residues in proteins that are relevant for intestinal enzymes such as alpha 1 antitrypsin. Variation in MSRA is postulated to alter digestion of intestinal contents contributing to the formation of viscous meconium that obstructs the intestine in CF fetuses and newborns. In a separate study, Dorfman and colleagues used parametric linkage methods to indentify a linkage signal on chromosome 12p13.3 (HLOD 2.935). Association analysis of the linkage region identified a SNP in the adiponectin receptor 2 (ADIPOR2) gene with risk for MI in the CGS population that replicated in the TSS sample.[87] Using a similar approach, a broad region on chromosome 4q13 was found to protect from MI. Causative variants have not been found in either *MSRA* or *ADIPOR2*, thus further evaluation of these genes may require functional studies. The CF mouse may be ideal for assess how knockout or hypomorphs of these modifier genes affect intestinal development and function given that this animal model develops intestinal obstruction.

Diabetes

Candidate gene studies for diabetes were informed by the observation that diabetes in CF may share genetic origins with type 2 diabetes.[37] SNPs near the transcription factor 7-like 2 *(TCF7L2)* gene[88] demonstrate the most consistent association with type 2 diabetes with an estimated odds ratio of ~1.5 per allele.[89] Transmission testing of the most highly associated SNP (rs7903146) in 539 families (998 patients) in the TSS revealed significant association with risk for diabetes ($P = 0.004$). The association was replicated in 802 unrelated patients from the GMS ($P = 0.02$; combined $P = 0.0002$).[37] Treatment with systemic glucocorticoids, an independent environmental risk factor for diabetes in CF,[90] obscured the association with *TCF7L2*. However, risk for diabetes associated with *TCF7L2* was substantially increased (HR 2.9 per allele; $P = 0.00011$) when patients who had not been treated with glucocorticoids in the past year were analyzed.[37] Together, these observations indicate that *TCF7L2* modifies risk for diabetes in CF patients who have not had recent or prolonged exposure to systemic steroids. *TCF7L2* is postulated to play a role in proliferation and function of the beta cells of pancreatic islets.[91,92] Thus, implication of *TCF7L2* as a modifier of risk in CF-related diabetes suggests that factors intrinsic to the beta cells of the pancreas contribute to the development of this common complication of CF.

Future directions and implications for other monogenic disorders

The candidate gene approach has been of moderate success in finding genetic modifiers of CF, and several associations have been replicated numerous times. These replicated modifiers account for a small fraction of the heritable contribution to each trait (<1%). The identification of loci and genes using genome-wide methods appears highly promising. The sample sizes accrued by the North American CF Modifier Consortium (~3,500 pts) have been sufficient to find several loci/genes responsible for variation in lung function and for risk of MI and diabetes. Initial screens found at least seven loci meeting or exceeding suggestive evidence of association. Thus, it is reasonable to predict that additional modifier loci and genes, each contributing small effect sizes, exist, and should be discoverable with larger sample sizes. Furthermore, environmental factors are likely to work in concert with genetic modifiers to amplify effect, as in the case of *TGFb1* and second-hand smoke, or diminish effect, as noted for steroid exposure and *TCF7L2*. Longitudinal studies of patients with single-gene disorders are likely to be needed to identify key exposures that contribute to gene–environment interactions. The unfortunate chronicity of monogenic conditions requires frequent medical evaluation, facilitating collection of detailed exposure histories. Furthermore, the near-universal screening for CF in newborns provides an ideal opportunity to recruit patients for longitudinal studies.

A primary reason to identify disease modifiers is to inform the development of therapies and to provide accurate prediction of disease course. Achieving these goals will require characterization of the mechanism of disease modification. One needs to know if improvement in function is afforded by increased activity of a modifier protein, or the reverse. The availability of various mouse strains carrying CFTR mutations associated with mild and severe disease provides an excellent foundation for functional studies and investigation of interactions between CFTR and specific modifier genes. The recent development of pig and ferret models of CF that more completely reflect human CF pathology should be useful for therapy evaluation.[43,93] Prognostication will be useful when the effect of genetic and nongenetic factors can be combined to accurately estimate a meaningful change in outcome. The latter may likely require prospective studies in sizable patient cohorts.

The experience in CF modifier research holds some useful insight for other single gene disorders. The importance of consistent phenotype measures and complementary study designs cannot be overemphasized. Family studies can help estimate the contribution of genetic and nongenetic factors. Sibling analysis should be considered if sufficient twins are not available for statistically robust concordance/correlation analyses. By their very nature, Mendelian studies lend themselves to recruitment of families that can be employed in linkage and association studies. Genotyping of parents provides accurate assessment of descent in linkage analysis and enables transmission disequilibrium methods. Ascertaining unrelated patients from the across the severity spectrum is essential for identification and replication studies using by association tools. Finally, recruitment of singleton patients is facilitated by their Mendelian condition as they can be verified to have disease due to the same etiologic factor.

Conflicts of interest

The author declares no conflicts of interest.

References

1. Antonarakis, S.E. & J.S. Beckmann. 2006. Mendelian disorders deserve more attention. *Nat. Rev. Genet.* **7:** 277–282.
2. Scriver, C.R. & P.J. Waters. 1999. Monogenic traits are not simple: lessons from phenylketonuria. *Trends Genet.* **15:** 267–272.
3. Nadeau, J.H. 2001. Modifier genes in mice and humans. *Nat. Rev. Genet.* **2:** 165–174.
4. Drumm, M. 2001. Modifier genes and variation in cystic fibrosis. *Respir. Res.* **2:** 125–128.
5. Cutting, G.R. 2005. Modifier genetics: cystic fibrosis. *Annu. Rev. Genomics Hum. Genet.* **6:** 237–260.
6. Welsh, M.J., B.W. Ramsey, F.J. Accurso & G.R. Cutting. 2001. Cystic fibrosis. In *The Metabolic and Molecular Bases of Inherited Disease*, Vol. III. C.R. Scriver, A.L. Beaudet, D. Valle & W.S. Sly, Eds.: 5121–5188. McGraw-Hill, Inc. New York.
7. Anderson, M.P., R.J. Gregory, S. Thompson, *et al.* 1991. Demonstration that CFTR is a chloride channel by alteration of its anion selectivity. *Science* **253:** 202–205.
8. Cutting, G.R. 2007. Cystic fibrosis. In *Emery and Rimoin's Principles and Practice of Medical Genetics*, Vol. 2. D.L. Rimoin, J.M. Connor, R.E. Pyeritz & B.R. Korf, Eds.: 1354–1394. Churchill Livingstone Elsevier. Philadelphia.
9. Cystic Fibrosis Foundation. 2005. *Cystic Fibrosis Foundation Patient Registry Annual Data Report 2005*. Cystic Fibrosis Foundation. Bethesda, MD.
10. Bobadilla, J.L., M. Macek, J.P. Fine & P.M. Farrell. 2002. Cystic fibrosis: a worldwide analysis of CFTR mutations—correlation with incidence data and application to screening. *Hum. Mutat.* **19:** 575–606.
11. The Cystic Fibrosis Genotype-Phenotype Consortium. 1993. Correlation between genotype and phenotype in patients with cystic fibrosis. *N. Engl. J. Med.* **329:** 1308–1313.
12. Wilschanski, M., A. Dupuis, L. Ellis, *et al.* 2006. Mutations in cystic fibrosis transmembrane regulator gene and in vivo transepithelial potentials. *Am. J. Respir. Crit Care Med.* **78:** 787–794.
13. Mickle, J.E. & G.R. Cutting. 2000. Genotype-phenotype relationships in cystic fibrosis. *Med. Clin. North Am.* **84:** 597–607.
14. Zielenski, J. 2000. Genotype and phenotype in cystic fibrosis. *Respiration* **67:** 117–133.

15. Stecenko, A.A. & A. Moran. 2010. Update on cystic fibrosis-related diabetes. *Curr. Opin. Pulm. Med.* **16:** 611–615.
16. Kerem, E., J. Reisman, M. Corey, *et al.* 1992. Prediction of mortality in patients with cystic fibrosis. *N. Engl. J. Med.* **326:** 1187–1191.
17. Kerem, E., M. Corey, B.-S. Kerem, *et al.* 1990. The relation between genotype and phenotype in cystic fibrosis—analysis of the most common mutation (deltaF508). *N. Engl. J. Med.* **323:** 1517–1522.
18. Gan, K.-H., H.J. Veeze, A.M.W. Van Den Ouweland, *et al.* 1995. A cystic fibrosis mutation associated with mild lung disease. *N. Engl. J. Med.* **333:** 95–99.
19. De Braekeleer, M., C. Allard, J.-P. Leblanc, *et al.* 1997. Genotype-phenotype correlation in cystic fibrosis patients compound heterozygous for the A455E mutation. *Hum. Genet.* **101:** 208–211.
20. Falconer, D.S. 1965. Inheritance of liability to certain diseases estimated from incidence among relatives. *Ann. Hum. Genet.* **29:** 51–71.
21. Spector, T.D. 2000. The history of twin and sibling-pair studies. In *Advances in Twin and Sib-Pair Analysis.* T.D. Spector, H. Snieder & A.J. MacGregor, Eds.: 1–10. Greenwich Medical Media Ltd. London.
22. Mekus, F., M. Ballmann, I. Bronsveld, *et al.* 2000. Categories of deltaF508 homozygous cystic fibrosis twin and sibling pairs with distinct phenotypic characteristics. *Twin. Res.* **3:** 277–293.
23. Vanscoy, L.L., S.M. Blackman, J.M. Collaco, *et al.* 2007. Heritability of lung disease severity in cystic fibrosis. *Am. J. Respir. Crit. Care Med.* **175:** 1036–1043.
24. O'Connor, G.T., H.B. Quinton, T. Kneeland, *et al.* 2003. Median household income and mortality rate in cystic fibrosis. *Pediatrics* **111:** e333–e339.
25. Schechter, M.S. 2004. Non-genetic influences on CF lung disease: the role of sociodemographic characteristics, environmental exposures and healthcare interventions. *Pediatr. Pulmonol. Suppl.* **26:** 82–85.
26. Rubin, B.K. 1990. Exposure of children with cystic fibrosis to environmental tobacco smoke. *N. Engl. J. Med.* **323:** 782–788.
27. Goss, C.H., S.A. Newsom, J.S. Schildcrout, *et al.* 2004. Effect of ambient air pollution on pulmonary exacerbations and lung function in cystic fibrosis. *Am. J. Respir. Crit. Care Med.* **169:** 816–821.
28. Collaco, J.M., S.M. Blackman, J. McGready, *et al.* 2010. Quantification of the relative contribution of environmental and genetic factors to variation in cystic fibrosis. *Lung Funct. J. Pediatr.* **157:** 802–807.
29. Stanke, F., T. Becker, V. Kumar, *et al.* 2010. Genes that determine immunology and inflammation modify the basic defect of impaired ion conductance in bystic fibrosis epithelia. *J. Med. Genet.* doi:10.1136/jmg.2010.080937.
30. Kristidis, P., D. Bozon, M. Corey, *et al.* 1992. Genetic determination of exocrine pancreatic function in cystic fibrosis. *Am. J. Hum. Genet.* **50:** 1178–1184.
31. Durie, P., G.G. Forstner, K.J. Gaskin, *et al.* 1986. Age-related alteration of immunoreactive pancreatic cationic trypsinogen in sera from cystic fibrosis patients with and without pancreatic insufficiency. *Pediatr. Res.* **20:** 209–213.
32. Sontag, M.K., M. Corey, J.E. Hokanson, *et al.* 2006. Genetic and physiologic correlates of longitudinal immunoreactive trypsinogen decline in infants with cystic fibrosis identified through newborn screening. *J. Pediatr.* **149:** 650–657.
33. Moran, A., J. Dunitz, B. Nathan, *et al.* 2009. Cystic fibrosis-related diabetes: current trends in prevalence, incidence, and mortality. *Diabetes Care* **32:** 1626–1631.
34. Milla, C.E., W.J. Warwick & A. Moran. 2000. Trends in pulmonary function in patients with cystic fibrosis correlate with the degree of glucose intolerance at baseline. *Am. J. Respir. Crit. Care Med.* **162:** 891–895.
35. Moran, A., D. Hardin, D. Rodman, *et al.* 1999. Diagnosis, screening and management of cystic fibrosis related diabetes mellitus: a consensus conference report. *Diabetes Res. Clin. Pract.* **45:** 61–73.
36. Blackman, S.M., S. Hsu, L.L. Vanscoy, *et al.* 2009. Genetic modifiers play a substantial role in diabetes complicating cystic fibrosis. *J. Clin. Endocrinol. Metab.* **94:** 1302–1309.
37. Blackman, S.M., S. Hsu, S.E. Ritter, *et al.* 2009. A susceptibility gene for type 2 diabetes confers substantial risk for diabetes complicating cystic fibrosis. *Diabetologia* **52:** 1858–1865.
38. Escobar, M.A., J.L. Grosfeld, J.J. Burdick, *et al.* 2005. Surgical considerations in cystic fibrosis: a 32-year evaluation of outcomes. *Surgery* **138:** 560–571.
39. Hamosh, A., T.M. King, B.J. Rosenstein, *et al.* 1992. Cystic fibrosis patients bearing the common missense mutation Gly->Asp at codon 551 and the deltaF508 are indistinguishable from deltaF508 homozygotes except for decreased risk of meconium ileus. *Am. J. Hum. Genet.* **51:** 245–250.
40. Blackman, S.M., R. Deering-Brose, R. McWilliams, *et al.* 2006. Relative contribution of genetic and nongenetic modifiers to intestinal obstruction in cystic fibrosis. *Gastroenterology* **131:** 1030–1039.
41. Rozmahel, R., M. Wilschanski, A. Matin, *et al.* 1996. Modulation of disease severity in cystic fibrosis transmembrane conductance regulator deficient mice by a secondary genetic factor. *Nat. Genet.* **12:** 280–287.
42. Stoltz, D.A., D.K. Meyerholz, A.A. Pezzulo, *et al.* 2010. Cystic fibrosis pigs develop lung disease and exhibit defective bacterial eradication at birth. *Sci. Transl. Med.* **2:** 29ra31.
43. Sun, X., H. Sui, J.T. Fisher, *et al.* 2010. Disease phenotype of a ferret CFTR-knockout model of cystic fibrosis. *J. Clin. Invest.* **120:** 3149–3160.
44. Blackman, S.M., L.L. Vanscoy, J.M. Collaco, *et al.* 2008. Variability in body mass index in cystic fibrosis is determined partly by a genetic locus on chromosome 5. *Pediatr. Pulmonol. Supp.* **31:** 271. 10–23-0008.
45. Bronsveld, I., F. Mekus, J. Bijman, *et al.* 2000. Residual chloride secretion in intestinal tissue of deltaF508 homozygous twins and siblings with cystic fibrosis. The European CF Twin and Sibling Study Consortium. *Gastroenterology* **119:** 32–40.
46. Bronsveld, I., F. Mekus, J. Bijman, *et al.* 2001. Chloride conductance and genetic background modulate the cystic fibrosis phenotype of Delta F508 homozygous twins and siblings. *J. Clin. Invest.* **108:** 1705–1715.

47. Knowles, M.R., A.M. Paradiso & R.C. Boucher. 1995. In vivo nasal potential difference: techniques and protocols for assessing efficacy of gene transfer in cystic fibrosis. *Hum. Gene Ther.* **6:** 445–455.
48. Wilschanski, M., J. Zielenski, D. Markiewicz, *et al.* 1995. Correlation of sweat chloride concentration with classes of the cystic fibrosis transmembrane conductance regulator gene mutations. *J. Pediatr.* **127:** 705–710.
49. Bartlett, J.R., K.J. Friedman, S.C. Ling, *et al.* 2009. Genetic modifiers of liver disease in cystic fibrosis. *JAMA* **302:** 1076–1083.
50. Castaldo, G., A. Fuccio, D. Salvatore, *et al.* 2001. Liver expression in cystic fibrosis could be modulated by genetic factors different from the cystic fibrosis transmembrane regulator genotype. *Am. J. Med. Genet.* **98:** 294–297.
51. Davies, J.C., U. Griesenbach & E. Alton. 2005. Modifier genes in cystic fibrosis. *Pediatr. Pulmonol.* **39:** 383–391.
52. Knowles, M.R. 2006. Gene modifiers of lung disease. *Curr. Opin. Pulm. Med.* **12:** 416–421.
53. Collaco, J.M. & G.R. Cutting. 2008. Update on gene modifiers in cystic fibrosis. *Curr. Opin. Pulm. Med.* **14:** 559–566.
54. Zielenski, J., M. Corey, R. Rozmahel, *et al.* 1999. Detection of a cystic fibrosis modifier locus for meconium ileus on human chromosome 19q13. *Nat. Genet.* **22:** 128–129.
55. Garred, P., T. Pressler, H.O. Madsen, *et al.* 1999. Association of mannose-binding lectin gene heterogeneity with severity of lung disease and survival in cystic fibrosis. *J. Clin. Invest.* **104:** 431–437.
56. Carlsson, M., A.G. Sjoholm, L. Eriksson, *et al.* 2005. Deficiency of the mannan-binding lectin pathway of complement and poor outcome in cystic fibrosis: bacterial colonization may be decisive for a relationship. *Clin. Exp. Immunol.* **139:** 306–313.
57. Drumm, M.L., M.W. Konstan, M.D. Schluchter, *et al.* 2005. Gene modifiers of lung disease in cystic fibrosis. *N. Engl. J. Med.* **353:** 1443–1453.
58. Choi, E.H., M. Ehrmantraut, C.B. Foster, *et al.* 2006. Association of common haplotypes of surfactant protein A1 and A2 (SFTPA1 and SFTPA2) genes with severity of lung disease in cystic fibrosis. *Pediatr. Pulmonol.* **41:** 255–262.
59. Faria, E.J., I.C. Faria, A.F. Ribeiro, *et al.* 2009. Association of MBL2, TGF-beta1 and CD14 gene polymorphisms with lung disease severity in cystic fibrosis. *J. Bras. Pneumol.* **35:** 334–342.
60. McDougal, K.E., D.M. Green, L.L. Vanscoy, *et al.* 2010. Use of a modeling framework to evaluate the effect of a modifier gene (MBL2) on variation in cystic fibrosis. *Eur. J. Hum. Genet.* **18:** 680–684.
61. Olesen, H.V., J.C. Jensenius, R. Steffensen, *et al.* 2006. The mannan-binding lectin pathway and lung disease in cystic fibrosis–disfunction of mannan-binding lectin-associated serine protease 2 (MASP-2) may be a major modifier. *Clin. Immunol.* **121:** 324–331.
62. Gabolde, M., M. Guilloud-Bataille, J. Feingold & C. Besmond. 1999. Association of variant alleles of mannose binding lectin with severity of pulmonary disease in cystic fibrosis: cohort study. *Br. Med. J.* **319:** 1166–1167.
63. Davies, J.C., M.W. Turner & N. Klein. 2004. Impaired pulmonary status in cystic fibrosis adults with two mutated MBL-2 alleles. *Eur. Respir. J.* **24:** 798–804.
64. Yarden, J., D. Radojkovic, K. De Boeck, *et al.* 2004. Polymorphisms in the mannose binding lectin gene affect the cystic fibrosis pulmonary phenotype. *J. Med. Genet.* **41:** 629–633.
65. Pulleyn, L.J., R. Newton, I.M. Adcock & P.J. Barnes. 2001. TGFbeta1 allele association with asthma severity. *Hum. Genet.* **109:** 623–627.
66. Wu, L., J. Chau, R.P. Young, *et al.* 2004. Transforming growth factor-beta1 genotype and susceptibility to chronic obstructive pulmonary disease. *Thorax* **59:** 126–129.
67. Silverman, E.S., L.J. Palmer, V. Subramaniam, *et al.* 2004. Transforming growth factor-beta1 promoter polymorphism C-509T is associated with asthma. *Am. J. Respir. Crit. Care Med.* **169:** 214–219.
68. Akhurst, R.J. 2004. TGF beta signaling in health and disease. *Nat. Genet.* **36:** 790–792.
69. Bremer, L.A., S.M. Blackman, L.L. Vanscoy, *et al.* 2008. Interaction between a novel TGFB1 haplotype and CFTR genotype is associated with improved lung function in cystic fibrosis. *Hum. Mol. Genet.*
70. Dorfman, R., A. Sandford, C. Taylor, *et al.* 2008. Complex two-gene modulation of lung disease severity in children with cystic fibrosis. *J. Clin. Invest.* **118:** 1040–1049.
71. Collaco, J.M., L. Vanscoy, L. Bremer, *et al.* 2008. Interactions between secondhand smoke and genes that affect cystic fibrosis lung disease. *JAMA* **299:** 417–424.
72. Brazova, J., K. Sismova, V. Vavrova, *et al.* 2006. Polymorphisms of TGF-beta1 in cystic fibrosis patients. *Clin. Immunol.* **121:** 350–357.
73. Arkwright, P.D., S. Laurie, M. Super, *et al.* 2000. TGF-beta(1) genotype and accelerated decline in lung function of patients with cystic fibrosis. *Thorax* **55:** 459–462.
74. Gu, Y., I.T. Harley, L.B. Henderson, *et al.* 2009. Identification of IFRD1 as a modifier gene for cystic fibrosis lung disease. *Nature* **458:** 1039–1042.
75. Hillian, A.D., D. Londono, J.M. Dunn, *et al.* 2008. Modulation of cystic fibrosis lung disease by variants in interleukin-8. *Genes Immun.* **9:** 501–508.
76. Bonfield, T.L., M.W. Konstan & M. Berger. 1999. Altered respiratory epithelial cell cytokine production in cystic fibrosis. *J. Allergy Clin. Immunol.* **104:** 72–78.
77. Chmiel, J.F., M. Berger & M.W. Konstan. 2002. The role of inflammation in the pathophysiology of CF lung disease. *Clin. Rev. Allergy Immunol.* **23:** 5–27.
78. Corvol, H., P.Y. Boelle, J. Brouard, *et al.* 2008. Genetic variations in inflammatory mediators influence lung disease progression in cystic fibrosis. *Ped. Pulm.* **43:** 1224–1232.
79. Darrah, R., E. McKone, C. O'Connor, *et al.* 2010. EDNRA variants associate with smooth muscle mRNA levels, cell proliferation rates, and cystic fibrosis pulmonary disease severity. *Physiol. Genomics* **41:** 71–77.
80. Knowles, M.R. 2009. CF genome-wide association study: Update from NA Consortium. *Pediatric Pulm.* **32** (Supplement): 165–166.
81. Cutting, G.R. & CF Modifier Consortium. 2010. A genome-wide association and linkage study (GWALS) identifies two

loci modifying lung disease severity in cystic fibrosis: North American CF Modifier Consortium. In *American Society of Human Genetics Annual Meeting-60th Annual Meeting Program Guide*. Available at www.ashg.org/2010meeting.

82. Knowles, M.R. 2010. Genetic Modifiers in cystic fibrosis. In *American Society of Human Genetics Annual Meeting-60th Annual Meeting Program Guide*. Available at www.ashg.org/2010meeting.
83. Haston, C.K. & L.C. Tsui. 2003. Loci of intestinal distress in cystic fibrosis knockout mice. *Physiol. Genomics* **12:** 79–84.
84. Norkina, O. & R.C. De Lisle. 2005. Potential genetic modifiers of the cystic fibrosis intestinal inflammatory phenotype on mouse chromosomes 1, 9, and 10 BMC. *Genet.* **6:** 29.
85. Young, F.D., S. Newbigging, C. Choi, *et al.* 2007. Amelioration of cystic fibrosis intestinal mucous disease in mice by restoration of mCLCA3. *Gastroenterology* **133:** 1928–1937.
86. Henderson, L.B., V. Doshi, S.M. Blackman, *et al.* 2009. A haplotype in the MSRA gene confers decreased risk of meconium ileus in cystic fibrosis. In *American Society of Human Genetics Annual Meeting—59th Annual Meeting Program Guide*. Available at www.ashg.org/2009meeting.
87. Dorfman, R., W. Li, L. Sun, *et al.* 2009. Modifier gene study of meconium ileus in cystic fibrosis: statistical considerations and gene mapping results. *Hum. Genet.* **126:** 763–778.
88. Grant, S.F., G. Thorleifsson, I. Reynisdottir, *et al.* 2006. Variant of transcription factor 7-like 2 (TCF7L2) gene confers risk of type 2 diabetes. *Nat. Genet.* **38:** 320–323.
89. Cauchi, S., Y. El Achhab, H. Choquet, *et al.* 2007. TCF7L2 is reproducibly associated with type 2 diabetes in various ethnic groups: a global meta-analysis. *J. Mol. Med.* **85:** 777–782.
90. Marshall, B.C., S.M. Butler, M. Stoddard, *et al.* 2005. Epidemiology of cystic fibrosis-related diabetes. *J. Pediatr.* **146:** 681–687.
91. Liu, Z. & J.F. Habener. 2008. Glucagon-like peptide-1 activation of TCF7L2-dependent Wnt signaling enhances pancreatic beta-cell proliferation. *J. Biol. Chem.* **283:** 8723–8735.
92. Wegner, L., M.S. Hussain, K. Pilgaard, *et al.* 2008. Impact of TCF7L2 rs7903146 on insulin secretion and action in young and elderly Danish twins. *J. Clin. Endocrinol. Metab.* **93:** 4013–4019.
93. Stoltz, D.A., D.K. Meyerholz, A.A. Pezzulo, *et al.* 2010. Cystic fibrosis pigs develop lung disease and exhibit defective bacterial eradication at birth. *Sci. Trans. Med.* **2:** 29ra31.
94. Trevisiol, C., M. Boniotto, L. Giglio, *et al.* 2005. MBL2 polymorphisms screening in a regional Italian CF Center. *J. Cyst. Fibros.* **4:** 189–191.
95. Doring, G., H. Krogh-Johansen, S. Weidinger & N. Hoiby. 1994. Allotypes of alpha 1-antitrypsin in patients with cystic fibrosis, homozygous and heterozygous for deltaF508. *Pediatr. Pulmonol.* **18:** 3–7.
96. Henry, M.T., S. Cave, J. Rendall, *et al.* 2001. An alpha(1)-antitrypsin enhancer polymorphism is a genetic modifier of pulmonary outcome in cystic fibrosis. *Eur. J. Hum. Genet.* **9:** 273–278.
97. Frangolias, D.D., J. Ruan, P.J. Wilcox, *et al.* 2003. Alpha 1-antitrypsin deficiency alleles in cystic fibrosis lung disease. *Am. J. Respir. Cell Mol. Biol.* **29:** 390–396.
98. de Faria, E.J., I.C. de Faria, A.E. Alvarez, *et al.* 2005. [Association between alpha 1 antitrypsin deficiency and cystic fibrosis severity]. *J. Pediatr. (Rio J)* **81:** 485–490.
99. Mahadeva, R., R.C. Westerbeek, D.J. Perry, *et al.* 1998. Alpha1-antitrypsin deficiency alleles and the Taq-I G->A allele in cystic fibrosis lung disease. *Eur. Respir. J.* **11:** 873–879.
100. Meyer, P., A. Braun & A.A. Roscher. 2002. Analysis of the two common alpha-1-antitrypsin deficiency alleles PiMS and PiMZ as modifiers of Pseudomonas aeruginosa susceptibility in cystic fibrosis. *Clin. Genet.* **62:** 325–327.
101. Arkwright, P.D., V. Pravica, P.J. Geraghty, *et al.* 2003. End-organ dysfunction in cystic fibrosis: association with angiotensin I converting enzyme and cytokine gene polymorphisms. *Am. J. Respir. Crit. Care Med.* **167:** 384–389.

Ann. N.Y. Acad. Sci. ISSN 0077-8923

ANNALS OF THE NEW YORK ACADEMY OF SCIENCES
Issue: *The Year in Human and Medical Genetics*

Mendelian genetics of rare—and not so rare—cancers

Charis Eng[1,2,3,4,5]

[1]Genomic Medicine Institute, Cleveland Clinic, Cleveland, Ohio. [2]Lerner Research Institute, Cleveland Clinic, Cleveland, Ohio. [3]Taussig Cancer Institute, Cleveland Clinic, Cleveland, Ohio. [4]Stanley Shalom Zielony Nursing Institute, Cleveland Clinic, Cleveland, Ohio. [5]Department of Genetics and CASE Comprehensive Cancer Center, Case Western Reserve University School of Medicine, Cleveland, Ohio

Address for correspondence: Charis Eng, M.D., Ph.D., Genomic Medicine Institute, Cleveland Clinic, 9500 Euclid Avenue, NE-50, Cleveland, OH 44195. engc@ccf.org

Mendelian genetics forms the basis for gene-informed risk assessment and management for the patient and family, and should be at the very foundation of 21st century personalization of healthcare. Yet this is an underutilized commodity. Identification and characterization of germline mutations in the *RET* proto-oncogene, encoding a receptor tyrosine kinase, as causing >90% of multiple endocrine neoplasia type 2 (MEN 2), an autosomal dominant disorder characterized by medullary thyroid cancer, pheochromocytoma, and hyperparathyroidism, heralded the era of evidence-based molecular diagnosis, predictive testing, genetic counseling, gene-informed cancer risk assessment, and preventative medicine. Since then, many syndromic endocrine neoplasias have proven to fall under this clinically utile and actionable model, such as those caused by mutations in *RET, VHL*, or SDHB-D. The familial risk associated with epithelial (nonmedullary) thyroid carcinoma is among the highest of all solid tumors, yet only a few highly penetrant heritable epithelial thyroid cancer syndrome exist. This is illustrated by Cowden syndrome, a difficult-to-recognize autosomal dominant disorder characterized by breast, thyroid, and other cancers, caused by germline mutations in *PTEN*, encoding a phosphatase, and minorly, *SDHB*/*SDHD* variants.

Keywords: personalized healthcare; molecular diagnostics; predictive testing; multiple endocrine neoplasia syndromes; *PTEN* hamartoma tumor syndrome; *SDH*-related pheochromocytoma/paraganglioma

Introduction

The recent fad and rallying cry of "personalized medicine" has led to a situation of trying to sprint at high speeds without even implementing and consolidating the infrastructure of how healthcare can walk in this field, that is, validated genetics-based personalized healthcare practices. The now too common cliché of personalizing for common diseases seems to have led to many instances of what we term the problem of misattributed equivalence.[1] For example, personalized medicine is viewed by many, even respected leaders in healthcare, to mean common variation for common disorders and pharmacogenomics, and indeed, is utilized to shun validated genetics to mean very rare (read: "useless") disorders. However, let us examine history: ABO blood typing is a superb example of widespread genetics-based personalized healthcare dating back to World War II, and yet has universal applicability and will continue for centuries to come. Consider a more recent example: common variation (e.g., SNP) associated with breast cancer, often misattributed as etiologic for common disorders, accounts for 1.5–3% of all breast cancers, whereas *BRCA1/2* (considered "rare") alone accounts for 10% of all breast cancers. There exist six additional Mendelian breast cancer predisposition genes. Together, these eight susceptibility genes could account for >11%, and perhaps up to 15%, of the genetic load of breast cancer.[2,3]

In taking a very practical (as opposed to theoretical) point of view, everyone, without exception, wants to use genetic and genomic information to benefit the patient. Yet, despite all good intent, how successful have we been? Our entire healthcare system saw 3.8 million visits (1.8 million new visits) in 2009. We know from decades' worth of clinical

doi: 10.1111/j.1749-6632.2010.05789.x

epidemiologic and clinical genetic studies that an average of 10% of all diseases have a high penetrance Mendelian genetic cause(s) and perhaps 20–50% more have a genomic component. Even at its most conservative, our healthcare system should have referred 190,000 (5%) individuals for genetic consultation. In reality, only 3,000 new referrals to genetics clinics were seen, without wait times or access being an issue, that year (CY 2009). At the national level, these figures are identical. Levy *et al.* surveyed 35,000 healthy individuals by taking cancer family histories.[4] By family history alone, 350 appeared to be at risk for hereditary breast and ovarian cancer syndrome due to germline *BRCA1/2* mutations, yet only 35 discussed their concern about family history with their healthcare providers, and only four reached appropriate *BRCA1/2* testing. Even more alarming, it is unknown how many received appropriate pre- and posttest genetic counseling, considered the standard of care. Thus, in a field, clinical cancer genetics, where evidence-based management can be changed by genetic information, only 1% of those who require such 21st century genetics-informed personalized healthcare are even identified and referred to genetics professionals.

The genetics of endocrine neoplasias has paved the way for accurate evidence-based practice of clinical cancer genetics and related fields.[5] When germline gain-of-function mutations of the *RET* proto-oncogene were identified in multiple endocrine neoplasia type 2 (MEN 2), mutation-informed molecular diagnostics, predictive testing, genetic counseling, surveillance, and prevention became possible.[6–9]

Heritable endocrine neoplasias

MEN 2, with an incidence of 1 in 200,000 live births, is an autosomal dominant heritable neuroendocrine neoplasia syndrome characterized by medullary thyroid carcinoma (MTC), pheochromocytoma, and hyperparathyroidism (HPT). MTC is derived from the calcitonin-producing C cells of the thyroid and pheochromocytoma from the adrenal medulla. In the MEN 2 setting, pheochromocytomas are usually benign. Because germline *RET* mutations are found in >95% of those with MEN 2, and in general, confer high penetrance predisposition such that risk assessment and validated evidence-based clinical practice are possible.[5,9] Shortly after identification of *RET* as the susceptibility gene for MEN 2, the

Table 1. Examples of heritable endocrine neoplasia syndromes and their high penetrance predisposition genes that are clinically valid and actionable in medical management

Syndrome name	Key component neoplasias	Gene(s)
MEN 2	MTC, phaeochromocytoma, HPT	RET
MEN 1	HPT, pituitary adenoma, pancreatic islet cell tumors	*MEN1, ?p27*
VHL	Central nervous system hemangioma/blastoma, retinal angioma, clear cell renal cell carcinoma, pheochromocytoma	VHL
Paraganglioma syndromes	Pheochromocytoma, Paraganglioma, (renal cell carcinoma, papillary thyroid carcinoma)	*SDHB, SDHC, SDHD; SDHAF2, TMEM127*
Carney complex	Spotty pigmentation (lentigenes), primary pigmented adrenocortinal nodular hyperplasia, cardiac myxoma, testicular germ cell tumors (epithelial thyroid neoplasias)	PRKAR1A
Cowden syndrome	Breast cancer and hamartoma, epithelial thyroid cancer, and benign neoplasia	PTEN; (*SDHB/SDHD*)

predisposing genes for von Hippel-Lindau syndrome (VHL), MEN 1 and other endocrine neoplasias followed (Table 1).[10] In general, these predisposition genes account for the great majority of these syndromes and have similar attributes to *RET* and MEN 2 such that clinical practice is empowered by genetic information which guides medical management.[9]

RET germline mutations in multiple endocrine neoplasia type 2

MEN 2 and *RET* genotype–phenotype associations and gene-informed practice are universally considered the paradigm for evidence-based practice of clinical cancer genetics and genetics-based personalized cancer care.[9]

The *RET* proto-oncogene, localized to 10q11.2, encodes a receptor tyrosine kinase expressed mainly in neural crest-derived tissues.[11] Germline gain-of-function mutations affecting specific hot-spot codons result in MEN 2 (Fig. 1).[6,12,13] MEN 2A, the most common clinical subtype of MEN 2, is characterized by the triad of MTC, pheochromocytoma, and HPT. Germline mutations in one of the extracellular cysteine codons in exon 10 or 11 have been identified in more than 98% of MEN 2A, and hence, has served as a highly sensitive molecular diagnostic test.[5,8] MEN 2B, the most fulminant and uncommon clinical subtype of MEN 2, is characterized by MTC and pheochromocytoma (with an onset an average of 10 or more years younger), and classic stigmata such as ganglioneuromatosis of the gut and marfanoid habitus.[14] Virtually, all MEN 2B individuals have been found to have a germline mutation in *RET*, either M918T (>97%) or A883F (>2%).[5,12,13,15] In rare instances of MEN 2B, the FMTC/MEN 2A-specific mutation V804M is associated with a second variant in proximity.[16–18] Familial MTC (FMTC) is a clinical subtype which we now believe to be a phenotypic consequence of decreased penetrance *RET* alleles. Indeed, FMTC rarely is associated with the most penetrant mutations in codon 634 and is most likely to occur with the least penetrant mutations in codons 609, 611, 618, and 620. Thus, the pre-*RET* clinical subdivision of MEN 2A and FMTC may seem artificial in the genetic and genomic medicine era: in fact, the clinical manifestations characterizing these two subtypes are not dichotomous but a continuum, thus representing allelic diversity with variable penetrance.

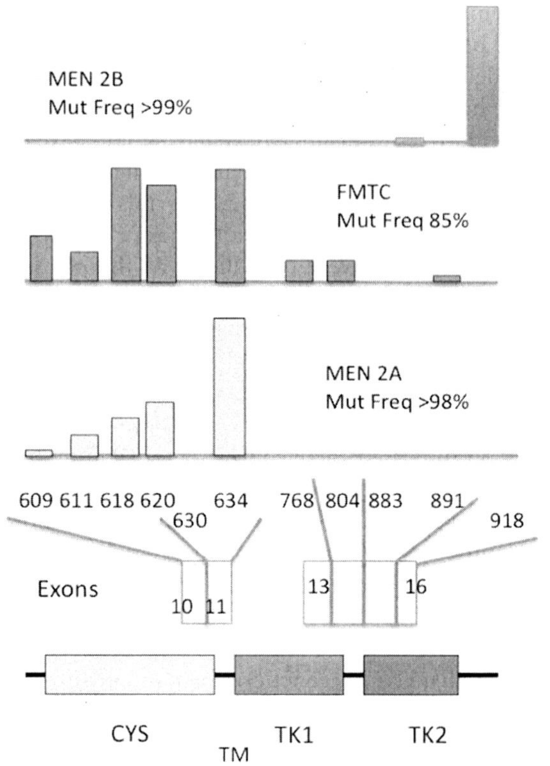

Figure 1. Germline *RET* mutations causing MEN 2 mainly affect hot-spot codons. Bottom panel is a schematic of the RET receptor tyrosine kinase showing the cysteine-rich domain (CYS), transmembrane domain (TM), and the two intracellular tyrosine kinase domains (TK1, TK2). There are mainly 11 hotspot codons housed within 6 of *RET*'s 21 total exons. Note the distinct mutational spectra found in individuals with MEN 2A, FMTC, and MEN 2B.

There also exist what was first thought to be lower penetrance germline *RET* mutations associated with MEN 2A/FMTC, such as those at codons 790 and 791.[19] Initially believed to be German founder mutations, their functional attributes came under suspicion when reports of these "mutations" seemed to mainly emanate from Austria and Germany.[19,20] Even founder mutations should be found elsewhere due to migration patterns. Recent systematic analysis of two population-based registries demonstrated that at least the codon 791 (Y791F) *RET* mutation could not be a pathogenic mutation for three main reasons.[21] First, Y791F was found not to segregate with the MEN 2 phenotype in registry families. Second, the frequency of the Y791F allele in MEN 2 individuals was identical to that in 1000 ancestry-matched population controls. Finally, *in silico* functional modeling of Y791F and wild-type Y791 gave similar results. Whether this variant can modify phenotype in germline *RET* (or other genes, e.g., *VHL*) pathogenic mutation positive individuals requires further investigation.

Heritable pheochromocytoma and paraganglioma syndromes

The genetic differential diagnosis of pheochromocytoma or paraganglioma includes syndromes caused by germline mutations in the *SDHx* genes, *VHL*, *RET* (MEN 2), and rarely in *MEN1* and *NF1* (type 1 neurofibromatosis).[22,23] VHL disease is an autosomal dominant disorder characterized by retinal angiomas, pheochromocytomas, central nervous system (CNS) hemangiomas and hemangioblastomas, renal cell carcinomas, and/or pancreatic endocrine neoplasias.[24] Germline mutations, including large deletions/rearrangements, in the *VHL* gene, linked to 3p25-p26, are etiologic for virtually all VHL disease.[25–27] Thus, *VHL* molecular diagnostic testing and predictive testing, in the setting of genetic counseling, are highly accurate resulting in well-established gene-informed medical management with published practice guidelines. The reader is referred to several classic and recent reviews in this regard.[28,29] There is a genotype–phenotype correlation, but these are not absolute such that the practice guidelines are not altered by specific genotype. For example, missense mutations within *VHL* are associated with pheochromocytoma and rarely, with renal cell carcinoma, whereas truncating mutations (including deletions) are associated with an increased risk of developing renal cell carcinoma.[30,31] Because of the small, but finite, possibility of developing renal cell carcinomas in individuals harboring missense mutations, most clinicians dare not stop renal surveillance even when a *VHL* missense mutation is revealed in their patient. From a practical point of view, however, the standard imaging for pheochromocytoma will also visualize the kidneys.

Autosomal dominant segregation of head and neck paragangliomas, showing a paternal inheritance (maternal imprinting) pattern, has been known for several decades before the identification of its predisposition gene, *SDHD*, an autosomal gene lying in a region (11q23) known to be maternally imprinted. It encodes one of the four subunits of mitochondrial complex II, also known as succinate dehydrogenase (SDH).[32] SDH sits at the cross-road of the electron transport chain and the Kreb's tribcarboxylic acid cycle. The so-called PGL1 families were shown to have a Dutch founder effect, and were linked to 11q23, where subsequently, heterozygous *SDHD* germline mutations were identified as segregating in these families.[32,33] Initially, only head and neck paragangliomas, especially carotid body tumors, were thought to be component to PGL, but subsequently such mutations were also found in familial pheochromocytoma.[34] Interestingly, a study analyzing individuals with apparently sporadic pheochromocytoma revealed the unexpected occurrence of germline *SDHD* mutations.[35] This anecdotal observation led to a systematic, population-based study examining the prevalence and clinical characteristics of germline mutations in the then known genes which cause heritable pheochromocytoma (*VHL, RET, SDHD, SDHB*) in registrants presenting with apparently sporadic symptomatic pheochromocytoma.[22,23] Of these registrants, ~30% were found to have germline mutations in one of these four genes, with almost half in *VHL* and the remaining distributed equally among *RET, SDHD,* and *SDHB*.[22,23] While early-onset, multifocal pheochromocytoma and extraadrenal disease were (not surprisingly) overrepresented in those with germline mutations, almost 20% presented after the age of 40, and 92% of those with mutations presented without any associated syndromic features. Interestingly, the age-related penetrance was 38 years for *SDHD* mutation and 30 for *SDHB* mutation.[23] The penetrance between head and neck PGL and pheochromocytoma between the two genes is different. For example, head and neck PGL penetrance approaches 100% by age 70 for *SDHD* and ~75% by age 70 for *SDHB*. In contrast, adrenal pheochromocytoma penetrance approaches 60% by age 50 and remains level after 50 for *SDHD*, and is 35% by age 50 (a penetrance that remains stable with age) for *SDHB*.[23] Typically, gene-specific mutation frequencies and penetrance are lower in population-based series compared to highly selected series. It is, therefore, important to note that even consortial series of highly selected cases revealed mutation frequencies and penetrances similar to those obtained in the population-based studies.[36,37]

In the population-based registry, *SDHD* mutations favor head and neck PGL whereas *SDHB* mutations intra-abdominal extra-adrenal pheochromocytomas.[23] It was also observed that prevalent malignant pheochromocytomas/PGL occurred in 1/3 of *SDHB* mutation positive individuals who had tumors and that perhaps renal cell carcinoma and papillary thyroid carcinoma were overrepresented in these mutation carriers as well.[23,38] In three series totaling 183 nonsyndromic renal cell carcinomas of various histologies, ages at diagnoses and family histories, five (2.7%) were found to have germline mutations, all in *SDHB*.[38–40] In examining all the known renal cell carcinoma patients with germline *SDHB* mutations, virtually all have mutations affecting arginine codons 5′ of codon 48, and except for one, were diagnosed before the age of 39.[39]

Germline mutations in *SDHC* are less common than those in *SDHB* or *SDHD* and were predominantly described in individuals and families with head and neck PGL.[41] In the population-based registries of the European-American study group, no germline mutations in *SDHC* were found in 371 registrants with pheochromocytoma.[42] Subsequently, only very rare cases of pheochromocytoma were found to carry *SDHC* mutations.[43] Of 121 head and neck PGL registrants, 4% were found to carry germline *SDHC* mutations.[42] Compared to those with *SDHD* mutations, *SDHC*-related head and neck PGLs had more of the characteristics of sporadic head and neck PGLs than *SDHD*-related PGL. For example, compared to *SDHD* or *SDHB*, *SDHC* mutations tended to confer unifocal PGL.[43] The mean age at diagnosis for *SDHC* disease was in between that of sporadic disease and of *SDHB/D* disease. *SDHC*-related tumors have never been found to be malignant nor have they been associated with extraparaganglial malignancies, as yet.

Because not all pheochromocytomas or PGLs carrying the red flags for heritability are accounted for by germline mutations or deletions in *VHL*, *RET*, *NF1*, *SDHB*, *SDHC*, and *SDHD*, there must be other predisposition genes. Recently, two such putative susceptibility genes, *Sdh5* and *TMEM127*, have been described.[44,45] Sdh5 or SDHAF2 interacts with SDHA and is required for flavination of SDH. Very little is known about TMEM127 except that it is a trans-endomembrane protein. However, no clinical outcomes studies have been performed, as yet, and so mutation frequencies, penetrance, and clinical phenotypes are currently unknown.

Apparently sporadic endocrine neoplasia

Because 10% of all individuals presenting with apparently sporadic MTC have germline *RET* mutations, the standard of care is to offer *RET* mutation analysis in the setting of genetic counseling to all presentations of MTC.[9] Patients found to carry germline *RET* mutations should be managed according to genotype, as detailed in recent practice guidelines, and family members should be offered predictive testing.[9]

Accumulating data over the last decade suggest that ~30% of apparently sporadic pheochromocytoma/PGL presentations carry germline mutations in *SDHB*, *SDHC*, *SDHD*, *VHL*, or *RET*.[22,23,36,42,46] Based on these observations, the standard of care is to offer all presentations of pheochromocytoma and/or PGL genetic testing for these genes in the setting of genetic counseling. Given the expense of bundled testing of all five genes, many have chosen to try to prioritize testing using the presence or absence of groups of demographic and clinical features (Fig. 2).[47–49] As an example, when a patient, without a family history, presents with head and neck PGL, features such as preceding pheochromocytoma/PGL, multiple tumors, malignant tumor, age < 40, and male gender. If any one of these features is present, then gene testing proceeds. If there are multiple tumors, testing begins with *SDHD*.[47] In contrast, if solitary, then testing begins with *SDHB*, followed by *SDHD* and *SDHC*. This type of algorithm reduces the costs of testing by 60% per proband. VHL should be suspected in very young (<10) bilateral pheochromocytoma presentations. It is usually unnecessary to test for *NF1* because other clinical features of this syndrome are obvious.[50]

Currently, there is no evidence-based data for surveillance of *SDH*-related pheochromocytoma/PGL. The European-American Pheochromocytoma-PGL Study Groups typically utilize annual MRI and FDG-PET.[48] There currently exists no effective treatment once pheochromocytoma or PGL become metastatic. Targeted therapies based on signaling pathways should be sought. Interestingly, the RET, VHL and SDH pathways may intersect downstream at the HIF1 signaling pathway, suggesting the importance of hypoxia and the electron transport chain.[51,52] Thus, therapies targeting the HIF1 and hypoxic-signaling pathways may be relevant.

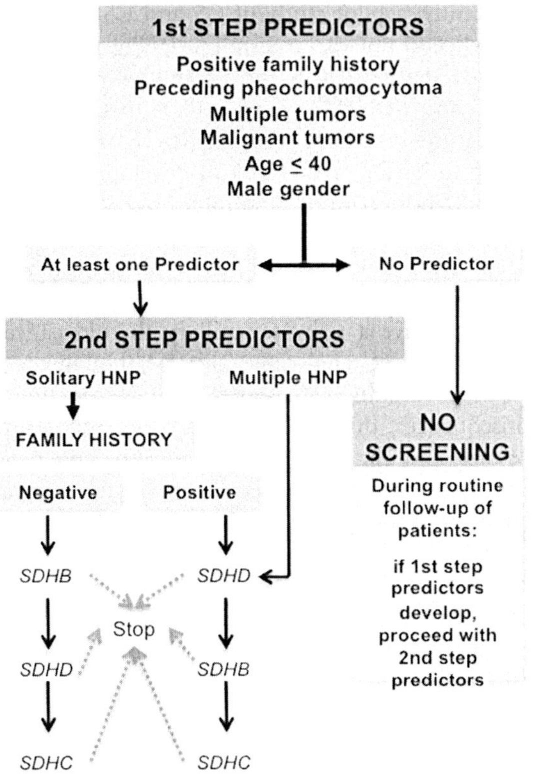

Figure 2. Illustration of algorithm for prioritizing gene testing and then prioritizing specific gene for head and neck PGL presentations.

Cowden syndrome and the *PTEN* hamartoma tumor syndromes

Although MTC carries only a single genetic differential diagnosis, MEN 2, the genetic differential diagnosis for epithelial thyroid carcinomas is broader. One important genetic differential diagnosis for epithelial thyroid carcinoma is Cowden syndrome (CS).[53] Other differential diagnoses include Carney complex, familial adenomatous polyposis, and Werner syndrome.[53]

Germline mutations in *PTEN*, a tumor suppressor gene on 10q23, cause ~80% of CS, an autosomal dominant disorder characterized by hamartomas and a high risk of beast and thyroid cancers, and that serves as the prototype of a *PTEN* hamartoma tumor syndrome (PHTS).[5] Other clinically distinct syndromes which belong to the PHTS include subsets of Bannayan–Riley–Ruvalcaba syndrome (BRRS), Proteus syndrome, and autism spectrum disorder with macrocephaly.[54–58] Thus, from a technical point of view, any clinical presentation found to harbor germline *PTEN* mutations should fall under the PHTS rubric, that is, molecular classification over clinical diagnosis.

CS occurs in 1 in 300,000 live births, however, it is likely more prevalent because many of the pathognomonic mucocutaneous features are commonly found within the general population and therefore are often overlooked resulting in underdiagnosis of CS.[5] It is characterized by multiple developmentally disorganized benign growths, or hamartomas, with an increased risk of both benign and malignant tumors. Mucocutaneous manifestations are the most prevalent phenotypic feature of CS and include trichilemmomas, papillomatous papules, and acral and plantar keratoses.

Individuals with CS are at increased risks of developing benign and malignant tumors of the breast, thyroid, and endometrium. Recently, a variety of gastrointestinal polyps and colorectal cancers may be added as component.[59] For this reason, accurate recognition of the CS phenotype is critical for both diagnosis and management of the proband and his/her family members.

Similar to other inherited cancer syndromes, affected individuals are more likely to develop bilateral disease in paired organs and multifocal disease. Women with CS have a 25–50% lifetime risk of developing malignant breast disease, with an average age of diagnosis between 38 and 46 years (range 17–65) and a 67% lifetime risk for developing benign breast disease.[60,61] Male breast cancers have also been reported in CS, but it is unclear how prevalent they are.[62] The second most frequently reported manifestation of CS is thyroid disease affecting between two-thirds and three-quarters of patients. The CS-associated benign thyroid abnormalities include multinodular goiter, adenomatous nodules, and follicular adenomas. Follicular and papillary thyroid cancer, but not medullary cancers, are also frequently observed with a 10% increased lifetime risk.[5] Women with CS are also more likely to develop endometrial cancer and uterine fibroids, but these components have not yet been documented in a systematic study. The lifetime risk of developing endometrial cancer is estimated to be between 5% and 10%, and roughly half of the women with CS will develop multiple large uterine fibroids. A recent systematic prospective study of germline *PTEN* mutation positive individuals revealed that gastrointestinal polyps occurred in >90% of such individuals that underwent at least one upper or lower endoscope.[59] Although hamartomatous polyps have been reported in CS, the high frequency and the varied histologies have not been documented in the past. Indeed, hyperplastic and mixed polyposis were not uncommon, and the age- and sex-adjusted standardized incidence ratio for colorectal cancer in PHTS is elevated to >200.[59] Other neoplasias and hamartomas, including skin cancers, renal cell cancer and brain tumors, are suspected to be associated with CS, however, their prevalence has not yet been systematically studied.

Clinical and molecular diagnosis of Cowden syndrome

Because it is difficult to recognize CS and because many more individuals possess combinations of CS-like features (CSL), operational criteria for clinical diagnosis of CS was initially developed by the International Cowden Consortium for purposes of the gene hunt.[63] These criteria were subsequently proven to be robust because >80% of individuals who meet the strict diagnostic criteria (Table 2) were found to carry germline *PTEN* mutations.[64] However, for purposes of clinical referral to genetics professionals, these operational diagnostic criteria are too stringent, because a broader net must be cast for clinical purposes. What clinical features or combinations of clinical features which would trigger this referral are currently unknown and the subject of systematic study.

Although a presumptive diagnosis of PHTS can be made from clinical observation, the molecular diagnosis of PHTS is only confirmed by identification of a germline *PTEN* mutation. Finding a germline *PTEN* mutation in an individual, irrespective of exhibited phenotype or family history, makes the definitive diagnosis of PHTS. However, it is important to emphasize that an individual who meets the strict Consortium clinical diagnostic criteria for CS (Table 2), where a *PTEN* mutation cannot be easily identified, should be managed as though he/she has CS. It is the >95% of CSL individuals with CS-like features, who do not have germline *PTEN* mutations, that pose challenges to risk assessment, genetic counseling, and medical management. In this situation, predictive testing of the family members is also not possible.

Recently, a pilot study of CS and CSL individuals who do not carry germline *PTEN* mutations revealed that ~10% harbor germline variants in *SDHB* and *SDHD*.[65] Individuals with CS or CS-like features who have germline *SDHB/SDHD* were shown to have a greater risk of developing breast, thyroid, and renal cell carcinomas over those with germline *PTEN* mutations.[65] These data are consistent with the observations in pheochromocytoma/PGL patients with *SDHB* mutations, who were found to have a risk of papillary thyroid cancer and early-onset renal cancers.[38,39] Although these missense variants were shown to have functional consequences, this pilot was based on small sample size and the observations must be independently validated before being used in the clinical setting.

PTEN mutation frequency in other PHTS

Although the diagnostic criteria for BRRS have not yet been systematically established, BRRS is clinically diagnosed in the presence of macrocephaly, lipomatosis, hamartomatous intestinal polyposis, and in males, pigmented macules of the glans penis.[66] Intragenic *PTEN* mutations have been identified in ~60% of BRRS cases.[54,67] Among those who remain mutation "negative," approximately 10% were found to carry large deletions of *PTEN*.[68] Approximately 20% of individuals with Proteus syndrome, 50% in Proteus-like syndrome, and 10–20% of ASD with macrocephaly harbor germline *PTEN* mutations are those subsets are considered PHTS and should be managed accordingly.[55–57,69–71]

Medical management of PHTS

PHTS-related mucocutaneous manifestations are rarely cause for concern and visual surveillance is considered to be sufficient, provided they are asymptomatic. Should they become symptomatic, temporary relief may be obtained through cryosurgery or laser surgery.[72] Surgical treatment is sometimes complicated by cheloid formation and recurrence of the lesions (Eng, unpublished data). Because PHTS-associated mucocutaneous lesions have a tendency to re-grow rapidly with cheloid formation, surgical resection should only be performed if malignancy is suspected or symptoms (e.g., pain, deformity) are significant. Both benign and malignant manifestations of PHTS are treated in a

Table 2. International Cowden Consortium Operational Criteria for Clinical Diagnosis (similar to NCCN 2006 version)

Pathognomic criteria
Adult Lhermitte–Duclos disease (LDD), defined as the presence of a cerebellar dysplastic gangliocytoma
Mucocutaneous lesions
• Trichilemmomas (facial)
• Acral keratoses
• Papillomatous lesions
• Mucosal lesions
Major criteria
• Breast cancer
• Epithelial thyroid cancer (nonmedullary), especially follicular thyroid cancer
• Macrocephaly (occipital frontal circumference $\geq$97th percentile)
• Endometrial carcinoma
Minor criteria
• Other thyroid lesions (e.g., adenoma, multinodular goiter)
• Mental retardation (IQ $\leq$ 75)
• Hamartomatous intestinal polyps
• Fibrocystic disease of the breast
• Lipomas
• Fibromas
• Genitourinary tumors (especially renal cell carcinoma)
• Genitourinary malformation
• Uterine fibroids
An operational diagnosis of Cowden syndrome is made if an individual meets any one of the following criteria
• Pathognomonic mucocutaneous lesions alone if there are
○ Six or more facial papules, of which three or more must be trichilemmoma
OR
○ Cutaneous facial papules and oral mucosal papillomatosis
OR
○ Oral mucosal papillomatosis and acral keratoses
OR
○ Six or more palmo-plantar keratoses
OR
• Two or more major criteria
OR
• One major and at least three minor criteria
OR
• At least four minor criteria
In a family in which one individual meets the diagnostic criteria for Cowden syndrome listed earlier, other relatives are considered to have a diagnosis of CS if they meet any of the following criteria:
• The pathognomonic criteria
OR
• Any one major criterion with or without minor criteria
OR
• Two minor criteria
OR
History of Bannayan–Riley–Ruvalcaba syndrome

similar manner to their sporadic counterparts at this time.

Because germline *PTEN* mutations are associated with breast and thyroid malignancies, and probably with endometrial, colorectal, and renal cancers, surveillance and primary prevention/early treatment are the cornerstones of clinical management in all PHTS. Comprehensive physical examinations are recommended yearly beginning at age 18, or 5 years prior to the family's earliest age of cancer diagnosis (whichever occurs first). Particular attention should be paid to breasts, thyroid, and mucocutaneous regions. Annual urinalysis with urine cytology and renal ultrasounds are recommended for individuals with a positive family history for renal cell cancer.

Breast cancer. Both men and women with CS or PHTS should conduct monthly breast self-examinations beginning at 18 years of age. In addition, women should receive annual clinical breast exams (age 25), mammography, and breast MRI imaging (age 30–35) at the indicated age, or 5–10 years prior to a family's earliest known breast cancer diagnosis (whichever is earlier).

Thyroid cancer. Baseline thyroid ultrasound examination should be done at age 18 (unless a younger age at diagnosis for thyroid cancer is documented in the family). Following the initial evaluation, individuals with CS/PHTS should consider yearly ultrasounds thereafter. Benign thyroid disease is prevalent in individuals with PHTS and often, the evaluation for malignancy is challenging, and should be managed in large academic medical centers. Should a thyroidectomy be indicated, for whatever reason, a total thyroidectomy should be performed because partial resections often result in regrowth of the remnant thyroid with hamartomas/neoplasias.

Endometrial cancer. Blind suction biopsies are recommended for premenopausal women on an annual basis starting at age 35–40, or 5 years prior to the earliest diagnosis of endometrial cancer in the family. Recommendations for postmenopausal women include annual transabdominal ultrasound examination with biopsy of suspicious areas. These recommendations are consistent with the NCCN guidelines up to the 2008 revisions, which do not advocate any endometrial cancer surveillance for women with PHTS. Experts in this field however, feel strongly that this surveillance should be continued. Therefore, as a compromise, endometrial surveillance as noted above is advocated in families having at least one relative with endometrial cancer.

Colorectal cancer. In the recent past, it was believed that colon cancer is not component to PHTS. Recently, a prospective series of individuals with PHTS revealed that component colorectal cancers occur prior to age 50 and as young as in the 30s.[59] Until further data, it may be prudent to begin routine colonoscopies in individuals carrying pathogenic germline *PTEN* mutations by the age of 35 years or 5 years young than the earliest age at diagnosis in the family, whichever occurs first.[59] In this series of PHTS patients, gastrointestinal polyps were found to occur in >90% of individuals with pathogenic germline *PTEN* mutations, suggesting that polyps are among the most common (top 2) features of PHTS.[59]

Experimental therapies. PTEN is a dual-specific phosphatase, being a lipid and protein phosphatase as well as a tyrosine and a serine–threonine phosphatase. Its lipid phosphatase activity downregulates the AKT pathway. When PTEN is nonfunctional, AKT is upregulated leading to upregulation of mTOR. Thus, mTOR inhibition would be a rational targeted therapy.[73] Currently, although mTOR inhibitors are promising candidates, they remain in clinical trials. Eligibility criteria for a rapamycin trial for PHTS includes > 18 years of age and CS/CSL with pathogenic germline *PTEN* mutation.

Validated practice of evidence-based cancer genetic healthcare

The power of *RET* testing and genotype-informed medical management in MEN 2 serves as the paradigm for validated cancer genetics practice.[9] Since the discovery of *RET* as the *MEN2* susceptibility gene in 1993, many other heritable cancer syndromes and their predisposition genes have been identified and characterized. This armamentarium of evidence allows for accurate molecular diagnosis, cancer risk assessment, genetic counseling and risk management for the proband.[5,9] Thus, after a medical and family history are obtained and a genetic differential diagnosis formed, the most likely predisposition gene is tested in an affected individual. Once the family-specific mutation is found, then predictive testing of as yet unaffected relatives can be offered in the setting of genetic counseling. Family members who test negative for the family-specific mutation are at population risk for cancer; those that test mutation positive will be at significantly increased risks of cancers (dependent on gene involved), and undergo genotype-informed surveillance and/or prophylaxis, thus saving lives from cancer and dramatically diminishing healthcare costs.

Accumulating genetics-clinical outcomes data suggest complementary strategies of identifying individuals at risk for cancer. For example, analysis of microsatellite instability (MSI) and immunohistochemistry (IHC) for the expression of mismatch repair proteins in colorectal cancers could be an effective way of identifying individuals at risk for Lynch syndrome, the most common heritable colon cancer syndrome in adults.[74,75] However, although the research evidence is present and the cost-effectiveness shown on paper, true clinical adoption remains a challenge. For instance, even in centers that pioneered this research, only a small minority of all individuals with colorectal cancers subjected to MSI/IHC and who are MSI-H/IHC null actually reach appropriate genetics care. At the Cleveland Clinic, passive response and adoption of MSI and IHC analyses of every colorectal cancer resected on main campus, with a special note placed in the pathology report, resulted in 14% of those with MSI-H/IHC null colorectal cancers reaching genetic healthcare.[76] In contrast, when we began to take an active concerted multidisciplinary effort, ~80% of those with MSI-H/IHC null tumors reach cancer genetics care.[76] In addition to a cost-efficient manner of identifying, and then managing, those at genetic risk for adult-onset colorectal cancer, and their families, the results of MSI and IHC even in sporadic cancers may help decisions whether adjuvant 5FU is warranted.

Even more so in the near future, germline mutation status will dictate drug choice, even in the setting of metastatic disease. For example, clinical trials have demonstrated great promise of PARP inhibitors for *BRCA1/2*-related metastatic ovarian and pancreatic cancers, and mTOR inhibition for *PTEN*-related cancers.[73,77]

Challenges and opportunities

It is now clear that the results from genome-wide association studies with SNP-associated risks for "common" cancers taken in isolation is not useful.[2] Believed to take into account the majority of common cancers and common diseases, these SNP-related risks take into account 1–3% of almost all common disorders. In contrast, *BRCA1/2* alone account for ~10% of all breast cancers; germline mismatch repair gene mutations account for ~3% of all colorectal cancers. While the latter substantially increases the risk of component cancers (e.g., from 5% lifetime risk for colorectal cancer in the population to up to 80% lifetime risk), SNP-associated cancer risks are only slightly increased, often in the RR1.01-1.3x range, that is, from a 5% lifetime risk to a 5.05–6.5% lifetime risk, certainly not enough to action.[2]

The integration of risk SNP's with family health history or the integration of SNP risks and germline high penetrance mutation to predict who is the 80% at risk for colon cancer and the 20% not at risk would be extremely useful to personalize surveillance. For example, do SNP's modify or complement family history? By themselves, prostate cancer-associated SNPs were not particularly helpful for identifying men who would go on to develop prostate cancer. However, the SNPs had value in specific situations. For example, it is standard clinical practice to biopsy the prostate when blood levels of prostate-specific antigen (PSA) are >4 ng/mL, but there is no consensus about how to handle borderline PSA levels often operationally defined as 2–3.9 ng/mL. Xu *et al.* examined the utility of these SNPs in such men, using the SNPs in conjunction with PSA levels to deliver a SNP-adjusted PSA score to help in the biopsy versus watchful-waiting-rebiopsy later decision. These researchers also examined the utility of combining SNP risks and family history of prostate cancer. In the absence of a family history of prostate cancer, they found that the presence of all 14 risk-SNPs raised the risk of a white individual developing prostate cancer between the ages of 55 and 72 years from 13% (the general population risk among whites) to ~20%.[78] In the presence of a family history of prostate cancer, however, having eight or more of these SNPs raised the risk of developing prostate cancer between the ages of 55 and 72 from 13% to 25%. In this scenario, having >8 SNPs delivered incremental risks beyond 25% and would be actionable, if these data can be replicated independently. In contrast, most studies examining a range of common diseases have shown that these disease-associated SNPs do not add to the standard clinical risk assessments.[79,80] Nonetheless, it is vital to investigate the clinical context whereby genome-wide SNPs and variations become useful to significantly inform clinical information and family health history as well as Mendelian genetic risk.

Finally, genomic variation is argued to be a powerful pharmacogenomic tool, that is, to choose drugs or to modulate dosing. Examples include the SNP's associated with Plavix metabolism and warfarin metabolism, although controversy remains. Ideally, the integration of high penetrance germline pathogenic variation with lower penetrance genomic variation and epigenomic alterations must be considered in the future practice and genetic and genomic medicine.

Acknowledgements

I am eternally grateful to all my patients who continue to teach me invaluable lessons and who have enrolled in my research protocols, without whom gene-informed cancer genetics practice would not be possible. I am deeply appreciative of my late mentors, Edward D. Garber and Robert J. Gorlin, who continue to inspire me from the other side, and of my longtime neuroendocrine neoplasia collaborators and friends, including Stan Lyonnet, Arnold Munnich, Lois M. Mulligan, and Hartmut P.H. Neumann. C.E. is the Sondra J. and Stephen R. Hardis Endowed Chair of Cancer Genomic Medicine at the Cleveland Clinic, and an American Cancer Society Clinical Research Professor, generously funded, in part, by the F.M. Kirby Foundation.

Conflicts of interest

The author declares no conflicts of interest.

References

1. Eng, C. & R.R. Sharp. 2010. Bioethical and clinical dilemmas of direct-to-consumer genomic testing: the problem of misattributed equivalence. *Science Transl. Med.* **2:** 1–5 (2:17cm5).
2. Khoury, M.J., C.M. McBride, S.D. Schully, *et al.* 2009. The scientific foundation for personal genomics: recommendations from a National Institutes of Health-Centers for Disease Control and Prevention multidisciplinary workshop. *Genet. Med.* **11:** 559–567.
3. Edelman, E. & C. Eng. 2009. A practical guide to the interpretation and clinical application of personal genomic screening. *Br. Med. J.* **339:** b4253.
4. Levy, D.E., J.E. Garber & A.E. Shields. 2009. Guidelines for genetic risk assessment of hereditary breast and ovarian cancer: early disagreements and low utilization. *J. Gen. Intern. Med.* **24:** 822–828.
5. Zbuk, K. & C. Eng. 2007. Cancer phenomics: RET and PTEN as illustrative models. *Nat. Rev. Cancer* **7:** 35–45.
6. Mulligan, L.M., J.B.J. Kwok, C.S. Healey, *et al.* 1993. Germline mutations of the *RET* proto-oncogene in multiple endocrine neoplasia type 2A. *Nature* **363:** 458–460.
7. Mulligan, L.M., C. Eng, C.S. Healey, *et al.* 1994. Specific mutations of the *RET* proto-oncogene are related to disease phenotype in MEN 2A and FMTC. *Nat. Genet.* **6:** 70–74.
8. Eng, C., D. Clayton, I. Schuffenecker, *et al.* 1996. The relationship between specific *RET* proto-oncogene mutations and disease phenotype in multiple endocrine neoplasia type 2: International *RET* Mutation Consortium analysis. *JAMA* **276:** 1575–1579.
9. Kloos, R.T., C. Eng, D.B. Evans, *et al.* 2009. Medullary thyroid carcinoma: Management Guidelines of the American Thyroid Association. *Thyroid* **19:** 565–612.
10. Maher, E.R. & C. Eng. 2002. The pressure rises: update on the genetics of phaeochromocytoma. *Hum. Mol. Genet.* **11:** 2347–2354.
11. Gardner, E., L. Papi, D.F. Easton, *et al.* 1993. Genetic linkage studies map the multiple endocrine neoplasia type 2 loci to a small interval on chromosome 10q11.2. *Hum. Mol. Genet.* **2:** 241–246.
12. Eng, C., D.P. Smith, L.M. Mulligan, *et al.* 1994. Point mutation within the tyrosine kinase domain of the *RET* proto-oncogene in multiple endocrine neoplasia type 2B and related sporadic tumours. *Hum. Mol. Genet.* **3:** 237–241.
13. Hofstra, R.M.W., R.M. Landsvater, I. Ceccherini, *et al.* 1994. A mutation in the *RET* proto-oncogene associated with multiple endocrine neoplasia type 2B and sporadic medullary thyroid carcinoma. *Nature* **367:** 375–376.
14. Gorlin, R.J., H.O. Sedano, R.A. Vickers & J. Cervenka. 1968. Multiple mucosal neuromas, phaeochromocytoma and medullary carcinoma of the thyroid—a syndrome. *Cancer* **22:** 293–299.
15. Gimm, O., D.J. Marsh, S.D. Andrew, *et al.* 1997. Germline dinucleotide mutation in codon 883 of the *RET* proto-oncogene in multiple endocrine neoplasia type 2B without codon 918 mutation. *J. Clin. Endocrinol. Metab.* **82:** 3902–3904.
16. Cranston, A.N., C. Carniti, K. Oakhill, *et al.* 2006. RET is actively constitutively activated by novel tandem mutations that later the active site resulting in multiple endocrine neoplasia type 2B. *Cancer Res.* **66:** 10179–10187.
17. Miyahuchi, A., H. Futami, N. Hai, *et al.* 1999. Two germline missense mutations at codons 804 and 806 of the *RET* proto-oncogene in the same allele in a patient with multiple endocrine neoplasia type 2B without codon 918 mutation. *Jpn. J. Cancer Res.* **90:** 1–5.
18. Iwashita, T., H. Murakami, K. Kurokawa, *et al.* 2000. A two-hit model for development of multiple endocrine neoplasia type 2B by *RET* mutations. *Biochem. Biophys. Res. Commun.* **268:** 804–808.
19. Berndt, I., M. Reuter, B. Saller, *et al.* 1998. A new hotspot for mutations in the *RET* proto-oncogene causing familial medullary thyroid carcinoma and multiple endocrine neoplasia. *J. Clin. Endocrinol. Metab.* **83:** 770–774.

20. Gimm, O., B.E. Niederle, T. Weber, *et al.* 2002. RET protooncogene mutations affecting codon 790/791: a mild form of multiple endocrine neoplasia type 2A syndrome. *Surgery* **132:** 952–959.
21. Erlic, Z., M.M. Hoffmann, M. Sullivan, *et al.* 2010. Pathogenicity of DNA variants and double mutations in multiple endocrine neoplasia type 2 and von Hippel-Lindau disease. *J. Clin. Endocrinol. Metab.* **95:** 308–313.
22. Neumann, H.P.H., B. Brauch, S.R. McWhinney, *et al.* 2002. Germ-line mutations in nonsyndromic pheochromocytoma. *N. Engl. J. Med.* **346:** 1459–1466.
23. Neumann, H.P.H., C. Pawlu, M. Pęczkowska, *et al.* 2004. Distinct clinical features characterize paraganglioma syndromes associated with SDHB and SDHD mutations. *JAMA* **292:** 943–951.
24. Neumann, H.P.H. 1987. Basic criteria for clinical diagnosis and genetic counseling in von Hippel-Lindau disease. *J. Vasc. Dis.* **16:** 220–226.
25. Latif, F., K. Tory, J. Gnarra, *et al.* 1993. Identification of the von Hippel-Lindau disease tumor suppressor gene. *Science* **260:** 1317–1320.
26. Stolle, C., G. Glenn, B. Zbar, *et al.* 1998. Improved detection of germline mutations in the von Hippel-Lindau disease tumor suppressor gene. *Hum. Mutat.* **12:** 417–423.
27. Zbar, B., T. Kishida, F. Chen, *et al.* 1996. Germline mutations in the von Hippel-Lindau disease (VHL) gene in families from North American, Europe and Japan. *Hum. Mutat.* **8:** 348–357.
28. Woodward, E.R. & E.R. Maher. 2006. von Hippel-Lindau disease and endocrine tumour susceptibility. *Endocr. Rel. Cancer* **13:** 415–425.
29. Nordstrom-OBrien, M., R.B. Van Der luijt, E. vonRooijen, *et al.* 2010. Genetic analysis of von Hippel-Lindau disease. *Hum. Mutat.* **31:** 521–527.
30. Maher, E.R., A.R. Webster, F.M. Richards, *et al.* 1996. Phenotypic expression in von Hippel-Lindau disease: correlations with germline *VHL* mutations. *J. Med. Genet.* **81:** 328–332.
31. Chen, F., T. Kishida, M. Yao, *et al.* 1995. Germline mutations in the Von Hippel-Lindau disease tumor suppressor gene: correlations with phenotype. *Hum. Mutat.* **5:** 66–75.
32. Baysal, B.E., R.E. Ferrell, J.E. Willett-Brozick, *et al.* 2000. Mutations in SDHD, a mitochondrial complex II gene, in hereditary paraganglioma. *Science* **287:** 848–851.
33. van Schothorst, E.M., M. Beekman, P. Torremans, *et al.* 1998. Founder effect of PGL1 in heredity head and neck paraganglioma families from the Netherlands. *Am. J. Hum. Genet.* **63:** 468–473.
34. Astuti, D., F. Douglas, T.W.J. Lennard, *et al.* 2001. Germline *SDHD* mutation in familial phaeochromocytoma. *Lancet* **357:** 1181–1182.
35. Gimm, O., M. Armanios, H. Dziema, *et al.* 2000. Somatic and occult germline mutations in *SDHD*, a mitochondrial complex II gene, in non-familial pheochromocytomas. *Cancer Res.* **60:** 6822–6825.
36. Benn, D.E., A.P. Gimenez-Roqueplo, J.R. Reilly, *et al.* 2006. Clinical presentation and penetrance of pheochromocytoma/paraganglioma syndromes. *J. Clin. Endocrinol. Metab.* **91:** 827–836.
37. Ricketts, C.J., J.R. Forman, E. Rattenberry, *et al.* 2010. Tumor risks and genotype-phenotype-proteotype analysis in 358 patients with germline mutations in SDHB and SDHD. *Hum. Mutat.* **31:** 41–51.
38. Vanharanta, S., S.R. Virta, R. Lehtonen, *et al.* 2004. Early onset renal cell carcinoma as novel extra-paraganglial component of SDHB-associated heritable paraganglioma. *Am. J. Hum. Genet.* **74:** 153–159.
39. Eng, C. 2008. Editorial: SDHB—a gene for all tumors? *J. Natl. Cancer Inst.* **100:** 1193–1195.
40. Ricketts, C., E.R. Woodward, P. Killick, *et al.* 2008. Germline SDHB mutations and familial renal cell carcinoma. *J. Natl. Cancer Inst.* **100:** 1260–1262.
41. Niemann, S. & U. Muller. 2000. Mutations in *SDHC* cause autosomal dominant paraganglioma. *Nat. Genet.* **26:** 141–150.
42. Schiavi, F., C.C. Boedeker, B. Bausch, *et al.* 2005. Predictors and prevalence of paraganglioma syndrome associated with mutations of the *SDHC* gene. *JAMA* **294:** 2057–2063.
43. Peczkowska, M., A. Cascon, A. Prejbisz, *et al.* 2007. Metachronous extraadrenal and adrenal pheochromocytomas associated with a germline succinate dehydrogenase subunit C (*SDHC*) mutation. *Nat. Clin. Pract. Endocrinol. Metab.* **4:** 111–115.
44. Hao, H.X., O. Khalimonchuk, M. Schraders, *et al.* 2009. SDH5, a gene required for flavination of succinate dehydrogease, is mutated in paraganglioma. *Science* **325:** 1139–1142.
45. Qin, Y., L. Yao, E.E. King, *et al.* 2010. Germline mutations in TMEM127 confer susceptibility to pheochromocytoma. *Nat. Genet.* **42:** 229–233.
46. Benn, D.E., T. Dwight, A.L. Richardson, *et al.* 2001. Sporadic and familial pheochromocytomas are associated with loss of at least two discrete regions on chromosome 1p. *Cancer Res.* **60:** 7048–7051.
47. Neumann, H.P.H., Z. Erlic, C.C. Boedeker, *et al.* 2009. Clinical predictors for germline mutations in head and neck paraganglioma patients: cost reduction strategy in genetic diagnostic process as fall-out. *Cancer Res.* **69:** 3650–3656.
48. Neumann, H.P.H. & C. Eng. 2009. The approach to the patient with paraganglioma. *J. Clin. Endocrinol. Metab.* **94:** 2677–2683.
49. Erlic, Z., L.A. Rybicki, M. Peczkowska, *et al.* 2009. Clinical predictors and algorithm for the genetic diagnosis of pheochromocytoma patients. *Clin. Cancer Res.* **15:** 6378–6385.
50. Bausch, B., A.-C. Koschker, M. Fassnacht, *et al.* 2006. Comprehensive mutation scanning of *NF1* in apparently sporadic cases of pheochromocytoma. *J. Clin. Endocrinol. Metab.* **91:** 3478–3481.
51. Eng, C., M. Kiuru, M.J. Fernandez & L.A. Aaltonen. 2003. A role for mitochondrial enzymes in inherited neoplasia and beyond. *Nat. Rev. Cancer* **3:** 193–202.
52. Maxwell, P.H. 2005. A common pathway for genetic events leading to pheochromocytoma. *Cancer Cell* **8:** 91–93.
53. Hodgson, S.V., W.D. Foulkes, C. Eng & E.R. Maher. 2006. *Practical Handbook of Human Cancer Genetics.* 3rd ed. Cambridge University Press. Cambridge.
54. Marsh, D.J., J.B. Kum, K.L. Lunetta, *et al.* 1999. *PTEN* mutation spectrum and genotype-phenotype correlations in

Bannayan-Riley-Ruvalcaba syndrome suggest a single entity with Cowden syndrome. *Hum. Mol. Genet.* **8:** 1461–1472.

55. Zhou, X.P., H. Hampel, H. Thiele, *et al.* 2001. Association of germline mutation in the *PTEN* tumour suppressor gene and a subset of Proteus sand Proteus-like syndromes. *Lancet* **358:** 210–211.
56. Butler, M.G., M.J. Dasouki, X.P. Zhou, *et al.* 2005. Subset of individuals with autism spectrum disorders and macrocephaly associated with germline mutations in the PTEN tumour suppressor gene. *J. Med. Genet.* **42:** 318–321.
57. Varga, E.A., M. Pastore, T. Prior, *et al.* 2009. The prevalence of PTEN mutations in a clinical pediatric cohort with autism spectrum disorders, developmental delay and macrocephaly. *Genet. Med.* **11:** 111–117.
58. Orloff, M.S. & C. Eng. 2008. Genetic and phenotypic heterogeneity in the *PTEN* hamartoma tumour syndrome. *Oncogene* **27:** 5387–5397.
59. Heald, B., J. Mester, L.A. Rybicki, *et al.* 2010. Frequent gastrointerestinal polyps and colorectal cancer in prospective series of *PTEN* mutation carriers. *Gastroenterology*. In press.
60. Brownstein, M.H., M. Wolf & J.B. Bikowski. 1978. Cowden's disease: a cutaneous marker of breast cancer. *Cancer* **41:** 2393–2398.
61. Starink, T.M., J.P. Van Der Veen, F. Arwert, *et al.* 1986. The Cowden syndrome: a clinical and genetic study in 21 patients. *Clin. Genet.* **29:** 222–233.
62. Fackenthal, J.D., D.J. Marsh, A.L. Richardson, *et al.* 2001. Male breast cancer in Cowden syndrome patients with germline PTEN mutations. *J. Med. Genet.* **38:** 159–164.
63. Nelen, M.R., G.W. Padberg, E.A.J. Peeters, *et al.* 1996. Localization of the gene for Cowden disease to 10q22–23. *Nat. Genet.* **13:** 114–116.
64. Marsh, D.J., V. Coulon, K.L. Lunetta, *et al.* 1998. Mutation spectrum and genotype-phenotype analyses in Cowden disease and Bannayan-Zonana syndrome, two hamartoma syndromes with germline *PTEN* mutation. *Hum. Mol. Genet.* **7:** 507–515.
65. Ni, Y., K.M. Zbuk, T. Sadler, *et al.* 2008. Germline mutations and variants in the succinate dehydrogenase genes in Cowden and Cowden-like syndromes. *Am. J. Hum. Genet.* **83:** 261–268.
66. Gorlin, R.J., M.M. Cohen, L.M. Condon & B.A. Burke. 1992. Bannayan-Riley-Ruvalcaba syndrome. *Am. J. Med. Genet.* **44:** 307–314.
67. Marsh, D.J., P.L.M. Dahia, Z. Zheng, *et al.* 1997. Germline mutations in *PTEN* are present in Bannayan-Zonana syndrome. *Nat. Genet.* **16:** 333–334.
68. Zhou, X.P., K.A. Waite, R. Pilarski, *et al.* 2003. Germline *PTEN* promoter mutations and deletions in Cowden/Bannayan-Riley-Ruvalcaba syndrome result in aberrant PTEN protein and dysregulation of the phosphoinositol-3-kinase/Akt pathway. *Am. J. Hum. Genet.* **73:** 404–411.
69. Smith, J.M., E.P.E. Kirk, G. Theodosopoulos, *et al.* 2002. Germline mutation of the tumour suppressor *PTEN* in Proteus syndrome. *J. Med. Genet.* **39:** 937–940.
70. Herman, G.E., N. Henninger, K. Ratliff-Schaub, *et al.* 2007. Genetic testing in autism: how much is enough? *Genet. Med.* **9:** 268–174.
71. Herman, G.E., E. Butter, B. Enrile, *et al.* 2007. Increasing knowledge of germline PTEN mutations: two additional patients with autism and macrocephaly. *Am. J. Med. Genet. A* **143:** 589–593.
72. Hildenbrand, C., W.H. Burgdorf & S. Lautenschlager. 2001. Cowden syndrome-diagnostic skin signs. *Dermatology* **202:** 362–366.
73. Marsh, D.J., T.N. Trahair, J.L. Martin, *et al.* 2008. Rapamycin treatment for a child with germline PTEN mutation. *Nat. Clin. Pract. Oncol.* **5:** 357–361.
74. Aaltonen, L.A., R. Salovaara, P. Kristo, *et al.* 1998. Incidence of hereditary nonpolyposis colorectal cancer and the feasibility of molecular screening for the disease. *N. Engl. J. Med.* **338:** 1481–1487.
75. Hampel, H., W.L. Frankel, E. Martin, *et al.* 2005. Screening for Lynch syndrome (Hereditary nonpolyposis colorectal cancer). *N. Engl. J. Med.* **352:** 1851–1860.
76. Leach, B.H., C. Eng, M.F. Kalady, *et al.* 2009. Sharing the responsibility: a multidisciplinary model improves colorecal cancer microsatellite instability testing. *InSIGHT* **25.06:** 50.
77. Fong, P.C., D.S. Boss, T.A. Yap, *et al.* 2009. Inhibition of poly(ADP-ribose) polymerase in tumors from BRCA mutation carriers. *N. Engl. J. Med.* **361:** 123–134.
78. Xu, J., J. Sun, A.K. Kader, *et al.* 2009. Estimation of absolute risk for prostate cancer using genetic markers and family history. *Prostate* **69:** 1562–1572.
79. Paynter, N.P., D.I. Chasman, J.E. Buring, *et al.* 2009. Cardiovascular disease risk prediction with and without knowledge of genetic variation at chromosome 9p21.3. *Ann. Intern. Med.* **150:** 65–72.
80. Lyssenko, V., A. Jonsson, P. Almgren, *et al.* 2008. Clinical risk factors, DNA variants, and the development of type 2 diabetes. *N. Engl. J. Med.* **359:** 2220–2232.

Ann. N.Y. Acad. Sci. ISSN 0077-8923

ANNALS OF THE NEW YORK ACADEMY OF SCIENCES
Issue: *The Year in Human and Medical Genetics*

Hereditary kidney diseases: highlighting the importance of classical Mendelian phenotypes

Geneviève Benoit,[1,2] Eduardo Machuca,[1,3] Laurence Heidet,[4] and Corinne Antignac,[1,5,6]

[1]Inserm, U983, Hôpital Necker-Enfants Malades, Paris, France. [2]CHU Sainte-Justine, Université de Montréal, Canada. [3]Pontificia Universidad Católica de Chile, Escuela de Medicina, Chile. [4]Service de Néphrologie Pédiatrique, Hôpital Necker-Enfants Malades, Assistance Publique-Hôpitaux de Paris (AP-HP), Paris, France. [5]Université Paris Descartes, Faculté de Médecine Paris Descartes, Paris, France. [6]Assistance Publique-Hôpitaux de Paris (AP-HP), Département de Génétique, Hôpital Necker-Enfants Malades, Paris, France

Address for correspondence: Corinne Antignac, M.D., Ph.D., Inserm, U983, 6e étage, Tour Lavoisier, Hôpital Necker-Enfants Malades, 149 rue de Sèvres, 75015 Paris, France.corinne.antignac@inserm.fr

A Mendelian inheritance underlies a nonnegligible proportion of hereditary kidney diseases, suggesting that the encoded proteins are essential for maintenance of the renal function. The identification of genes involved in congenital anomalies of the kidney and in familial forms of nephrotic syndrome significantly increased our understanding of the renal development and kidney filtration barrier physiology. This review will focus on the classical phenotype and clinical heterogeneity observed in the monogenic forms of these disorders. In addition, the role of susceptibility genes in kidney diseases with a complex inheritance will also be discussed.

Keywords: renal development disorders; nephrotic syndrome; podocyte; Mendelian inheritance

Introduction

The essential function of the kidney is to preserve constancy of body fluid and electrolytes. During normal renal development, interactions between the branching utereteric bud and the metanephric mesenchyme generate the nephrons, which are the basic structural and functional unit of the kidney. Each nephron has an initial filtering component (the glomerulus) and a tubule specialized for reabsorption and secretion. In the last 10 years, the discovery of several genes involved in different nephropathies represented a major breakthrough in the field of renal medicine. These genes encode proteins having a role in all nephron segments and leading to clinically different disorders (Fig. 1); their identification often provided a starting point for the elucidation of the related disease pathogenesis. Indeed, before the identification of the first podocyte gene,[1] the crucial role of podocytes (visceral epithelial cells that wrap around the capillaries of the glomerulus) in the maintenance of the kidney filtration barrier was far underestimated (Fig. 1). In addition, strategies of positional cloning and candidate gene analysis identified genes involved in various types of tubulopathies, thereby showing that primary genetic defects in transporter or channel proteins operating along the renal tubules underlie these diseases. Finally, the finding of mutations in ciliary genes in cases with cystic kidney diseases generated new insights on the mechanisms of these disorders, revealing that the primary cilium is a key structure necessary for the regulation and maintenance of cellular/tissue homeostasis; the pathogenesis of ciliopathies is detailed in another section of this volume.

It has been shown that mutation in one single gene is often sufficient in itself to cause these diseases. However, the demonstration of a great genetic and clinical heterogeneity has confirmed that a significant proportion of "simple" monogenic disorders are far more complex than previously thought. In the next sections, the genetics aspects of renal developmental disorders and inherited glomerular diseases will be discussed, highlighting the importance of classical Mendelian phenotypes in

doi: 10.1111/j.1749-6632.2010.05817.x

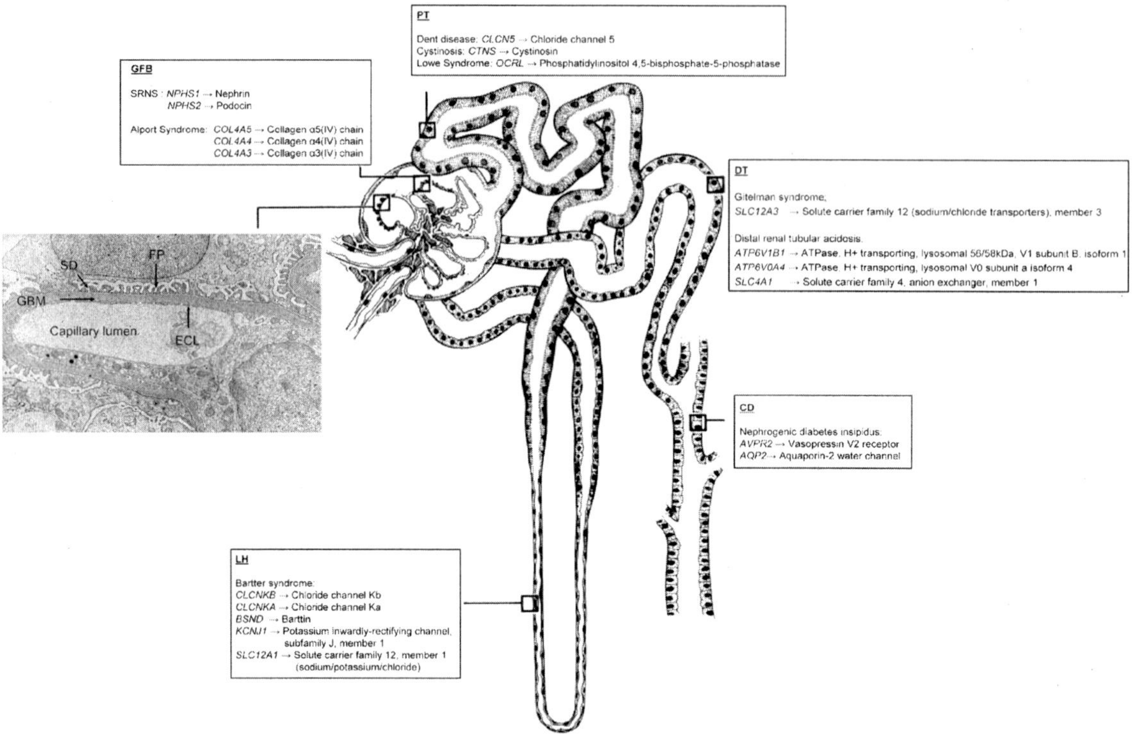

Figure 1. Various monogenic renal diseases due to mutations in genes affecting different nephron segments. The list of genes, proteins, and disorders is not exhaustive. Mutations in the genes encoding renin, angiotensinogen, angiotensin converting enzyme or angiotensin II receptor type 1 may cause autosomal recessive renal tubular dysgenesis, a rare disorder histologically characterized by absence or paucity of differentiated proximal tubules. Dominant mutations in the first exon of the renin gene have also been found in patients with early-onset hyperuricemia, anemia, and progressive renal failure. Mutated genes involved in cystic kidney disease are expressed at the primary cilia and centrosome complex and malorientation of the mitotic spindle during kidney morphogenesis might cause dilatation of the renal tubules, resulting in renal cysts. These genes are not shown in Figure 1. In addition, monogenic forms of hypertension are not represented here. The glomerular filtration barrier structure on electron microscopy is illustrated in the left part of the figure. In the glomerulus, the plasma is ultrafiltered through a semipermeable glomerular capillary wall composed of the fenestrated endothelial cell layer, the glomerular basement membrane, and the podocytes. These latter cells have foot processes, which are interconnected by the slit diaphragm, a signal transduction unit composed of several proteins that contribute to the regulation of podocyte function. CD, collecting duct; DT, distal tubule; ECL, fenestrated endothelial cell layer; FP, podocyte foot process; GBM, glomerular basement membrane; GFB, glomerular filtration barrier; LH, loop of Henle; PT, proximal tubule; SD, slit-diaphragm; SRNS, steroid-resistant nephrotic syndrome. Modified from Miner J.H. 1999. Renal basement components. Kidney Int. 56: 2016–2024, with permission. The electron microscopy is a courtesy of Dr. Marie-Claire Gubler.

hereditary kidney diseases but also showing the role of susceptibility genes and complex inheritance in some cases.

Renal development disorders

Due to the uncertainty about the clinical classification of the various kidney and urinary tract malformations, these entities are usually represented under the acronym CAKUT (Congenital Anomalies of the Kidney and Urinary Tract). CAKUT are major causes of chronic renal failure in childhood.[2] Until recently, providing a precise genetic counseling about the risk of disease transmission in families with renal developmental disorders was often not possible, as almost no specific genetic and/or environmental etiologies were identified. Functional animal experiments provided invaluable inputs for the identification of genes playing a role in renal anomalies in humans and it is now known that several multiorgan malformation syndromes and some isolated CAKUT cases have a Mendelian inheritance.

About two decades ago, *Pax2* loss-of-function mutations in mice have been shown to be associated with renal hypoplasia and retinal defect.[3]

This gene encodes a transcriptional factor expressed in the mouse developing kidney, optic cup, and other parts of the central nervous system.[4,5] Based on the similarity between the *Pax2* mutant mice phenotype and the association of renal hypoplasia, vesicoureteral reflux, and optic nerve coloboma observed in humans,[6,7] heterozygous *PAX2* mutations have been found in 1995 to cause the human autosomal-dominant Renal Coloboma syndrome.[8] To date, about 20 different inactivating mutations have been described, the most common being a single-base pair insertion (G619-insertion mutation within a string of seven guanidines) in the second exon of the gene. Mice studies showed that heterozygous mutations of *Pax2* are associated with increased apoptosis and reduced branching of the ureteric bud, due to reduced PAX2 dosage during a critical window in kidney development.[9–11] By inducing a delay in urinary tract maturation, PAX2 may also influence the structure of the vesicoureteral junction;[12] this is concordant with the finding that vesicoureteral reflux is noted in roughly one-third of patients with mutations in this gene.[13] Interestingly, a relatively common *PAX2* haplotype has been associated with a 10% reduction in newborn kidney volume (a surrogate of nephron number).[14] Considering that suboptimal number of nephrons has been observed in patients with primary hypertension compared to age-matched control subjects,[15] one may also hypothesize a role for *PAX2* in the pathogenesis of essential hypertension.

Mutations in several other genes may lead to autosomal-dominant syndromic forms of CAKUT. Among them, mutations in *EYA1* and *SIX1/SIX5* are involved in branchio-oto-renal (BOR) syndrome[16–18] while mutations in *SALL1* are implicated in Townes-Brocks syndrome.[19] However, probably the most interesting discovery in the field of nephrogenesis anomalies has been the identification of mutations in the *HNF1B* gene, encoding the transcription factor hepatocyte nuclear factor 1β. This factor is a homeoprotein expressed in several tissues including pancreatic, hepatic, and renal epithelia. *Hnf1b*-null mice die around embryonic day 7 due to a defect in visceral endoderm differentiation.[20,21] A Cre-LoxP strategy has therefore been used to assess the roles of this gene during renal development. It has been shown that *Hnf1b* deletion in mouse collecting ducts causes a renal cystic phenotype and that renal cyst formation is accompanied by a drastic transcription downregulation of the *Umod*, *Pkhd1*, and *Pkd2* genes (genes responsible for distinct cystic kidney diseases).[22,23] Surprisingly, in humans, the description of mutations in the *HNF1B* gene (also known as *TCF2*) as a cause of renal developmental disorders arose from observations in the study of monogenic diabetes. Indeed, *HNF1B* mutations were first implicated in maturity-onset diabetes of the young (MODY) 5,[24] although all the mutations carriers in the initial family described also had renal disease. Thereafter, the identification of further families with *HNF1B* mutations showed that severe nondiabetic renal manifestations, particularly renal cystic disease, was the major feature of the phenotype while diabetes was confirmed in only 58% of cases.[25] This led to name this entity "renal cysts and diabetes syndrome" (MIM # 137920);[25] this syndrome may also include nonrenal features such as pancreatic atrophy,[26] abnormal liver function tests,[27] and genital tract malformations.[28] The major role of this gene in CAKUT needs to be emphasized, as heterozygous *HNF1B* mutations have been found in 20–30% of patients with various renal phenotypes (hyperechogenic kidneys, multicystic kidney disease, renal agenesis, renal hypoplasia, cystic dysplasia, and hyperuricemic tubulointerstitial nephropathy).[29–31] About 50% of the mutations are *de novo*,[29,30,32,33] explaining the sporadic appearance of this disease; the identification of "renal cysts and diabetes syndrome" as a Mendelian disorder was therefore a complex task and represents a major breakthrough in the field of genetic renal diseases.

Although the above-mentioned genes are usually associated with syndromic forms of renal hypodysplasia, they should also be considered as a potential etiology in cases with isolated renal development disorders. Indeed, among 77 patients with renal hypodysplasia and no obvious extrarenal phenotype, almost 12% had a detectable sequence variant in *HNF1B*, *PAX2*, or *SALL1* genes.[32] In addition, significant phenotype variability, including intrafamilial variability, has been observed in patients bearing mutations in these genes. Indeed, in two families bearing an heterozygous *EYA1* mutation, a complete BOR syndrome was present in the index cases whereas one of their parents, carrying the same mutation, presented deafness, ear tags, and cervical fistulas without kidney anomalies.[32] In some patients with *PAX2* mutation, the optic nerve abnormality is barely detectable or even absent.[32,34–36] Moreover, an additional *SIX1* variant has been identified in two siblings presenting Renal Coloboma syndrome (early-onset rapidly progressing chronic renal insufficiency and optic disc dysplasia) and *PAX2* mutation, while the father, bearing the same *PAX2* mutation without the *SIX1* variant, had only isolated renal hypodysplasia with late-onset moderate alteration of the renal function.[32] This suggests that the presence of modifiers may contribute to the phenotype variability observed in these patients. It has also been shown that disruption of noncoding elements with site- and stage-specific enhancer function surrounding master developmental genes may be involved in some human congenital malformations;[37] one can speculate that this mutational mechanism may also have a role in the clinical heterogeneity observed in some cases.

Another domain that has been well studied in animal experiments concerned the key players of ureteric bud branching morphogenesis during kidney development. Gdnf, a glial cell line-derived neurotrophic factor produced by the metanephric mesenchyme cells, signals through the receptor tyrosine kinase Ret and the coreceptor Gdnf family receptor-α1, which are coexpressed in the Wolffian duct and later at the tips of the ureteric bud.[38] In animal models, inactivation of the genes encoding these proteins leads to failure of primary ureteric bud growth, which clinically translates into renal agenesis.[39–42] Although the control of animal kidney formation by Ret, Gdnf, and Gfra1 is known since more than one decade, the implication of these components in human renal diseases remains unclear. One study suggested that *RET* mutations may be involved in more than one-third of fetuses with bilateral agenesis and 20% of those with unilateral agenesis,[43] but this has not yet been confirmed in further reports. Mutations in other genes encoding proteins participating in the reciprocal interactions of ureteric budding have recently been found in patients with kidney development anomalies.[44] Indeed, the similarity between murine knockout phenotypes of *Six2* and *Bmp4* and human renal malformations suggested a role of these genes is human renal dysplasia; Six2 is a transcription factor whose expression in the renal mesenchyme is required for Gdnf synthesis,[45] while Bmp4, a member of the TGF-β superfamily, is one of the key regulators of ureteric budding and has an inhibitory effect on Gdnf function.[46] Several other molecules (including Etv4, Etv5) required for kidney branching morphogenesis have been identified in mice experiments, but their role in human diseases need to be determined.[47] Nevertheless, animal models contributed significantly to our understanding of nephrogenesis, thereby continuously suggesting novel pathways that may potentially be of importance in the occurrence of CAKUT in humans.

Hereditary disorders of the renin–angiotensin system

The renin–angiotensin system (RAS) has widespread and diverse roles, including maintenance of renal sodium homeostasis, modulation of vascular tone, and erythropoiesis. While several publications showed the effect of the angiotensin converting enzyme gene polymorphisms on renal function, the crucial role of RAS in human kidney development has been highlighted in 2005 by the identification of homozygous or compound heterozygous mutations in the genes encoding renin, angiotensinogen, angiotensin converting enzyme or angiotensin II receptor type 1 in individuals with autosomal recessive renal tubular dysgenesis (RTD).[48] RTD is a rare lethal disorder histologically characterized by absence or paucity of differentiated proximal tubules; it has been suggested that the absence or poor development of proximal tubules, as well as renal vascular changes, may be attributable to renal hypoperfusion. Interestingly, dominant mutations in the first exon of the renin gene, which alter the signal sequence of the protein, have been found few years later in patients with a completely different phenotype, including early-onset hyperuricemia, anemia, and progressive renal failure.[49] The mutations identified prevented endoplasmic reticulum translocation/processing of nascent preprorenin which are necessary steps for prorenin/renin synthesis, secretion, and activity. It was hypothesized that the expression of the mutant proteins reduced the viability of renin-expressing cells, which may explain, at least partly, the clinical manifestations observed. Indeed, reduced renin activity is associated with volume depletion with secondary increased proximal tubular reabsorption of uric acid, resulting in hyperuricemia. Anemia has previously been described in patients with decreased RAS activity.[50] However, although apoptosis of the renin-expressing cells may perturb kidney autoregulation, the exact pathophysiological mechanisms leading to progressive renal failure in these patients need to be better characterized. The functional impact of mutations in the other exons of the renin gene also needs to be determined.

Inherited glomerular diseases

In contrast to renal developmental disorders, resulting in a reduced number of nephrons or an abnormal patterning of the renal tissue, patients with genetic glomerular diseases have structural defects in a specific part of the kidney, the glomerular filtration barrier, which clinically leads to the development of various renal diseases including Alport syndrome (inherited glomerular basement membrane [GBM] disease) and nephrotic syndrome (NS) (inherited podocyte disease). NS is a heterogeneous group of disorders characterized by heavy urinary proteins losses with hypoalbuminemia, edema, and dyslipidemia. The prevalence of idiopathic NS has been estimated to be 11 per 100,000 individuals. While most patients have steroid-sensitive NS (SSNS), approximately 10–20% of children and 40% of adult cases do not achieve sustained remission after steroid therapy;[51–54] they are therefore at higher risk of reaching end-stage kidney disease (ESKD). In contrast to SSNS which usually presents minimal glomerular changes on kidney biopsy, steroid-resistant nephrotic syndrome (SRNS) typically manifests histologically as focal segmental glomerulosclerosis (FSGS), a lesion characterized by sclerosis and podocyte foot process effacement in some of the glomeruli and a part of each entire glomerulus.

SSNS is believed to have an immune pathogenesis; the involvement of a circulating permeability factor, whose production seems to follow T cell dysfunction, has been suggested.[55] Recently, a humanized mouse model of NS has been developed by injecting $CD34^+$ stem cells or $CD34^-$ peripheral blood mononuclear cells from patients with FSGS or controls into immunocompromised mice.[56] This model confirmed the immune origin of idiopathic NS and suggested that the cells responsible for the pathogenesis are more likely to be immature differentiating cells rather than mature peripheral T cells. Familial forms of SSNS have been described; the exact incidence is unknown, but it may represent up to 3% of cases.[57] At least two attempts to identify a putative disease locus have been performed.[58,59] A genome-wide scan (GWS) analysis performed in a consanguineous SSNS kindred allowed the identification of a locus on chromosome 2p12-p13.2.[59] More recently, an extended SSNS Bedouin family (11 affected individuals) with a high rate of consanguinity has been studied;[58] however, the GWS analysis did not support linkage to any chromosomal locus, thereby suggesting that either all the affected patients may not have mutations in the same gene or that a more complex inheritance is involved.

On the other hand, it has clearly been shown that inherited structural defects of the glomerular filtration barrier are responsible of a large proportion of SRNS. Indeed, genetic linkage studies have identified several genes involved in the development of SRNS, coding for novel proteins (nephrin, podocin) or already known proteins for which their role in the kidney was unexpected (PLCε1, CD2AP, ACTN4, TRPC6, INF2) (Fig. 2 and Table 1). These findings completely changed our understanding of the GBM physiology, clearly emphasizing the pivotal role of the podocyte. Indeed, podocytes play a crucial role in the pathogenenesis of glomerulopathies; these cells have foot processes, which are interconnected by the slit diaphragm (SD), a signal transduction unit composed of several proteins that contribute to the regulation of podocyte function. The SD is linked to the actin-based contractile apparatus of foot processes and tightly controls its dynamics.

The first major step in the field of hereditary NS occurred in 1998 with the cloning of nephrin, encoded by *NPHS1*, together with its localization to the SD of podocytes.[1] Nephrin is an integral membrane protein of the immunoglobulin superfamily located at opposing sites of the foot processes formed by podocytes. *NPHS1* mutations were initially found in patients with congenital NS of the Finnish type (CNF),[1] a recessively inherited disorder typically characterized by massive proteinuria detectable at birth, a large placenta, and characteristic microcystic dilatation of the proximal tubules. Although CNF is observed worldwide, it is particularly common in Finland, where two founder mutations, a two-base pair deletion in exon 2 (Fin major: p.L41fsX91) and a nonsense mutation in exon 26 (Fin minor: p.R1109X) are present. To date, more than 165 different *NPHS1* mutations have been identified and most non-Finnish patients have private *NPHS1* mutations, including nonsense, missense, frameshift insertion/deletion, and splicing mutations spanning over the whole gene.[68–79] Overall, *NPHS1* mutations

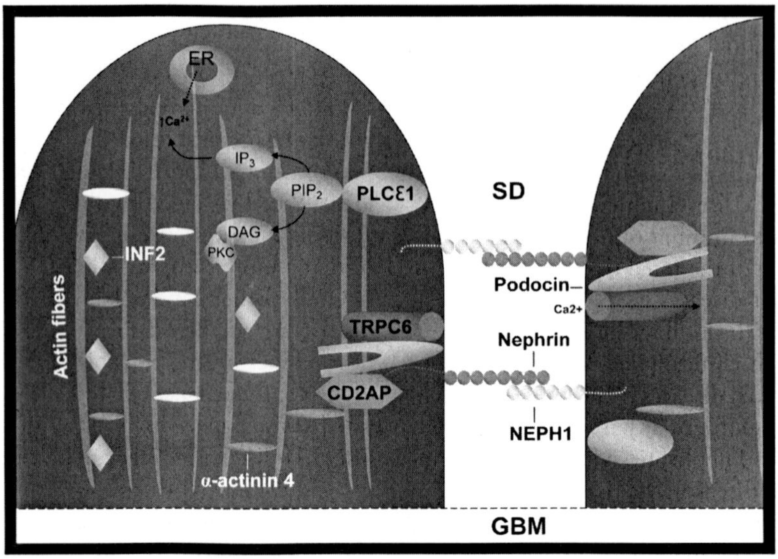

Figure 2. Overview of the podocyte and slit-diaphragm proteins involved in hereditary nephrotic syndrome. *NPHS1* encodes nephrin, a transmembrane protein of the immunoglobulin superfamily accounting for one the major components of the slit membrane.[1] Nephrin extracellular domain forms homodimers and heterodimerizes with NEPH1.[60] Nephrin also interacts through its C-terminal part with podocin,[61] encoded by *NPHS2*, and with CD2AP.[62] While CD2AP is a cytoplasmic protein playing a role in cytoskeletal remodeling, podocin is a hairpin-like protein that localizes to the insertion site of the slit diaphragm of podocytes. In addition, podocin is a lipid raft protein, which recruits nephrin in these specialized microdomains of the plasma membrane.[63] TRPC6 has found to be regulated by the podocin-lipid complex and also interacts with nephrin.[63,64] PLCε1 belongs to the phospholipase family of proteins that catalyze hydrolysis of phosphoinositides,[65] thereby generating IP_3 and DAG; DAG activates protein kinase C and also regulates TRPC channels. Finally, *ACTN4* encodes the actin-bundling protein α-actinin-4[66] while the newly identified *INF2* gene encodes a member of the formin family of actin-regulating protein;[67] the FSGS phenotype associated with mutations in these two genes confirms the critical importance of an intact podocyte actin cytoskeleton. DAG, diacylglycerol; ER, endoplasmic reticulum; GBM, glomerular basement membrane; IP_3, inositol trisphosphate; PIP_2, phosphatidyl inositol-bisphosphate; PKC, protein kinase C; SD, slit-diaphragm.

are found in 39–56% of patients presenting NS in the first 3 months of life.[76,77,79] Soon thereafter, in 2000, positional cloning allowed identification of mutations in the *NPHS2* gene, encoding podocin.[80] Since then, our knowledge on the podocyte function has continuously increased, and the role of podocin and *NPHS2* variants in human glomerular diseases has been better defined. Podocin is a raft-associated component of the SD, which interacts with the cytoplasmic tail of nephrin.[63] Homozygous or compound heterozygous mutations in this gene have been found in about 40% of familial and 10–28% of sporadic SRNS cases.[81–85] More than one-third of patients with NS onset between 4 months and 1 year of age have *NPHS2* mutations.[76] So far, about 110 different pathogenic mutations have been reported, most of which are missense and nonsense mutations.[80,83,84] Small insertion/deletions and splicing mutations also occur in some cases.

Although mutations in these two genes represented the first evidence of NS with a Mendelian inheritance, recent researches confirmed that a great genetic and clinical heterogeneity exists among NS cases and that the gap between "simple monogenic" and complex disorders is probably thinner than previously thought. Indeed, our knowledge of the phenotype spectrum associated with *NPHS1* mutations has increased dramatically in the very last years, by the identification of disease-causing nephrin mutations in patients presenting NS later in childhood; in one study, *NPHS1* mutations were found in cases with age at disease onset varying from 6 months to 8 years,[86] while other authors found compound heterozygous mutations (p.R827X and p.R979S) in a patient with an adult-onset glomerular disorder.[87] While nephrin mutants are usually retained in the endoplasmic reticulum (probably secondary to misfolding of the protein) in patients with early onset NS,[88,89] the presence of at least one mild mutation allowing partial maintenance of the protein function may explain the later onset of disease in other cases. Similarly, although *NPHS2* is the main gene involved in infantile NS, it has also been implicated in congenital[76,79] and even adult forms of the disease.[90,91] Interestingly, the p.R229Q variant, which is the most frequently reported nonsynonymous *NPHS2* variant in Caucasians, has been shown to lead to late-onset NS when found in association with one pathogenic *NPHS2* mutation.[90,91] *In vitro* studies demonstrated decreased binding of the p.R229Q mutant protein to nephrin, probably explaining its pathogenic role.[91] Although concomittant mutations in both *NPHS1* and *NPHS2* have been found in some patients,[72,93] there is so far no convincing evidence for oligogenic inheritance in cases with SRNS. Nevertheless, the difference in the mutation detection rate between familial and sporadic cases indicates that all SRNS cases are not hereditary with a Mendelian inheritance and points to a more complex inheritance for a significant proportion of them. Indeed, while monogenic diseases are more prevalent in patients presenting clinical manifestations early in life,[76] multifactorial etiologies have a more predominant role in those with late-onset disorder. In addition, rarely, identical mutations leading to significantly different phenotypes, from early to late onset NS, have been found,[86,87] also suggesting that modifying genes or environmental factors may influence the renal presentation and outcome. Interestingly, in a mouse model in which the podocin gene was constitutively inactivated, it has been shown that maternal lactational environment interacts with modifier genes to influence the renal disease progression.[94]

Probably one of the most intriguing findings of the recent years in the field of hereditary NS has been the identification of asymptomatic adult individuals carrying in the homozygous state the mutations found in symptomatic patients of their respective families.[95,96] This is a particularly unusual feature for autosomal recessive disorders which, again, emphasizes the role of modifier genes or environmental factors in the disease phenotype variability, even in diseases initially considered as monogenic. This situation has been found in cases bearing mutations in the *PLCE1* gene, coding for the phospholipase C epsilon 1 (PLCε1) protein, and causing early-onset NS and ESKD. Phospholipase C (PLC) is a key enzyme that generates the second messengers inositol 1,4,5-triphosphate (IP_3) and diacylglycerol (DAG), thereby initiating intracellular pathways of cell growth and differentiation; it also plays a role in regulating the calcium flux.[97] In the kidney, PLCε1 is expressed in the podocytes.[65] To explain the nonpenetrance seen in some cases bearing recessive *PLCE1* mutations, one may hypothesize a protective effect arising from other members of the phospholipase C family in the asymptomatic individuals or, alternatively, a deleterious additional effect caused by mutations or variants in other podocyte genes in symptomatic patients; these modifiers hypotheses have however not been confirmed.[96]

Nevertheless, it is tempting to speculate that modifiers may play a significant role in the phenotypic variability observed in SRNS cases bearing mutations in the same gene and even carrying the same mutation. Another growing interest resides in the determination of genes that may predispose individuals to develop kidney diseases. While recessive mutations in the *CD2AP* gene has been identified in a young nephrotic child, thereby providing so far the only clear evidence for a direct pathogenic role of this gene in NS,[98] heterozygous *CD2AP* mutations have also been found in patients with FSGS as well as in completely asymptomatic individuals.[93,98–100] Moreover, in the *CD2AP* heterozygous mouse model, proteinuria is noted only after injection of nephrotoxic antibodies, leading to consider that these mice have a decreased capacity to sustain glomerular damage.[99] The above observations have therefore suggested *CD2AP* as a determinant of human susceptibility to glomerular disease. Interestingly, the CD2-associated protein is an adapter molecule initially found as a ligand for the T cell membrane protein CD2. Before genetics studies identified the role of *CD2AP* in the kidney, no one could have imagined that its encoded protein was one of the main components of the SD and an interacting partner of nephrin and podocin.[63]

Another gene, initially identified as directly responsible of a Mendelian disorder, has thereafter been considered as having a role in the susceptibility to develop renal diseases. Indeed, *MYH9*, encoding the podocyte-expressed nonmuscle myosin heavy chain type II isoform A protein, has originally been found as the disease-causing gene of the rare giant-platelet disorders, which may include renal manifestations.[101] It has thereafter

Table 1A. Genes involved in steroid-resistant nephrotic syndrome (autosomal recessive inheritance)

Gene	Protein	Type/ localization of protein	Typical phenotype	Phenotypic variability
Autosomal recessive inheritance				
NPHS1	Nephrin	SD	Congenital NS and infantile SRNS	Homozygous mutations have been found in cases with childhood[86,87] and even adult-onset (one case reported)[87] SRNS
NPHS2	Podocin	SD	Infantile and childhood-onset SRNS May also cause congenital NS	The association of a pathogenic *NPHS2* mutation with the p.R229Q variant may lead to juvenile and adult-onset SRNS[90,91]
CD2AP	CD2 associated protein	SD	Infantile SRNS (only one case reported)	
PLCE1	Phospholipase C epsilon 1	SD	Infantile and early-childhood SRNS with DMS	Asymptomatic individuals carrying in the homozygous state the *PLCE1* mutations found in symptomatic patients of their respective families have been reported (see Ref. 92)[95,96]
LAMB2	Laminin β2	GBM	Pierson syndrome: congenital NS, microcoria and other ocular anomalies, neuromuscular features	Isolated congenital NS has been reported in one family [123]
ITGB4	Integrin β4	GBM	Epidermolysis bullosa with pyloric atresia may be associated with NS	
SCARB2	Scavenger receptor class B, member 2 (LIMP2, Lysosome membrane protein 2)	Lysosomal protein	Action myoclonus-renal failure	

Continued

Table 1A. *Continued*

Gene	Protein	Type/ localization of protein	Typical phenotype	Phenotypic variability
COQ2	Para-hydroxybenzoate-polyprenyl transferase	Mitochondrial protein	CoQ_{10} deficiency, early-onset SRNS with or without neurological anomalies	
PDDS2	Decaprenyl diphosphate synthase 2	Mitochondrial protein	CoQ_{10} deficiency, Leigh syndrome, and SRNS	
MTTL1	Mitochondrial tRNA for leucine (UUR)	Mitochondrial protein	MELAS Diabetes, deafness, and glomerular lesions	

DMS, diffuse mesangial sclerosis; GBM, glomerular basement membrane; MELAS, mitochondrial myopathy, encephalopathy, lactic acidosis, and stroke; NS, nephrotic syndrome; SD, slit-diaphragm; SRNS, steroid-resistant nephrotic syndrome.

been proposed that common *MYH9* genetic variants, although not directly pathogenic, may contribute to more than 40% of ESKD in the African–American population, playing major roles in idiopathic FSGS, HIV-associated nephropathy, and "hypertension-associated" ESKD.[102–104] However, it has very recently been demonstrated that variants in the *APOL1* gene (located immediately centromeric of *MYH9*) were more strongly associated with FSGS and ESKD among African Americans than previously reported *MYH9* variants; risk variants in *MYH9* and *APOL1* were in strong linkage disequilibrium.[105–107] The *APOL1* gene encodes apolipoprotein L-1 (ApoL1), which is a serum factor that lyses trypanosomes. Genovese *et al.* showed that only the kidney-associated *APOL1* variants (common in African chromosomes but absent in European chromosomes) lyse *Trypanosoma brucei rhodesiense*, the agent of African sleeping sickness in Eastern and Southern Africa;[105] these variants may thereby provide a selective advantage to carriers against African sleeping sickness at the cost of increased susceptibility to kidney disease. The results were consistent with a heterozygous advantage model: the protective effect against *Trypanosoma brucei rhodesiense* was dominant while the observed mode of inheritance of the *APOL1* kidney risk variants was fully recessive in the FSGS cohort. *APOL1* variants may have broader protective effects, as a recent report demonstrated that the trypanosome lytic factor also ameliorates Leishmania infection.[108] The preceding observations may explain why *APOL1* mutations were maintained in the African population and why high rates of renal disease are found among this group. The mechanisms by which *APOL1* variants induce kidney injury are unclear; based on the known functions of ApoL1 in autophagic cell death,[109] lipid metabolism, and vascular integrity,[110,111] several mechanisms can however be hypothesized.

Additional genetic causes of nephrotic syndrome

The above observations showed that major progresses have been made in unraveling the genetic etiology of NS and ESKD. While a significant number of causative genes are yet to be discovered, the continuous identification of novel gene loci is encouraging. In this regards, a third gene involved in the autosomal dominant, juvenile, or adult onset form of FSGS has been identified in late 2009.[67] Indeed, mutations in *INF2*, encoding a member of the formin family of actin-regulating proteins, have been found in families presenting moderate proteinuria and FSGS lesions in early adolescence or adulthood (see Ref. 92).[67] *INF2* mutations explained about 12–17% of familial FSGS cases with

Table 1B. Genes involved in steroid-resistant nephrotic syndrome (autosomal dominant inheritance and complex inheritance/susceptibility genes)

Gene	Protein	Type/ localization of protein	Typical phenotype	Phenotypic variability
Autosomal-dominant inheritance				
TRPC6	Transient receptor potential cation channel, subfamily C, member 6	SD	Juvenile and adult-onset FSGS	*TRPC6* missense mutations have also been identified in two FSGS cases with disease onset at 7 and 9 years of age[113,124]
ACTN4	α-actinin-4	Actin-binding component	Juvenile and adult-onset FSGS	*de novo ACTN4* mutations have been reported in 3 children (2 families) from 3 to 5 years of age which presented rapid progression to chronic kidney disease/ESKD[112,125]
INF2	Inverted formin (FH2 and WH2 domain containing)	Actin-binding component	Juvenile and adult-onset FSGS	
WT1	Wilms tumor 1	Nuclear protein	Denys-Drash syndrome (male pseudohermaphroditism, progressive glomerulopathy, and Wilm's tumor) Frasier syndrome (male pseudohermaphroditism, progressive glomerulopathy, and gonadoblastoma)	Can also cause isolated SRNS[126]
LMX1B	LIM homeobox transcription factor 1 beta	Nuclear protein	Nail-patella syndrome: Dysplasia of the nails and absent or hypoplastic patellae are the cardinal features, nephropathy may be present	
SMARCAL1	SWI/SNF related, matrix associated, actin dependent regulator of chromatin, subfamily a-like 1	Nuclear protein	Schimke immuno-osseous dysplasia: Spondyloepiphyseal dysplasia, nephropathy, and T cell deficiency	

Continued

Table 1B. *Continued*

Gene	Protein	Type/ localization of protein	Typical phenotype	Phenotypic variability
Complex inheritance/susceptibility genes				
CD2AP	CD2 associated protein	SD	Carrying a heterozygous *CD2AP* mutation has been reported to be a predisposing factor toward developing FSGS[92,98,99]	
MYH9	Non muscle myosin heavy chain type II isoform A	Component of the podocyte cytoskeleton	Polymorphisms in *MYH9* exhibit strong association with nondiabetic chronic kidney disease in African Americans[101–103]	
APOL1	Apolipoprotein L1	mRNA encoding ApoL1 are expressed in cultured podocytes[126]	The genetic risk of nondiabetic nephropathy in African Americans previously attributed to *MYH9* may reside, in part or in whole, in *APOL1*[104–106]	

ESKD, end-stage kidney disease; FSGS, focal segmental glomerulosclerosis; SD, slit-diaphragm; SRNS, steroid-resistant nephrotic syndrome.

an apparent autosomal dominant inheritance (see Ref. 92)[67] while mutations in *ACTN4*, encoding α-actinin-4, and *TRPC6*, encoding the transient receptor potential cation channel, subfamily C, member 6, were known to account for about 4–6% of familial FSGS.[112,113] Similar to the phenotype associated with mutations in *ACTN4* and *TRPC6*, disease in patients carrying *INF2* mutations was progressive and often lead to ESKD (age at ESKD: 13–70 years) (see Ref. 92).[67] INF2 is highly expressed in the kidney, including in glomerular podocytes. It has been shown that formins accelerate actin filament assembly;[114] this is particularly relevant as components of the actin cytoskeleton play a crucial role in the regulation of podocyte structure and function.[115] Therefore, screening for *INF2* mutations needs to be strongly considered in patients with an autosomal dominant familial history of FSGS, even prior to *ACTN4* and *TRPC6*.

Other recent interesting findings have been the identification of the genetic cause of some syndromic forms of NS, thereby showing that defects in structures nonspecific to the kidney, such as mitochondrial and lysosomal organelles, may also lead to podocyte dysfunction. Coenzyme Q_{10} (CoQ_{10}) is an essential component of the mitochondrial electron-transport and primary CoQ_{10} deficiency typically includes a group of rare autosomal recessive disorders characterized by neurological and muscular symptoms. However, mutations in the *COQ2* gene, which encodes the parahydroxybenzoate-polyprenyl-transferase enzyme of the CoQ_{10} pathway, have also been identified in patients presenting early-onset glomerular lesions with or without neuromuscular symptoms.[116] In addition, mutations in the *PDSS2* gene, encoding a subunit of decaprenyl diphosphate synthase—the first enzyme of the CoQ_{10} biosynthetic pathway—have been described in a patient with Leigh syndrome and NS[117] while mutations in the $tRNA^{LeuUUR}$ gene have been associated with the MELAS subgroup of mitochondrial encephalomyopathies[118] and FSGS.[119] Finally, the identification of *SCARB2* mutations in patients presenting with action myoclonus-renal failure syndrome,[120] an autosomal recessive disease characterized by FSGS and progressive myoclonic epilepsy, added a lysosomal protein, LIMP2, to the list of molecules involved in the development of NS. Table 1 summarizes the genes known to be involved in SRNS with the associated phenotypes.

Conclusions

A nonnegligible proportion of diseases in the field of nephrology are hereditary. In the last years, genetic studies have allowed identification of causative mutations in a wide variety of rare Mendelian renal disorders. These discoveries have tremendously increased our understanding of, among others, the nephrogenesis process and the role of the podocyte. However, for some patients, who likely have an inherited disorder, genetic testing fails to detect mutations. Further strategies, including exome sequencing,[121] may lead to identification of novel genes in the near future. On the other hand, monogenic diseases account for less than 15% of the burden of ESKD[122] and success in identifying susceptibility genes in renal diseases with a multifactorial etiology has been more limited. Nevertheless, recent studies provided important insights in the complex genetic determinants of glomerular diseases, notably by identifying *APOL1* as a major-effect risk gene for nondiabetic nephropathy in African Americans.[105–107] Most studies on kidney diseases inherited in a complex manner are based on the "common variant–common disease" hypothesis, but rare alleles or multiple variants may also contribute to disease in some cases; therefore, analysis of very large populations and the use of highly efficient resequencing techniques will be required in order to lead to their identification. In addition, as structural rearrangements may confer a risk of developing complex diseases, copy number variation analyses represent an additional promising strategy, particularly in the field of renal development anomalies and syndromic disorders. Finally, it is now clear that clinical heterogeneity is part of many presumably monogenic disorders and further studies searching for genetic and environmental modifiers are noteworthy.

Conflicts of interest

The authors declare no conflicts of interest.

References

1. Kestila, M., U. Lenkkeri, M. Mannikko, *et al.* 1998. Positionally cloned gene for a novel glomerular protein—nephrin: is mutated in congenital nephrotic syndrome. *Mol. Cell* **1:** 575–582.
2. Woolf, A.S. 2000. A molecular and genetic view of human renal and urinary tract malformations. *Kidney Int.* **58:** 500–512.
3. Keller, S.A., J.M. Jones, A. Boyle, *et al.* 1994. Kidney and retinal defects (Krd), a transgene-induced mutation with a deletion of mouse chromosome 19 that includes the Pax2 locus. *Genomics* **23:** 309–320.
4. Dressler G.R., U. Deutsch, K. Chowdhury, *et al.* 1990. Pax2, a new murine paired-box-containing gene and its expression in the developing excretory system. *Development* **109:** 787–795.
5. Nornes, H.O., G.R. Dressler, E.W. Knapik, *et al.* 1990. Spatially and temporally restricted expression of Pax2 during murine neurogenesis. *Development* **109:** 797–809.
6. Karcher, H. 1979. The morning glory syndrome. *Klin. Monbl. Augenheilkd.* **175:** 835–840.
7. Weaver, R.G., L.F. Cashwell, W. Lorentz, *et al.* 1988. Optic nerve coloboma associated with renal disease. *Am. J. Med. Genet.* **29:** 597–605.
8. Sanyanusin, P., L.A. Schimmenti, L.A. McNoe, *et al.* 1995. Mutation of the PAX2 gene in a family with optic nerve colobomas, renal anomalies and vesicoureteral reflux. *Nat. Genet.* **9:** 358–364.
9. Porteous, S., E. Torban, N.P. Cho, *et al.* 2000. Primary renal hypoplasia in humans and mice with PAX2 mutations: evidence of increased apoptosis in fetal kidneys of Pax2(1Neu) +/- mutant mice. *Hum. Mol. Genet.* **9:** 1–11
10. Torban, E., M.R. Eccles, J. Favor, P.R. Goodyer. 2000. PAX2 suppresses apoptosis in renal collecting duct cells. *Am. J. Pathol.* **157:** 833–842.
11. Dziarmaga, A., M. Eccles & P. Goodyer. 2006. Suppression of ureteric bud apoptosis rescues nephron endowment and adult renal function in Pax2 mutant mice. *J. Am. Soc. Nephrol.* **17:** 1568–1575.
12. Murawski, I.J., D.B. Myburgh, J. Favor & I.R. Gupta. 2007. Vesico-ureteric reflux and urinary tract development in the Pax2 1Neu+/– mouse. *Am. J. Physiol. Renal Physiol.* **293:** F1736–F1745.
13. Eccles, M.R. & L.A. Schimmenti. 1999. Renal-coloboma syndrome: a multi-system developmental disorder caused by PAX2 mutations. *Clin. Genet.* **56:** 1–9.
14. Quinlan, J., M. Lemire, T. Hudson, *et al.* 2007. A common variant of the PAX2 gene is associated with reduced newborn kidney size. *J. Am. Soc. Nephrol.* **18:** 1915–1921.
15. Keller, G., G. Zimmer, G. Mall, *et al.* 2003. Nephron number in patients with primary hypertension. *N. Engl. J. Med.* **348:** 101–108.
16. Abdelhak, S., V. Kalatzis, R. Heilig, *et al.* 1997. A human homologue of the Drosophila eyes absent gene underlies branchio-oto-renal (BOR) syndrome and identifies a novel gene family. *Nat. Genet.* **15:** 157–164.
17. Hoskins, B.E., C.H. Cramer, D. Silvius, *et al.* 2007. Transcription factor SIX5 is mutated in patients with branchio-oto-renal syndrome. *Am. J. Hum. Genet.* **80:** 800–804.

18. Kochhar, A., D.J. Orten, J.L. Sorensen, *et al.* 2008. SIX1 mutation screening in 247 branchio-oto-renal syndrome families: a recurrent missense mutation associated with BOR. *Hum. Mutat.* **29:** 565.
19. Kohlhase, J. 2000. SALL1 mutations in Townes-Brocks syndrome and related disorders. *Hum. Mutat.* **16:** 460–466.
20. Coffinier, C., D. Thepot, C. Babinet, *et al.* 1999. Essential role for the homeoprotein vHNF1/HNF1beta in visceral endoderm differentiation. *Development* **126:** 4785–4794.
21. Barbacci, E., M. Reber, M.O. Ott, *et al.* 1999. Variant hepatocyte nuclear factor 1 is required for visceral endoderm specification. *Development* **126:** 4795–4805.
22. Fischer, E., E. Legue, A. Doyen, *et al.* 2006. Defective planar cell polarity in polycystic kidney disease. *Nat. Genet.* **38:** 21–23.
23. Gresh, L., E. Fischer, A. Reimann, *et al.* 2004. A transcriptional network in polycystic kidney disease. *EMBO J.* **23:** 1657–1668.
24. Horikawa, Y., N. Iwasaki, M. Hara, *et al.* 1997. Mutation in hepatocyte nuclear factor-1 beta gene (TCF2) associated with MODY. *Nat. Genet.* **17:** 384–385.
25. Bingham, C. & A.T. Hattersley. 2004. Renal cysts and diabetes syndrome resulting from mutations in hepatocyte nuclear factor-1beta. *Nephrol. Dial. Transplant.* **19:** 2703–2708.
26. Bellanne-Chantelot, C., D. Chauveau, J.F. Gautier, *et al.* 2004. Clinical spectrum associated with hepatocyte nuclear factor-1beta mutations. *Ann. Intern. Med.* **140:** 510–517.
27. Montoli, A., G. Colussi, O. Massa, *et al.* 2002. Renal cysts and diabetes syndrome linked to mutations of the hepatocyte nuclear factor-1 beta gene:description of a new family with associated liver involvement. *Am. J. Kidney Dis.* **40:** 397–402.
28. Lindner, T.H., P.R. Njolstad, Y. Horikawa, *et al.* 1999. A novel syndrome of diabetes mellitus, renal dysfunction and genital malformation associated with a partial deletion of the pseudo-POU domain of hepatocyte nuclear factor-1beta. *Hum. Mol. Genet.* **8:** 2001–2008.
29. Heidet, L., S. Decramer, A. Pawtowski, *et al.* 2010. Spectrum of HNF1B mutations in a large cohort of patients who harbor renal diseases. *Clin. J. Am. Soc. Nephrol.* **5:** 1079–1090.
30. Ulinski, T., S. Lescure, S. Beaufils, *et al.* 2006. Renal phenotypes related to hepatocyte nuclear factor-1beta (TCF2) mutations in a pediatric cohort. *J. Am. Soc. Nephrol.* **17:** 497–503.
31. Adalat, S., A.S. Woolf, K.A. Johnstone, *et al.* 2009. HNF1B mutations associate with hypomagnesemia and renal magnesium wasting. *J. Am. Soc. Nephrol.* **20:** 1123–1131.
32. Weber, S., V. Moriniere, T. Knuppel, *et al.* 2006. Prevalence of mutations in renal developmental genes in children with renal hypodysplasia: results of the ESCAPE study. *J. Am. Soc. Nephrol.* **17:** 2864–2870.
33. Decramer, S., O. Parant, S. Beaufils, *et al.* 2007. Anomalies of the TCF2 gene are the main cause of fetal bilateral hyperechogenic kidneys. *J. Am. Soc. Nephrol.* **18:** 923–933.
34. Ford, B., R. Rupps, D. Lirenman, *et al.* 2001. Renal-coloboma syndrome: prenatal detection and clinical spectrum in a large family. *Am. J. Med. Genet.* **99:** 137–141.
35. Nishimoto, K., K. Iijima, T. Shirakawa, *et al.* 2001. PAX2 gene mutation in a family with isolated renal hypoplasia. *J. Am. Soc. Nephrol.* **12:** 1769–1772.
36. Salomon, R., A.L. Tellier, T. Attie-Bitach, *et al.* 2001. PAX2 mutations in oligomeganephronia. *Kidney Int.* **59:** 457–462.
37. Benko, S., J.A. Fantes, J. Amiel, *et al.* 2009. Highly conserved non-coding elements on either side of SOX9 associated with Pierre Robin sequence. *Nat. Genet.* **41:** 359–364.
38. Takahashi, M. 2001. The GDNF/RET signaling pathway and human diseases. *Cytokine Growth Factor Rev.* **12:** 361–373.
39. Sanchez, M.P., I. Silos-Santiago, J. Frisen, *et al.* 1996. Renal agenesis and the absence of enteric neurons in mice lacking GDNF. *Nature* **382:** 70–73.
40. Schuchardt, A., V. D'Agati, V. Pachnis & F. Costantini. 1996. Renal agenesis and hypodysplasia in ret-k- mutant mice result from defects in ureteric bud development. *Development* **122:** 1919–1929.
41. Pichel, J.G., L. Shen, H.Z. Sheng, *et al.* 1996. Defects in enteric innervation and kidney development in mice lacking GDNF. *Nature* **382:** 73–76.
42. Cacalano, G., I. Farinas, L.C. Wang, *et al.* 1998. GFRalpha1 is an essential receptor component for GDNF in the developing nervous system and kidney. *Neuron* **21:** 53–62.
43. Skinner, M.A., S.D. Safford, J.G. Reeves, *et al.* 2008. Renal aplasia in humans is associated with RET mutations. *Am. J. Hum. Genet.* **82:** 344–351.
44. Weber, S., J.C. Taylor, P. Winyard, *et al.* 2008. SIX2 and BMP4 mutations associate with anomalous kidney development. *J. Am. Soc. Nephrol.* **19:** 891–903.
45. Brodbeck, S., B. Besenbeck & C. Englert. 2004. The transcription factor Six2 activates expression of the Gdnf gene as well as its own promoter. *Mech. Dev.* **121:** 1211–1222.
46. Miyazaki, Y., K. Oshima, A. Fogo, *et al.* 2000. Bone morphogenetic protein 4 regulates the budding site and elongation of the mouse ureter. *J. Clin. Invest.* **105:** 863–873.
47. Lu, B.C., C. Cebrian, X. Chi, *et al.* 2009. Etv4 and Etv5 are required downstream of GDNF and Ret for kidney branching morphogenesis. *Nat. Genet.* **41:** 1295–1302.
48. Gribouval, O., M. Gonzales, T. Neuhaus, *et al.* 2005. Mutations in genes in the renin-angiotensin system are associated with autosomal recessive renal tubular dysgenesis. *Nat. Genet.* **37:** 964–968.
49. Zivna, M., H. Hulkova, M. Matignon, *et al.* 2009. Dominant renin gene mutations associated with early-onset hyperuricemia, anemia, and chronic kidney failure. *Am. J. Hum. Genet.* **85:** 204–213.
50. Hubert, C., K. Savary, J.M. Gasc & P. Corvol. 2006. The hematopoietic system: a new niche for the renin-angiotensin system. *Nat. Clin. Pract. Cardiovasc. Med.* **3:** 80–85.
51. Children ArotISoKDi. 1981. The primary nephrotic syndrome in children. Identification of patients with minimal change nephrotic syndrome from initial response to prednisone. *J. Pediatr.* **98:** 561–564.
52. Cattran, D.C. & P. Rao. 1998. Long-term outcome in children and adults with classic focal segmental glomerulosclerosis. *Am. J. Kidney Dis.* **32:** 72–79.

53. Mendoza, S.A., V.M. Reznik, W.R. Griswold, *et al.* 1990. Treatment of steroid-resistant focal segmental glomerulosclerosis with pulse methylprednisolone and alkylating agents. *Pediatr. Nephrol.* **4:** 303–307.
54. Mekahli, D., A. Liutkus, B. Ranchin, *et al.* 2009. Long-term outcome of idiopathic steroid-resistant nephrotic syndrome: a multicenter study. *Pediatr. Nephrol.* **24:** 1525–1532.
55. Bruneau, S. & J. Dantal. 2009. New insights into the pathophysiology of idiopathic nephrotic syndrome. *Clin. Immunol.* **133:** 13–21.
56. Sellier-Leclerc, A.L., A. Duval, S. Riveron, *et al.* 2007. A humanized mouse model of idiopathic nephrotic syndrome suggests a pathogenic role for immature cells. *J. Am. Soc. Nephrol.* **18:** 2732–2739.
57. White, R.H. 1973. The familial nephrotic syndrome. I. A European survey. *Clin. Nephrol.* **1:** 215–219.
58. Landau, D., T. Oved, D. Geiger, *et al.* 2007. Familial steroid-sensitive nephrotic syndrome in Southern Israel: clinical and genetic observations. *Pediatr. Nephrol.* **22:** 661–669.
59. Ruf, R.G., A. Fuchshuber, S.M. Karle, *et al.* 2003. Identification of the first gene locus (SSNS1) for steroid-sensitive nephrotic syndrome on chromosome 2p. *J. Am. Soc. Nephrol.* **14:** 1897–1900.
60. Liu, G., B. Kaw, J. Kurfis, *et al.* 2003. Neph1 and nephrin interaction in the slit diaphragm is an important determinant of glomerular permeability. *J. Clin. Invest.* **112:** 209–221.
61. Huber, T.B., M. Kottgen, B. Schilling, *et al.* 2001. Interaction with podocin facilitates nephrin signaling. *J. Biol. Chem.* **276:** 41543–41546.
62. Huber, T.B., B. Hartleben, J. Kim, *et al.* 2003. Nephrin and CD2AP associate with phosphoinositide 3-OH kinase and stimulate AKT-dependent signaling. *Mol. Cell Biol.* **23:** 4917–4928.
63. Schwarz, K., M. Simons, J. Reiser, *et al.* 2001. Podocin, a raft-associated component of the glomerular slit diaphragm, interacts with CD2AP and nephrin. *J. Clin. Invest.* **108:**1621–1629.
64. Reiser, J., K.R. Polu, C.C. Moller, *et al.* 2005. TRPC6 is a glomerular slit diaphragm-associated channel required for normal renal function. *Nat. Genet.* **37:** 739–744.
65. Hinkes, B., R.C. Wiggins, R. Gbadegesin, *et al.* 2006. Positional cloning uncovers mutations in PLCE1 responsible for a nephrotic syndrome variant that may be reversible. *Nat. Genet.* **38:** 1397–1405.
66. Kaplan, J.M., S.H. Kim, K.N. North, *et al.* 2000. Mutations in ACTN4, encoding alpha-actinin-4, cause familial focal segmental glomerulosclerosis. *Nat. Genet.* **24:** 251–256.
67. Brown, E.J., J.S. Schlondorff, D.J. Becker, *et al.* 2010. Mutations in the formin gene INF2 cause focal segmental glomerulosclerosis. *Nat. Genet.* **42:** 72–76.
68. Aya, K., H. Tanaka & Y. Seino. 2000. Novel mutation in the nephrin gene of a Japanese patient with congenital nephrotic syndrome of the Finnish type. *Kidney Int.* **57:** 401–404.
69. Beltcheva, O., P. Martin, U. Lenkkeri & K. Tryggvason. 2001. Mutation spectrum in the nephrin gene (NPHS1) in congenital nephrotic syndrome. *Hum. Mutat.* **17:** 368–373.
70. Gigante, M., P. Greco, V. Defazio, *et al.* 2005. Congenital nephrotic syndrome of Finnish type: detection of new nephrin mutations and prenatal diagnosis in an Italian family. *Prenat. Diagn.* **25:** 407–410.
71. Gigante, M., F. Monno, R. Roberto, *et al.* 2002. Congenital nephrotic syndrome of the Finnish type in Italy: a molecular approach. *J. Nephrol.* **15:** 696–702.
72. Koziell, A., V. Grech, S. Hussain, *et al.* 2002. Genotype/phenotype correlations of NPHS1 and NPHS2 mutations in nephrotic syndrome advocate a functional inter-relationship in glomerular filtration. *Hum. Mol. Genet.* **11:** 379–388.
73. Sako, M., K. Nakanishi, M. Obana, *et al.* 2005. Analysis of NPHS1, NPHS2, ACTN4, and WT1 in Japanese patients with congenital nephrotic syndrome. *Kidney Int.* **67:** 1248–1255.
74. Shih, N.Y., J. Li, V. Karpitskii, *et al.* 1999. Congenital nephrotic syndrome in mice lacking CD2-associated protein. *Science* **286:** 312–315.
75. Frishberg, Y., Z. Ben-Neriah, M. Suvanto, *et al.* 2007. Misleading findings of homozygosity mapping resulting from three novel mutations in NPHS1 encoding nephrin in a highly inbred community. *Genet. Med.* **9:** 180–184.
76. Hinkes, B.G., B. Mucha, C.N. Vlangos, *et al.* 2007. Nephrotic syndrome in the first year of life: two thirds of cases are caused by mutations in 4 genes (NPHS1, NPHS2, WT1, and LAMB2). *Pediatrics* **119:** e907–e919.
77. Heeringa, S.F., C.N. Vlangos, G. Chernin, *et al.* 2008. Thirteen novel NPHS1 mutations in a large cohort of children with congenital nephrotic syndrome. *Nephrol. Dial. Transplant.* **23:** 3527–3533.
78. Ismaili, K., A. Pawtowski, O. Boyer, *et al.* 2009. Genetic forms of nephrotic syndrome: a single-center experience in Brussels. *Pediatr. Nephrol.* **24:** 287–294.
79. Machuca, E., G. Benoit, F. Nevo, *et al.* 2010. Genotype–phenotype correlations in non-Finnish congenital nephrotic syndrome. *J. Am. Soc. Nephrol.* **21:** 1209–1217.
80. Boute, N., O. Gribouval, S. Roselli, *et al.* 2000. NPHS2, encoding the glomerular protein podocin, is mutated in autosomal recessive steroid-resistant nephrotic syndrome. *Nat. Genet.* **24:** 349–354.
81. Caridi, G., R. Bertelli, M. Di Duca, *et al.* 2003. Broadening the spectrum of diseases related to podocin mutations. *J. Am. Soc. Nephrol.* **14:** 1278–1286.
82. Hinkes, B., C. Vlangos, S. Heeringa, *et al.* 2008. Specific podocin mutations correlate with age of onset in steroid-resistant nephrotic syndrome. *J. Am. Soc. Nephrol.* **19:** 365–371.
83. Karle, S.M., B. Uetz, V. Ronner, *et al.* 2002. Novel mutations in NPHS2 detected in both familial and sporadic steroid-resistant nephrotic syndrome. *J. Am. Soc. Nephrol.* **13:** 388–393.
84. Ruf, R.G., A. Lichtenberger, S.M. Karle, *et al.* 2004. Patients with mutations in NPHS2 (podocin) do not respond to standard steroid treatment of nephrotic syndrome. *J. Am. Soc. Nephrol.* **15:** 722–732.
85. Weber, S., O. Gribouval, E.L. Esquivel, *et al.* 2004. NPHS2 mutation analysis shows genetic heterogeneity of

steroid-resistant nephrotic syndrome and low post-transplant recurrence. *Kidney Int.* **66:** 571–579.
86. Philippe, A., F. Nevo, E.L. Esquivel, *et al.* 2008. Nephrin mutations can cause childhood-onset steroid-resistant nephrotic syndrome. *J. Am. Soc. Nephrol.* **19:** 1871–1878.
87. Santin, S., R. Garcia-Maset, P. Ruiz, *et al.* 2009. Nephrin mutations cause childhood- and adult-onset focal segmental glomerulosclerosis. *Kidney Int.* **76:** 1268–1276.
88. Liu, L., S.C. Done, J. Khoshnoodi, *et al.* 2001. Defective nephrin trafficking caused by missense mutations in the NPHS1 gene:insight into the mechanisms of congenital nephrotic syndrome. *Hum. Mol. Genet.* **10:** 2637–2644.
89. Roselli, S., I. Moutkine, O. Gribouval, *et al.* 2004. Plasma membrane targeting of podocin through the classical exocytic pathway: effect of NPHS2 mutations. *Traffic* **5:** 37–44.
90. Machuca, E., A. Hummel, F. Nevo, *et al.* 2009. Clinical and epidemiological assessment of steroid-resistant nephrotic syndrome associated with the NPHS2 R229Q variant. *Kidney Int.* **75:** 727–735.
91. Tsukaguchi, H., A. Sudhakar, T.C. Le, *et al.* 2002. NPHS2 mutations in late-onset focal segmental glomerulosclerosis: R229Q is a common disease-associated allele. *J. Clin. Invest.* **110:** 1659–1666.
92. Boyer, O., G. Benoit, O. Gribouval, *et al.* 2010. INF2, a major gene of autosomal dominant focal segmental glomerulosclerosis. *J. Am. Soc. Nephrol.* In press.
93. Lowik, M., E. Levtchenko, D. Westra, *et al.* 2008. Bigenic heterozygosity and the development of steroid-resistant focal segmental glomerulosclerosis. *Nephrol. Dial. Transplant.* **23:** 3146–3151.
94. Ratelade, J., T.A. Lavin, A.O. Muda, *et al.* 2008. Maternal environment interacts with modifier genes to influence progression of nephrotic syndrome. *J. Am. Soc. Nephrol.* **19:** 1491–1499.
95. Gilbert, R.D., C.L. Turner, J. Gibson, *et al.* 2009. Mutations in phospholipase C epsilon 1 are not sufficient to cause diffuse mesangial sclerosis. *Kidney Int.* **75:** 415–419.
96. Boyer, O., G. Benoit, O. Gribouval, *et al.* 2010. Mutational analysis of the PLCE1 gene in steroid resistant nephrotic syndrome. *J. Med. Genet.* **47:** 445–452.
97. Wing, M.R., D.M. Bourdon & T.K. Harden. 2003. PLC-epsilon: a shared effector protein in Ras-, Rho-, and G alpha beta gamma-mediated signaling. *Mol. Interv.* **3:** 273–280.
98. Lowik, M.M., P.J. Groenen, I. Pronk, *et al.* 2007. Focal segmental glomerulosclerosis in a patient homozygous for a CD2AP mutation. *Kidney Int.* **72:** 1198–1203.
99. Kim, J.M., H. Wu, G. Green, *et al.* 2003. CD2-associated protein haploinsufficiency is linked to glomerular disease susceptibility. *Science* **300:** 1298–1300.
100. Gigante, M., P. Pontrelli, E. Montemurno, *et al.* 2009. CD2AP mutations are associated with sporadic nephrotic syndrome and focal segmental glomerulosclerosis (FSGS). *Nephrol. Dial. Transplant.* **24:** 1858–1864.
101. Seri, M., R. Cusano, S. Gangarossa, *et al.* 2000. Mutations in MYH9 result in the May-Hegglin anomaly, and Fechtner and Sebastian syndromes. The May-Heggllin/Fechtner Syndrome Consortium. *Nat. Genet.* **26:** 103–105.
102. Kao, W.H., M.J. Klag, L.A. Meoni, *et al.* 2008. MYH9 is associated with nondiabetic end-stage renal disease in African Americans. *Nat. Genet.* **40:** 1185–1192.
103. Kopp, J.B., M.W. Smith, G.W. Nelson, *et al.* 2008. MYH9 is a major-effect risk gene for focal segmental glomerulosclerosis. *Nat. Genet.* **40:** 1175–1184.
104. Freedman, B.I., P.J. Hicks, M.A. Bostrom, *et al.* 2009. Polymorphisms in the non-muscle myosin heavy chain 9 gene (MYH9) are strongly associated with end-stage renal disease historically attributed to hypertension in African Americans. *Kidney Int.* **75:** 736–745.
105. Genovese, G., D.J. Friedman, M.D. Ross, *et al.* 2010. Association of trypanolytic ApoL1 variants with kidney disease in African Americans. *Science* **329:** 841–845.
106. Genovese, G., S.J. Tonna, A.U. Knob, *et al.* 2010. A risk allele for focal segmental glomerulosclerosis in African Americans is located within a region containing APOL1 and MYH9. *Kidney Int.* **78:** 689–704.
107. Tzur, S., S. Rosset, R. Shemer, *et al.* 2010. Missense mutations in the APOL1 gene are highly associated with end stage kidney disease risk previously attributed to the MYH9 gene. *Hum. Genet.* **128:** 345–350.
108. Samanovic, M., M.P. Molina-Portela, A.D. Chessler, *et al.* 2009. Trypanosome lytic factor, an antimicrobial high-density lipoprotein, ameliorates Leishmania infection. *PLoS Pathog.* **5:** e1000276.
109. Zhaorigetu, S., G. Wan, R. Kaini, *et al.* 2008. ApoL1, a BH3-only lipid-binding protein, induces autophagic cell death. *Autophagy.* **4:** 1079–1082.
110. Monajemi, H., R.D. Fontijn, H. Pannekoek & A.J. Horrevoets. 2002. The apolipoprotein L gene cluster has emerged recently in evolution and is expressed in human vascular tissue. *Genomics* **79:** 539–546.
111. Vanhollebeke, B. & E. Pays. 2006. The function of apolipoproteins L. *Cell Mol. Life Sci.* **63:** 1937–1944.
112. Weins, A., P. Kenlan, S. Herbert, *et al.* 2005. Mutational and biological analysis of alpha-actinin-4 in focal segmental glomerulosclerosis. *J. Am. Soc. Nephrol.* **16:** 3694–3701.
113. Santin, S., E. Ars, S. Rossetti, *et al.* 2009. TRPC6 mutational analysis in a large cohort of patients with focal segmental glomerulosclerosis. *Nephrol. Dial. Transplant.* **24:** 3089–3096.
114. Higgs, H.N. 2005. Formin proteins: a domain-based approach. *Trends Biochem. Sci.* **30:** 342–353.
115. Faul, C., K. Asanuma, E. Yanagida-Asanuma, *et al.* 2007. Actin upregulation of podocyte structure and function by components of the actin cytoskeleton. *Trends Cell Biol.* **17:** 428–437.
116. Diomedi-Camassei, F., S. Di Giandomenico, F.M. Santorelli, *et al.* 2007. COQ2 nephropathy: a newly described inherited mitochondriopathy with primary renal involvement. *J. Am. Soc. Nephrol.* **18:** 2773–2780.
117. Lopez, L.C., M. Schuelke, C.M. Quinzii, *et al.* 2006. Leigh syndrome with nephropathy and CoQ10 deficiency due to decaprenyl diphosphate synthase subunit 2 (PDSS2) mutations. *Am. J. Hum. Genet.* **79:** 1125–1129.
118. Goto, Y., I. Nonaka & S. Horai. 1990. A mutation in the tRNA(Leu)(UUR) gene associated with the MELAS subgroup of mitochondrial encephalomyopathies. *Nature* **348:** 651–653.

119. Kurogouchi, F., T. Oguchi, E. Mawatari, *et al.* 1998. A case of mitochondrial cytopathy with a typical point mutation for MELAS, presenting with severe focal-segmental glomerulosclerosis as main clinical manifestation. *Am. J. Nephrol.* **18:** 551–556.
120. Berkovic, S.F., L.M. Dibbens, A. Oshlack, *et al.* 2008. Array-based gene discovery with three unrelated subjects shows SCARB2/LIMP-2 deficiency causes myoclonus epilepsy and glomerulosclerosis. *Am. J. Hum. Genet.* **82:** 673–684.
121. Ng S.B, K.J Buckingham, C. Lee, *et al.* Exome sequencing identifies the cause of a mendelian disorder. *Nat. Genet.* **42:** 30–35.
122. Ansell D.F.J., D. Fogarty, D. Ford, *et al.* 2009. UK Renal Registry 11th Annual Report (December 2008). *Nephron. Clin. Pract.* **111:** c43–c68.
123. Hasselbacher, K., R.C. Wiggins, V. Matejas, *et al.* 2006. Recessive missense mutations in LAMB2 expand the clinical spectrum of LAMB2-associated disorders. *Kidney Int.* **70:** 1008–1012.
124. Heeringa, S.F., C.C. Moller, J. Du, *et al.* 2009. A novel TRPC6 mutation that causes childhood FSGS. *PLoS One* **4:** e7771.
125. Choi, H.J., B.H. Lee H.Y. Cho, *et al.* 2008. Familial focal segmental glomerulosclerosis associated with an ACTN4 mutation and paternal germline mosaicism. *Am. J. Kidney Dis.* **51:** 834–838.
126. Mucha, B., F. Ozaltin, B.G. Hinkes, *et al.* 2006. Mutations in the Wilms' tumor 1 gene cause isolated steroid resistant nephritic syndrome and occur in exons 8 and 9. *Pediatr. Res.* **59:** 325–331.
127. Freedman, B.I., J.B. Kopp, C.D. Langefeld, *et al.* 2010. The apolipoprotein L1 (APOL1) gene and nondiabetic nephropathy in African Americans. *J. Am. Soc. Nephrol.* **21:** 1422–1426.

Ann. N.Y. Acad. Sci. ISSN 0077-8923

ANNALS OF THE NEW YORK ACADEMY OF SCIENCES
Issue: *The Year in Human and Medical Genetics*

Disorders of dysregulated signal traffic through the RAS-MAPK pathway: phenotypic spectrum and molecular mechanisms

Marco Tartaglia[1] and Bruce D. Gelb[2]

[1]Dipartimento di Ematologia, Oncologia e Medicina Molecolare, Istituto Superiore di Sanità, Rome, Italy. [2]Child Health and Development Institute, Mount Sinai School of Medicine, New York, New York

Address for correspondence: Marco Tartaglia, Ph.D., Dipartimento di Ematologia, Oncologia e Medicina Molecolare, Istituto Superiore di Sanità, Viale Regina Elena 299, 00161 Rome, Italy. mtartaglia@iss.it

RAS GTPases control a major signaling network implicated in several cellular functions, including cell fate determination, proliferation, survival, differentiation, migration, and senescence. Within this network, signal flow through the RAF-MEK-ERK pathway—the first identified mitogen-associated protein kinase (MAPK) cascade—mediates early and late developmental processes controlling morphology determination, organogenesis, synaptic plasticity, and growth. Signaling through the RAS-MAPK cascade is tightly controlled; and its enhanced activation represents a well-known event in oncogenesis. Unexpectedly, in the past few years, inherited dysregulation of this pathway has been recognized as the cause underlying a group of clinically related disorders sharing facial dysmorphism, cardiac defects, reduced postnatal growth, ectodermal anomalies, variable cognitive deficits, and susceptibility to certain malignancies as major features. These disorders are caused by heterozygosity for mutations in genes encoding RAS proteins, regulators of RAS function, modulators of RAS interaction with effectors, or downstream signal transducers. Here, we provide an overview of the phenotypic spectrum associated with germline mutations perturbing RAS-MAPK signaling, the unpredicted molecular mechanisms converging toward the dysregulation of this signaling cascade, and major genotype–phenotype correlations.

Keywords: RAS signaling; Noonan syndrome; LEOPARD syndrome; cardiofaciocutaneous syndrome; Costello syndrome; neurofibromatosis type 1; Legius syndrome; Noonan-like syndrome with loose anagen hair

Introduction

RAS proteins are small guanosine triphosphate (GTP)/guanosine diphosphate (GDP)–binding GTPases that function as molecular switches controlling a major intracellular signaling network that, depending on the cellular context, guides diverse biological functions such as proliferation, migration, survival, cell fate determination, differentiation, and senescence.[1,2] RAS signaling is switched on upon stimulation by numerous cytokines, hormones, and growth factors, and mediates the appropriate cell response to external stimuli through the regulation of transcription, cytoskeletal rearrangement, and metabolism. Within this network, signal flow through the RAF-MEK-ERK protein kinase pathway (Fig. 1), the first described mitogen-associated protein kinase (MAPK) cascade, controls early and late developmental processes, including determination of morphology, organogenesis, synaptic plasticity, and growth.[3,4] Signal transmission via this cascade is initiated by the activation of cell surface receptors by extracellular ligands, which creates intracellular docking sites for adaptor molecules and signal-relay proteins recruiting and activating guanine nucleotide-exchange factors (GEFs). In turn, GEF proteins promote the release of GDP from RAS, favoring binding to the more prevalent GTP and activation of the GTPase. Activated GTP-bound RAS interacts with the RAF serine/threonine kinases (RAF1, BRAF, and ARAF) favoring their catalytic activation, which results in the phosphorylation and

doi: 10.1111/j.1749-6632.2010.05790.x

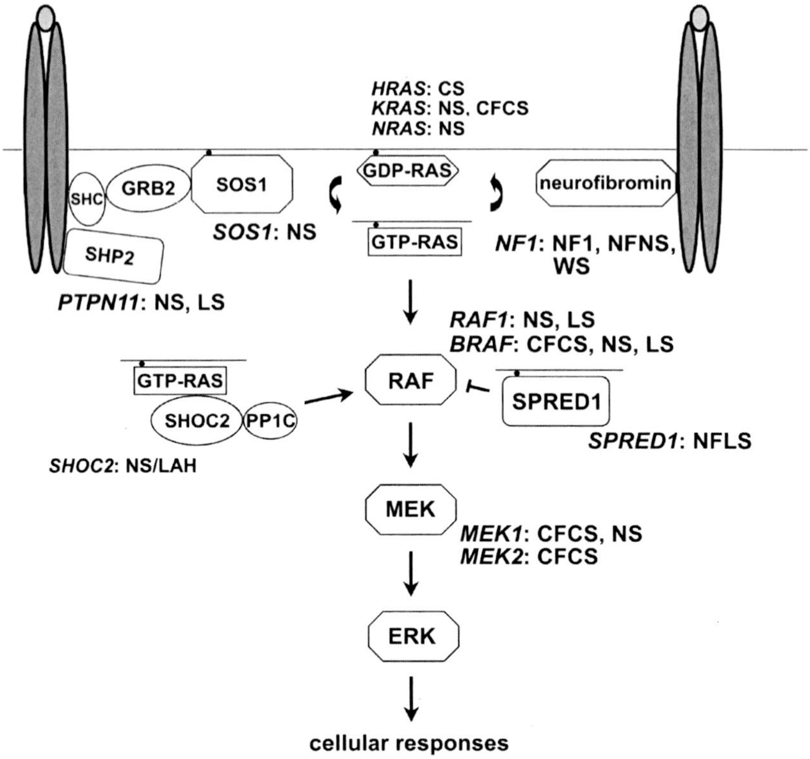

Figure 1. Schematic diagram showing the RAS-MAPK signal transduction pathway and affected disease genes in disorders of the neuro-cardio-facial-cutaneous syndrome family. The double ovals in dark gray and the light gray ovals represent generic dimerized cell-surface receptors binding to their ligand. Abbreviations: CFCS, cardiofaciocutaneous syndrome; CS, Costello syndrome; LS, LEOPARD syndrome; NF1, neurofibromatosis type 1; NFLS, neurofibromatosis type 1–like syndrome (also termed Legius syndrome); NFNS, Neurofibromatosis-Noonan syndrome; NS, Noonan syndrome; NS/LAH, Noonan-like syndrome with loose anagen hair; WS, Watson syndrome.

activation of their substrates, the MAPK/ERK kinases (MEK1 and MEK2). Upon activation, MEKs act as dual specificity kinases to phosphorylate regulatory residues of the extracellular signal-regulated kinases (ERK1 and ERK2), which function as serine/threonine proteine kinases to modulate the activity of a large number of cytoplasmic and nuclear substrates, including proteins implicated in the control of gene expression. Several feedback mechanisms negatively control signal flow through the RAS-MAPK pathway, including the functional inactivation of RAS by GTPase activating proteins (GAPs), which stimulate their intrinsically low GTPase activity.

Due to RAS's nodal role in signal transduction and cell function, signal traffic through it is tightly controlled. Loss of that control leading to enhanced flow contributes significantly to oncogenesis.[5–7] Activating somatic *RAS* gene mutations occur in approximately 30% of human cancers, but the upregulation of this signaling pathway can also result from enhanced function of upstream signal transducers or RAS effectors, as well as from inefficient function of feedback mechanisms. The recent, unpredicted discovery of germline mutations in a number of genes coding for proteins participating in RAS signaling, however, has established a novel scenario in which aberrant signal flow through RAS is causally linked to a group of clinically related developmental disorders characterized by facial dysmorphism, a wide spectrum of cardiac disease, reduced growth, variable cognitive deficits, ectodermal and musculoskeletal anomalies, and increased risk for certain malignancies (Fig. 1).[8,9] This group of diseases can now be viewed as a family of disorders of upregulated RAS-MAPK signaling. Based on the shared pathogenic mechanism and clinical overlap, these genetic diseases have been tentatively grouped as neuro-cardio-facial-cutaneous syndromes (NCFCS) or RAS-opathies.[10,11]

In this review, we outline the phenotypic spectrum associated with perturbed RAS-MAPK signaling, the molecular mechanisms converging toward the dysregulation of this pathway, and major genotype–phenotype correlations.

The RAS-opathies

Noonan syndrome

Noonan syndrome (NS, OMIM 163950) is a disorder characterized by particular facial dysmorphic features, congenital heart defects (CHD), hypertrophic cardiomyopathy (HCM), short stature, skeletal anomalies, and webbing of the neck. Other relatively common features are bleeding diathesis, ectodermal anomalies, lymphatic dysplasias, cryptorchidism, and cognitive deficits.[12–14] NS is characterized by marked variability, which is, in part, explained by the underlying molecular lesion (see below). Facial features and signs include high forehead, hypertelorism, downslanting palpebral fissures, epicantal folds, ptosis, low-set and/or posteriorly rotated ears (Fig. 2). Besides the short and/or webbed neck, a low posterior hairline commonly occurs. Affected individuals exhibit a wide spectrum of cardiac disease.[15,16] Pulmonic stenosis (PS) and HCM occur most commonly, but other lesions are also observed. Feeding problems are noted in the majority of affected infants and can cause failure to thrive.[17] While feeding difficulties resolve in most patients by around age 18 months, the inadequacy in oral intake may be severe enough to necessitate placement of gastrostomy tubes in some.

Developmental delay and learning problems are quite common. Some motor delay can be attributed to the hypotonia often observed in affected infants. An increased prevalence of attention deficit/hyperactivity disorder and frank mental retardation are also observed. Available data indicate that the heterogeneity in cognitive abilities observed in NS is at least partially ascribed to the specifics of the causative gene mutation.[18,19] Skeletal anomalies most frequently consist of pectus deformities, cubitus valgus, and vertebral defects. Of note, two childhood leukemias, juvenile myelomonocytic leukemia (JMML, OMIM 607785) and acute lymphoblastic leukemia (ALL), arise at increased prevalence in NS although they affect only a small percentage of patients.[20–22] The myeloproliferative disorder in children with NS (NS/JMML) may regress without treatment, follow an aggressive clinical course, or evolve to acute myeloid leukemia. This condition is strictly associated with a narrow spectrum of mutations affecting the *PTPN11* gene (see below).[23,24]

NS is thought to be relatively common, although its prevalence has not been determined accurately to date. Most authors cite the figure of 1 in 1,000–2,500 live births, but that estimate is not based on a population study.[25] NS is a Mendelian trait transmitted in an autosomal dominant manner, and, typical for that genetic mechanism, one-third to one-half of cases arise from spontaneous mutations. Thus far, seven genes (*PTPN11*, *SOS1*, *KRAS*, *NRAS*, *RAF1*, *BRAF*, and *MEK1*) have been causally related to this disorder. Mutations in these genes account for a large percentage of cases (approximately 70–75%, in our experience) but are highly dependent on the diagnostic criteria used to make the diagnosis. A recent review provides an up-to-date and thorough survey of the clinical issues and pathogenetic mechanisms for this disorder.[26]

LEOPARD syndrome

LEOPARD syndrome (LS, OMIM 151100) is an autosomal dominant trait that overlaps phenotypically with NS. It is also allelic with NS, with a restricted spectrum of mutations in *PTPN11* accounting for the vast majority of affected individuals (see below).[27,28] In a small proportion of cases, LS has been causally linked to mutations in *RAF1* or *BRAF*.[29–31] The acronymic name refers to the major features: Lentigines, ECG conduction abnormalities, Ocular hypertelorism, PS, Abnormal genitalia, Retardation of growth, and sensorineural Deafness.[32] Similar to NS, there are age-dependent aspects of the phenotype. Facial dysmorphism is similar to that of NS but is usually milder,[33] and the neck is short but not webbed. Multiple lentigines present as dispersed flat, black-brown macules, mostly on the face, neck, and upper part of the trunk, but sparing the mucosae. In general, lentigines appear at the age of 4–5 years and increase until puberty into the thousands. Café-au-lait spots are also observed, alone or in association with lentigines, in up to 70–80% of the patients,[33] and usually precede the appearance of lentigines, being present from the first months of life. Growth retardation (height below the third centile) is observed in 25% of affected individuals,

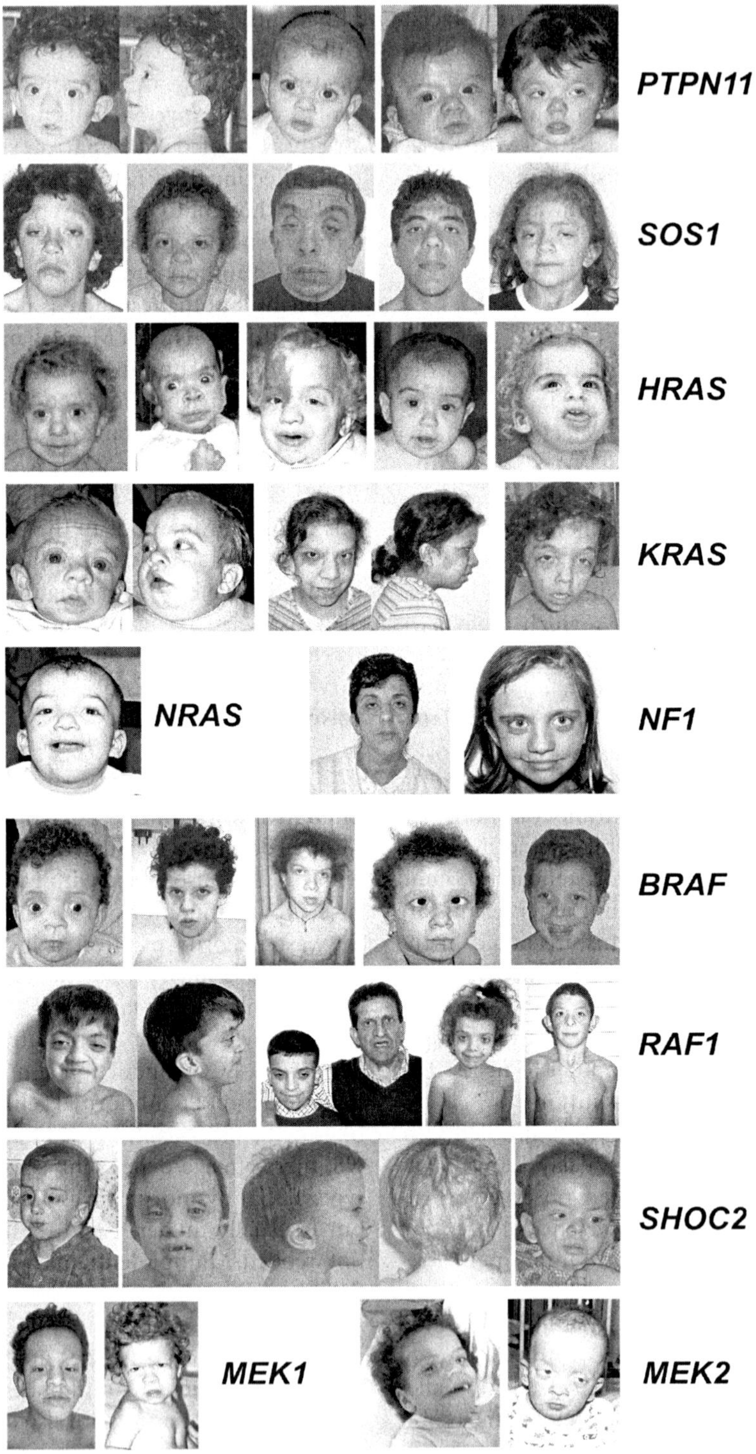

Figure 2. Dysmorphic facial features in disorders of the neuro-cardio-facial-cutaneous syndrome family caused by RAS-MAPK signaling dysregulation. Series of affected individuals heterozygous for mutations in different disease genes are shown [courtesy of G. Zampino (Università Cattolica del Sacro Cuore, Rome, Italy), B. Dallapiccola and M.C. Digilio (Ospedale "Bambino Gesù," Rome, Italy), G. Mancini (Erasmus Medical Center, Rotterdam, the Netherlands), G.B. Ferrero (Università di Torino, Turin, Italy), and M. Zenker (University Hospital, Magdeburg, Germany)].

and final height in 85% of subjects is below the 25th centile.[34,35] Approximately half of affected individuals have heart defects, which are similar to those recurring in NS but recur with different frequencies. ECG anomalies, progressive conduction defects, and HCM are the most frequent features. Of note, HCM is detected in up to 80% of LS patients with heart defects, most commonly appearing during infancy. It is progressive and paralleled by the appearance of lentigines.[36] Diagnostic criteria for LS have been outlined by Voron and colleagues.[34] A clinical diagnosis is not always feasible, particularly in young patients with no lentigines, due to the overlap with NS and neurofibromatosis-NS (see below). In these patients, the clinical differentiation from NS may be difficult during infancy and early childhood. The occurrence of distinct associated signs, including HCM, sensorineural deafness, and café-au-lait spots represents an important diagnostic handle in these young patients.[33] Sarkozy *et al.*[37] provide an updated review of the clinical and molecular genetics aspects of this disorder.

Noonan syndrome-like with loose anagen hair

Subjects with NS-like with loose anagen hair (NS/LAH, OMIM 607721) show features that are at first view reminiscent of NS (Fig. 2). The phenotype of these subjects, however, is notable for reduced growth associated with proven GH deficiency, cognitive deficits, distinctive hyperactive behavior, and easily pluckable, sparse, thin, slow-growing hair in the anagen phase but lacking an inner and outer root sheaths.[38,39] Such hair anomalies fit a well-known condition termed loose anagen hair (LAH) syndrome, which is clinically heterogeneous and has been reported to occur typically in young children and, in most cases, spontaneously improves clinically as early as adolescence.[40] In these subjects, LAH can be confirmed by microscopic examination of plucked hairs.[41] Most affected individuals exhibit hairless and darkly pigmented skin with eczema or ichthyosis, and a tendency to pruritus. Ectodermal anomalies also include sparse eyebrows and dystrophic or thin nails. The voice is characteristically hoarse or hypernasal. Cardiac anomalies are observed in the majority of cases, with dysplasia of the mitral valve and septal defects significantly overrepresented compared with the general NS population. So far, NS/LAH appears to be genetically homogeneous, as all affected individuals share the same 4A>G missense change (Ser2Gly) in *SHOC2*, which encodes a scaffold protein required for the efficient transmission of information from RAS to the MAPK cascade.[39]

Noonan-like/multiple giant cell lesion syndrome

NS is rarely accompanied by multiple giant cell lesions (MGCLs) of the jaw and/or other bone/soft tissues. This finding was originally introduced as a distinct nosologic entity named Noonan-like/multiple giant cell lesion syndrome (NL/MGCLS) or NS with cherubism (OMIM 163955).[42] Now, it is apparent that MGCLs are an associated feature of NS, LS, and cardiofaciocutaneous syndrome and that NL/MGCLS is not a separate disorder. This view is supported by the observation that this feature has been documented to occur in different conditions and be caused by mutations affecting *PTPN11*, *SOS1*, and *MEK1* genes (see below) previously reported in subjects with those syndromes without any evidence of bony involvement.[43–50] Based on these considerations, the use of NL/MGCLS should be avoided.

Cardiofaciocutaneous syndrome

Cardiofaciocutaneous syndrome (CFCS, OMIM 115150) is a rare, sporadic multiple congenital anomalies/mental retardation syndrome characterized by failure to thrive, severe feeding problems, developmental delay, reduced growth, distinctive dysmorphic facial features, abnormalities of the skin, gastrointestinal tract, and central nervous system, and cardiac defects.[51,52] CFCS has considerable clinical overlap with NS, and "borderline" cases are commonly observed, which justified the long debated question of whether it was a separate nosologic entity or an extreme phenotype/variant of NS.[53–55]

Recurrent facial/head features include relative/absolute macrocephaly, which is usually associated with high forehead, bitemporal narrowing, and facial dysmorphia that is coarser compared to NS (Fig. 2). The cutaneous involvement includes dry and hyperkeratotic skin (face, arms, and legs), ichthyosis, eczema, sparse, friable, and curly hair, absent/sparse eyebrows and eyelashes, pigmentary changes (such as café-au-lait spots, *naevi*, or lentigines), and hemangiomata. Heart defects occur in the majority of affected individuals and consist of PS most commonly, HCM and septal defects.[52]

Cognitive deficits are usually moderate to severe. Hypotonia results in marked motor delay. Structural brain abnormalities include cerebral atrophy, frontal lobe, corpus callosum or cerebellar vermis hypoplasia, and hydrocephalus.[52] Seizures are also frequently observed. As with Costello syndrome (see below), life expectancy is shortened on average due to the early death of affected individuals with severe cardiac involvement.

CFCS is genetically heterogeneous, with mutations in the *KRAS*, *BRAF*, *MEK1*, and *MEK2* genes occurring in approximately 60–90% of affected individuals.[56–60]

Costello syndrome

Costello syndrome (CS, OMIM 218040) is the eponymous name for the disorder originally described in 1971, and further delineated by the same author in 1977, as a condition characterized by prenatal overgrowth followed by postnatal feeding difficulties and severe failure to thrive, distinctive coarse facial features (Fig. 2), mental retardation, short stature, cardiac defects (most commonly HCM, septal defects, valve thickening, and/or dysplasia, and arrhythmias), and musculoskeletal and skin abnormalities.[61–64] Children and young adults with CS are predisposed to malignancies.[65] Solid tumors, including rhabdomyosarcoma and less frequently neuroblastoma and bladder carcinoma, have been estimated to occur in approximately 15% of affected individuals throughout childhood, while benign cutaneous papillomata in the perinasal or/and perianal region(s) represent a distinctive and common feature associated with the disorder.

CS is caused by germline missense mutations in the *HRAS* proto-oncogene (OMIM 190020).[66] Data from published surveys indicate that *HRAS* mutations account for 83–100% of subjects with a confirmed diagnosis of CS.[66–70] This finding has been explained as either locus heterogeneity or by the difficulties of diagnosis in cases exhibiting overlapping features with clinically related disorders or a phenotype that evolves with time. In our opinion, CS should be used only for those subjects who are heterozygous for a *HRAS* mutation. The absence of a mutation in the gene in a subject with phenotype apparently fitting CS would be predictive of a clinically related but nosologically distinct condition. Clinical reevaluation of *HRAS* mutation-negative cases is often helpful diagnostically.

Neurofibromatosis type I, including Neurofibromatosis-NS and Watson syndrome

Neurofibromatosis type 1 (NF1, OMIM 162200) is one of the most common autosomal dominant diseases (1 in 3,000–4,000 live births). Its main clinical characteristics include cutaneous, subcutaneous, and/or plexiform neurofibromas, café-au-lait spots, axillary, and/or inguinal freckling, Lisch nodules in the iris, skeletal deformities, vascular defects, learning disabilities, and behavioral problems, short stature, macrocephaly, and a predisposition for developing benign and malignant neoplasias, most commonly myeloid malignancies, including JMML, optic gliomas, pheochromocytomas, rhabdomyosarcomas, and neuroblastomas.[71,72] The diagnosis of NF1 is based on clinical criteria established at the NIH Consensus Development Conference in 1988.[73] It should be noted, however, that subjects diagnosed as having NF1 using those criteria may carry heterozygous mutations in *SPRED1*, which underlie Legius syndrome (see below). NF1 is caused by heterozygous loss-of-function mutations or deletions of the *NF1* gene, encoding the RAS-specific GTPase, neurofibromin.[74] Loss-of-function *NF1* mutations occur in the vast majority of affected individuals (>90%). NF1 is essentially fully penetrant, but there is considerable variation in the clinical features of the disease, even within families transmitting the trait. Fernel[75] and Lee and Stephenson[76] provide updated reviews of the clinical and molecular genetic aspects of this disorder.

Neurofibromatosis-NS (NFNS, OMIM 601321) is a clinical trait first noted in 1985 by Allanson and coworkers who described subjects with features of both NF1 and NS (Fig. 2).[77] It has been long questioned whether NFNS was a variant of either NF1 or NS, a chance association, or a distinct disorder.[78,79] Recent reports have provided some insights but have not completely resolved that issue. *NF1* mutation analysis performed in three independent cohorts of patients with features fitting NFNS documented occurrence of heterozygous *NF1* mutations in 23 of 28 subjects.[80–82] Lesions included nonsense mutations, out-of-frame deletions, missense changes (mostly affecting the GAP domain), and small in-frame deletions. Those groups also searched for *PTPN11* mutations in their subjects but failed to find any. In contrast, two cases of NFNS with *NF1* and *PTPN11* mutations have been reported.[83,84]

Finally, *NF1* mutations have cosegregated with the NFNS trait in a few kindreds large enough to make second mutations highly unlikely. Taken together, the current evidence indicates that most individuals with NFNS harbor an *NF1* mutation and that a single mutation appears to be sufficient to engender the trait. Double heterozygosity for *NF1* and *PTPN11* mutations rarely causes NFNS, and it is not yet clear how frequently mutations in other NS disease genes co-occur with *NF1* defects in the disorder. These findings support the view that NFNS is genetically distinct from NS and emphasize the extreme phenotypic variability associated with lesions in the *NF1* gene. The identification of specific *NF1* alleles recurring in NFNS, the fact that these alleles cosegregate in families with the condition, and the observation of a peculiar mutational spectrum strongly suggest that the term *NFNS* primarily characterizes a phenotypic variant of NF1, which manifests with a lower incidence of plexiform neurofibromas, skeletal anomalies, and internal tumors, in association with hypertelorism, ptosis, low-set ears, and CHDs.[79–82,85] It should be noted, however, that some of the mutations identified in patients with NFNS have also been reported in NF1 without any feature suggestive of NS.

A second trait closely related to NF1 is Watson syndrome (WS, OMIM 193520), an autosomal dominant condition reported by Watson in 1967,[86] characterized by PS, café-au-lait spots, reduced growth, and decreased intellectual ability.[86–89] This phenotype was originally considered distinct from NF1 and LS based on the absence of neurofibromata, Lisch nodules, lentigines, and deafness.[86,87] Allanson and coworkers, however, made observations expanding the clinical phenotype to include relative macrocephaly, Lisch nodules, and neurofibromata as recurrent features of the disorder.[88,89] Consistent with the clinical overlap with NF1, linkage to the *NF1* locus was documented,[89,90] and allelism between these two traits was proven by the identification of a 80-kb *NF1* deletion in a patient,[91] and an in-frame tandem duplication of 42 bases in exon 28 of the gene in an additional family.[92] The available records indicate that subjects with WS show only a mild expression of NF1 and lack the typical facial appearance of NS. As in the case of NFNS, WS should be considered as part of the broad clinical spectrum associated with *NF1* mutations.

Legius syndrome

Legius syndrome (OMIM 611431), previously termed neurofibromatosis type 1-like syndrome, is an autosomal dominant condition characterized by multiple axillary freckling, café-au-lait spots, macrocephaly, and a NS-like facial dysmorphism in some individuals.[93–96] Some patients have learning difficulties and/or hyperactive behavior. Despite the clinical overlap with NF1, affected individuals do not have Lisch nodules, cutaneous or plexiform neurofibromata, typical NF1 osseous lesions, or symptomatic optic gliomas, while lipomas represent a relatively common feature.[93,94,96] Additional studies are required, however, to establish susceptibility of affected individuals to malignancies.

Legius syndrome is caused by loss-of-function mutations of the *SPRED1* gene (see below).[93] Messiaen and coworkers observed that, among 42 *SPRED1* mutation-positive individuals, approximately half of the cases fulfilled NIH's diagnostic criteria for NF1.[96] In the same study, mutational analysis of a large cohort of patients fulfilling NIH criteria for diagnosis of NF1 revealed a *SPRED1* mutation in 2% of cases (i.e., approximately 8% of *NF1* mutation-negative cases).[96]

RAS-MAPK signaling dysregulation and phenotypic consequences

PTPN11 gene

The *PTPN11* gene (OMIM 176876) encodes SHP2, a widely expressed cytoplasmic protein tyrosine phosphatase that positively modulate RAS signaling.[97–100] It is characterized by two tandemly arranged *N*-terminal src-homology 2 domains (N-SH2 and C-SH2), a single catalytic domain (PTP) and a *C*-terminal tail of uncharacterized function. SHP2 binds to phosphotyrosyl-containing signaling partners through the SH2 domains, which control both SHP2's subcellular localization and function.[101]

We discovered *PTPN11* as a major NS disease gene using a positional candidacy approach in 2001.[102] *PTPN11* mutations account for approximately 50% of NS, and are more prevalent among subjects with pulmonary valve stenosis and short stature, and less common in individuals with HCM and/or severe cognitive deficits.[43,103] These mutations are almost always missense changes and perturb SHP2's function through distinct mechanisms.[104–106] Most mutations affect residues involved in the N-SH2/PTP

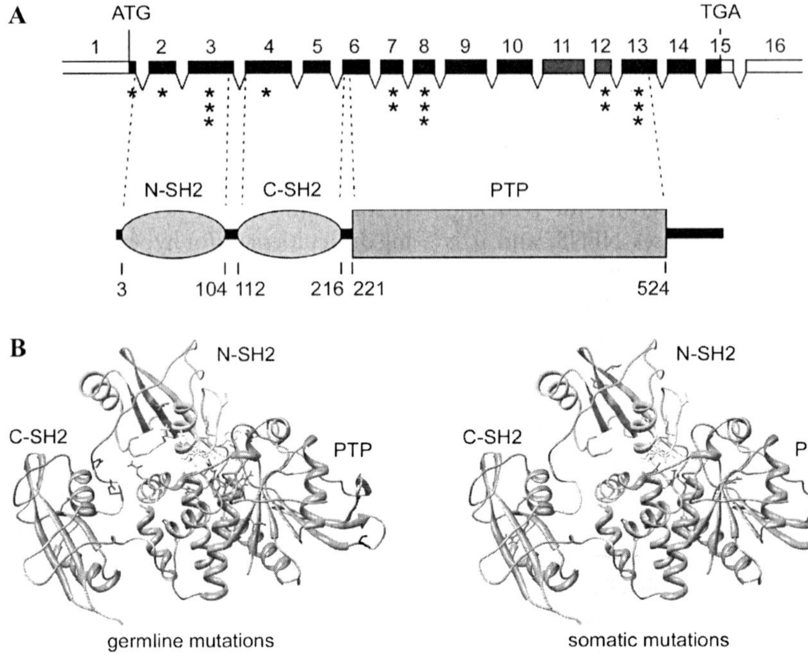

Figure 3. *PTPN11* gene organization, SHP2 domain structure, and location of affected residues in human disease. (A) The *PTPN11* gene and its encoded protein product. The numbered, filled boxes at the top indicate the coding exons; the positions of the ATG and TGA codons are shown. Exon location of disease-associated mutations is indicated (the number of asterisks is an index of the relative prevalence of mutations within each exon). SHP2's functional domains, consisting of two tandemly arranged SH2 domains at the N-terminus (N-SH2 and C-SH2) followed by a protein tyrosine phosphatase (PTP) domain, are shown below. The numbers below that cartoon indicate the amino acid boundaries of those domains. (B) Location of mutated residues in the three-dimensional structure of SHP2 in its catalytically inactive conformation.[94] Residues affected by germline (left-hand side) or somatically acquired (right-hand side) mutations are shown with their lateral chains (yellow, mutations affecting the N-SH2/PTP interaction; green, mutations affecting the N-SH2/PTP interaction and possibly substrate specificity and/or catalysis; red, mutations promoting increased SH2 phosphopeptide binding affinity or affecting specificity; blue, mutations affecting SH2 orientation or mobility; black, unclassified). The identity of affected residues and amino acid substitutions, and the prevalence of mutations are listed in Tartaglia *et al.*[98]

interdomain binding network that stabilizes SHP2 in its catalytically inactive conformation (Fig. 3). These mutations are predicted to upregulate SHP2's function by impairing the switch between the active and inactive conformations, favoring a shift in the equilibrium toward the former. There are, however, changes affecting residues located in the phosphopeptide-binding pocket of the N-SH2 and C-SH2 domains or in the linker stretch connecting these domains. There is evidence supporting the idea that these substitutions promote SHP2 gain of function by increasing binding affinity, altering the binding specificity for the signaling partners, or altering the flexibility of the N-SH2 domain in a manner that inhibits the N-SH2/PTP interaction.

A distinct class of *PTPN11* mutations has been identified to underlie LS.[27,28] Tyr279Cys and Thr468Met represent the most common defects, although additional mutations have been documented.[37] An impaired catalytic activity has been established as the biochemical behavior shared by these SHP2 mutants.[105–108] Their role in SHP2 functional dysregulation still requires further study, since they do not appear to perturb intracellular signaling by a merely dominant negative effect,[106,109,110] as supposed in the past.[108]

Children with NS are predisposed to a spectrum of hematologic abnormalities and malignancies, including JMML, a clonal myeloproliferative disorder of childhood,[111] and children with NS/JMML are heterozygous for *PTPN11* germline mutations in the majority of cases, with one mutation rarely observed in NS (218C>T, Thr73Ile) occurring in a large percentage of cases.[23,24,112] Remarkably, a distinct class of mutations in this gene, acquired as a somatic event, occurs in approximately one-third of children with JMML without NS as well as variable proportions of other myeloid and lymphoid malignancies of childhood.[23,24,112–115] As observed in NS, the vast majority of *PTPN11* lesions identified in this heterogeneous group of hematologic malignancies are missense changes that alter residues located at the interface between the N-SH2 and PTP domains (Fig. 3), but appear to be more activating.[104,105,112,116,117]

De novo duplications of the chromosomal region encompassing *PTPN11* (12q24) have been documented in two subjects with clinical features suggestive of NS without mutations in any known NS genes.[118,119] These findings suggest that increased dosage of SHP2 may have dysregulating effects on intracellular signaling and consequences on developmental processes. The screening of a relatively large cohort of NS cases negative for mutations in previously identified genes, however, indicate that *PTPN11* gene duplication represents an uncommon cause of NS.

SOS1 gene

Cell-surface tyrosine kinase receptors activate RAS proteins by recruiting GEFs to the cytoplasmic side of the plasma membrane. These proteins catalyze the release of GDP from RAS, facilitating the conversion of the inactive GDP-bound form of RAS to its active GTP-bound state.[120] Among the GEFs, two members of the son of sevenless (SOS) family, SOS1 and SOS2, promote guanine nucleotide exchange on RAS proteins, but not on RAS-related family members, such as the RAP and RHO proteins.[121,122]

Based on their modulatory role in RAS signaling, we and others discovered that *SOS1* (OMIM 182530) is mutated in a relatively large percentage of subjects with NS.[123,124] This gene encodes a large multidomain protein (Fig. 4).[125] The *N*-terminal portion of the protein contains a histone domain (HD) that is characterized by two

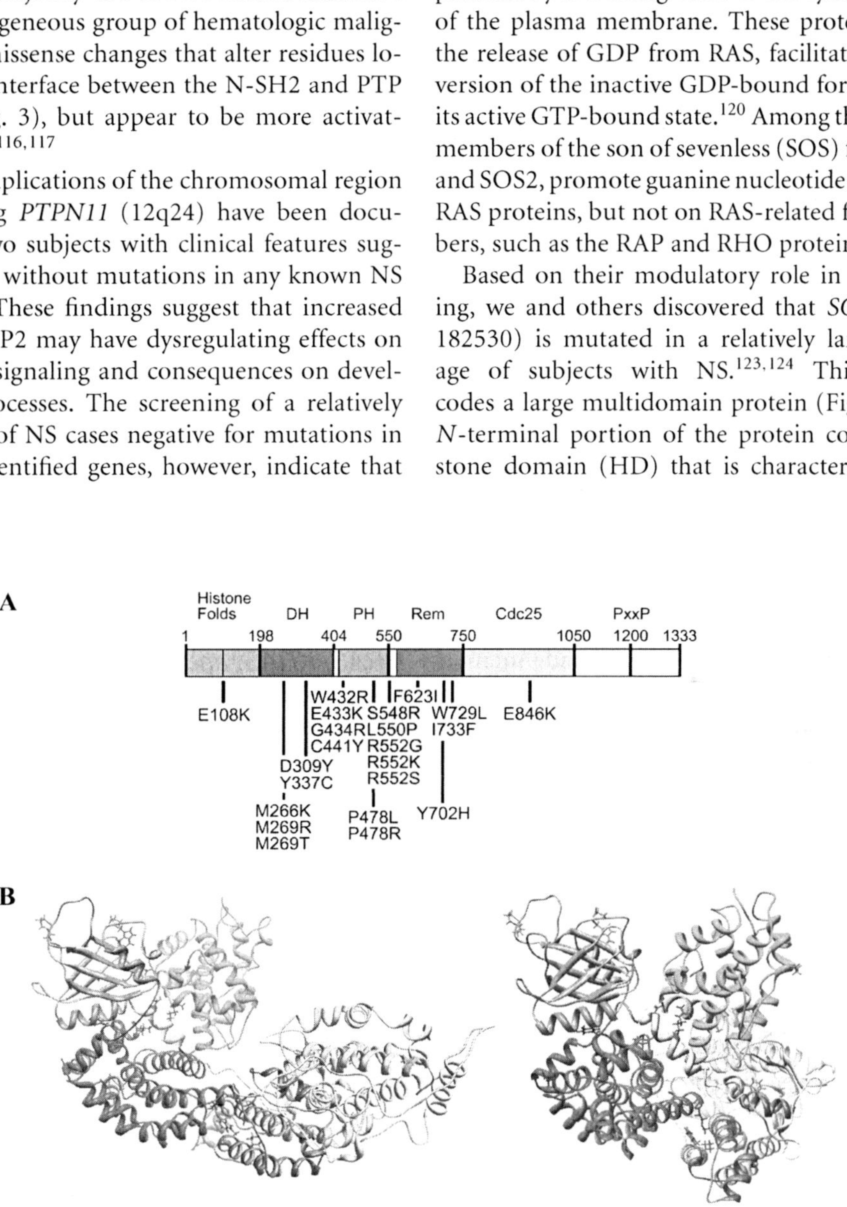

Figure 4. SOS1 domain structure and location of affected residues in NS. (A) *SOS1* missense mutations are positioned below the cartoon of the SOS1 protein with its functional domains indicated above. Abbreviations: DH, Dbl homology domain; PH, plekstrin homology domain; Rem, RAS exchanger motif; PxxP, proline-rich region. (B) Location of the mutated residues on the three-dimensional structure of SOS1. The functional domains are color coded as follows: Histone folds, cyan; DH, magenta; PH, orange; Rem, green; Cdc25, yellow. Residues affected by mutations are indicated with their lateral chains (histone folds, violet; DH, blue; PH, green; helical linker, red; Rem, orange; Cdc25, cyan). Based on Sondermann *et al.*,[121] which utilized structural data and computational modeling.

tandemly arranged histone folds, and is followed by Dbl homology (DH) and pleckstrin homology (PH) domains, which are implicated in the activation of RAC, a small GTPase of the RHO/CDC42 family.[126] The *C*-terminal half of the protein contains the RAS exchanger motif (REM) domain and the Cdc25 domain, which are required for the RAS-specific nucleotide exchange activity of SOS1. Finally, the region at the *C*-terminus contains recognition sites for SH3 domains and mediates interaction of SOS1 with SH3 domain-containing adaptor proteins that deliver SOS1 to the membrane upon receptor activation. Additional anchorage sites on the membrane are provided by the phosphatidylinositol phosphate-binding site within the PH domain,[127] and an extended positively charged surface of the HD domain.[128]

The available data indicate that *SOS1* is the second most frequently mutated NS disease gene, accounting for approximately 10% of subjects.[123,124,129] The vast majority of the mutations identified are missense changes and affect multiple domains (Fig. 4). Approximately half of the *SOS1* defects affect residues located in the short helical linker connecting the PH and REM domains, with substitutions of residue Arg552 accounting for approximately 30% of total mutations. A second mutation cluster is located within the PH domain (residues 432 to 434), while a third functional cluster resides at the interacting regions of the DH (Thr266 and Met269) and REM (Trp729 and Ile733) domains. A single amino acid change (Glu846Lys) within the Cdc25 domain accounts for more than 10% of defects.

Subjects heterozygous for a mutated *SOS1* allele tend to exhibit a distinctive phenotype characterized by ectodermal abnormalities generally associated with an absence of cognitive deficits and normal growth.[124,129] A few *SOS1* mutation-positive individuals with ectodermal manifestations and distinctive facial dysmorphia that might be suggestive of CFCS have been reported (Fig. 2).[130,131] In these subjects, cognitive deficits were generally absent or minor.

The GEF activity of SOS1 is controlled by an allosteric site, involving the Cdc25 and REM domains, that stimulates SOS1 exchange activity through the binding of nucleotide-bound RAS.[132] Basally, the interaction between the DH and REM domains stabilizes SOS1 in its catalytically inactive conformation by masking the allosteric binding site for RAS. Following SOS1's translocation to the membrane, the inhibitory effect of the DH domain is relieved by still undefined events allowing RAS binding to the allosteric site, which in turn promotes a conformational change of the REM and Cdc25 domains, and RAS binding to the catalytic site.[133] Most NS-causing *SOS1* mutations reside in regions within the molecule that are predicted to contribute structurally to the maintenance of the catalytically autoinhibited conformation (Fig. 4). Biochemical data confirmed such predictions, demonstrating that NS-causing *SOS1* mutations promote gain of function and enhance RAS and ERK activation.[123,124]

A possible contribution of activating *SOS1* mutations in cancer has been evaluated (our unpublished data).[134,135] Mutational analysis performed on genomic DNA from more than 1,000 solid tumors (pancreatic, lung, breast, colon, thyroid cancers) and hematological malignancied (myeloproliferative disorders and acute myeloid and lymphoblastic leukemias) indicate that *SOS1* is not mutated in most cancers, suggesting that NS patients heterozygous for a *SOS1* mutation are not at increased risk of developing cancer.

HRAS, KRAS, and NRAS genes

The three members of the RAS subfamily, HRAS (OMIM 190020), KRAS (OMIM 190070), and NRAS (OMIM 164790), are monomeric GTPases that use GDP/GTP-regulated molecular switching to control intracellular signal flow.[1,136] They exhibit high affinity for GDP and GTP, low GTPase activity, and cycle from a GDP-bound inactive state to a GTP-bound active state, the latter favoring their interaction with multiple downstream transducers. As anticipated, GDP/GTP cycling is controlled by GEFs, such as SOS1, which promote release of GDP, and GAPs, which accelerate the intrinsic GTPase activity. RAS proteins share a structure that includes a conserved domain (residues 1 to 165), known as the G domain, which is required for its signaling function, and a less conserved *C*-terminal tail, called the hypervariable region, that guides posttranslational processing and plasma membrane anchoring (Fig. 5). Within the G domain, conserved sequence elements direct GTP/GDP binding and GTP hydrolysis. Furthermore, two tracts, denoted as Switch I and Switch II, undergo major conformational changes upon GTP/GDP exchange and mediate

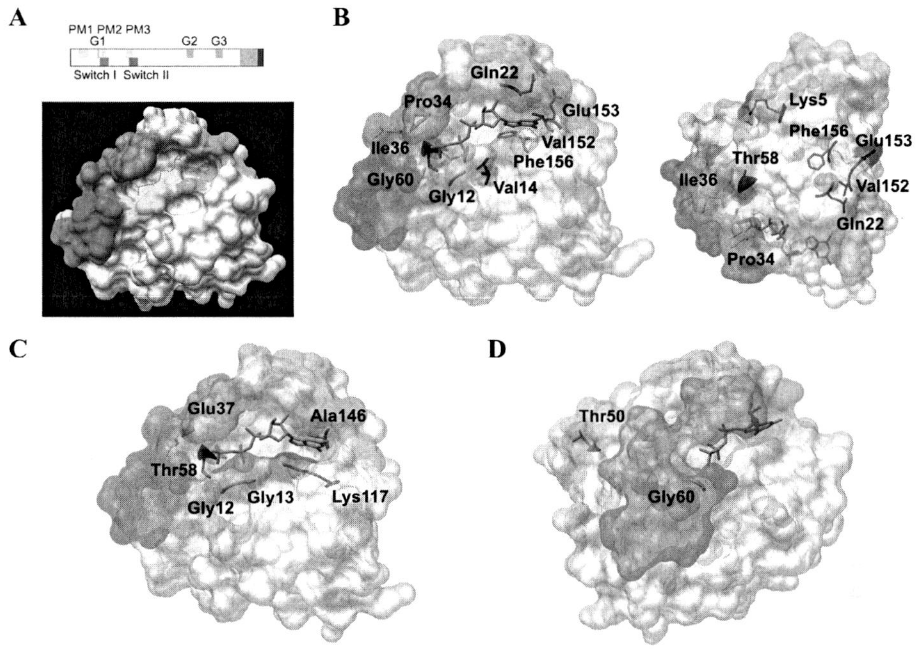

Figure 5. Domain structure of RAS proteins and location of affected residues in NS, CFCS, and CS. (A) Schematic diagram and three-dimensional representation of the structural and functional domains defined within RAS proteins. The motifs required for signaling function (PM1 to PM3 indicate residues involved in binding to the phosphate groups, while G1 to G3 are those involved in binding to the guanine base) are indicated. The hypervariable region is shown in gray, together with the *C*-terminal motifs that direct posttranslational processing and plasma membrane anchoring (dark gray). The GTP/GDP binding pocket is shown in cyan (guanine ring binding surface) and yellow (triphosphate group binding surface) together with the Switch I (forest green) and Switch II (magenta) domains, according to the GTP-bound RAS conformation. (B) Location of the KRAS residues mutated in NS and CFCS. Residues affected by mutations are indicated with their lateral chains (triphosphate group binding surface, orange; guanine ring binding surface, blue; Switch I, green; other regions; black). (C) Location of the HRAS residues mutated in CS. Residues affected by mutations are indicated with their lateral chains as reported above. (D) Location of the NRAS residues mutated in NS. Residues affected by mutations are indicated with their lateral chains as reported above.

binding to RAS effectors, GAPs and GEFs.[1,137] Heterozygous germline mutations in each of the three members of the RAS subfamily have been identified to be causally related with NS, CFCS, or CS.

Based on the link between the functions of SHP2 and RAS, Aoki and coworkers used a candidate gene approach to discover that a small number of missense mutations affecting residues Gly12 and Gly13 in *HRAS* underlie CS.[66] These residues represent hot spots for oncogenic somatic mutations.[138] Defects affecting those residues render the protein insensitive to GAP-promoted GTP hydrolysis, resulting in a constitutively GTP-bound, active state.[5,138] Other less common amino acid substitutions involving residues in the purine ring binding pocket of the protein (His117 and Ala146) are predicted to affect the stability of HRAS binding to GDP/GTP,[69,70] favoring a shift in the equilibrium toward the GTP-bound, active form of RAS, which bypasses the requirement for a GEF.[139–142] Recently, two different three-nucleotide insertions resulting in the duplication of Glu37 within the Switch I region have been reported.[143] This small in-frame insertion promotes a weak hyperactivation of signaling resulting from a balancing effect between a profound GAP insensitivity and inefficient binding to effector proteins.

In CS, the Gly12Ser substitution in HRAS accounts for approximately 85% of affected individuals. This change recurs with considerably lower prevalence in human cancers (approximately 5% of total HRAS amino acid substitutions) (COSMIC; http: //www.sanger.ac.uk/genetics/CGP/cosmic/). Experimental data indicate that Gly12Ser has a lower transformation potential than the Gly12Val mutant,[144] which is the most common somatic HRAS defect in human tumors (about 45% of total somatic *HRAS* gene mutations). The latter has been documented in only two individuals with CS who exhibited a relatively severe phenotype (approximately 2% of cases with an *HRAS* gene mutation).[66,145] Similarly, no *HRAS* mutation affecting residue Gln61, which constitute approximately 33% of total *HRAS* lesions in human cancer, has been documented in CS to date. Such a striking difference in prevalence and spectrum of HRAS amino acid substitutions suggests that CS-causative mutations have less potency for deregulating protein function than cancer-contributing ones and that the latter might exert a stronger perturbing effect on development when inherited.

Based on the identification of *HRAS* gene mutations in CS, we and others discovered that germline *KRAS* mutations can underlie NS or CFCS (Fig. 2).[56,146,147] Mutations account for a relatively small percentage of affected subjects in both disorders (<2–5%), and, in NS, are generally associated with a highly variable but generally severe phenotype. The diversity of mutations associated with these developmental disorders as well as their phenotypic spectrum have been investigated further, refining the picture of a clustered distribution of germline disease-associated *KRAS* defects, and confirming the high clinical variability.[59,148,149] The *KRAS* gene (OMIM 190070) produces two transcripts through alternative splicing, resulting in two proteins called KRASA and KRASB,[150] which differ at the *C*-terminus. Most *KRAS* mutations cluster in the first and second coding exons of the gene, which are shared by the two KRAS isoforms. Mutations affecting exon 6, however, have been documented in one-third of NS or CFCS subjects with a *KRAS* germline mutation. This exon codes for the *C*-terminal tail of KRASB but not KRASA. These findings document that isolated KRASB gain of function is sufficient for disease pathogenesis, which is also consistent with the available evidence about the relative roles of the KRAS isoforms on development from knock-out animal models.[151–153]

As observed for the somatically acquired oncogenic *RAS* gene mutations and germline transmitted *HRAS* lesions, the NS/CFCS-causing *KRAS* defects were found to upregulate protein function by either affecting the intrinsic and/or GAP-stimulated GTPase activity and, consequently, impairing the switch between the active and inactive conformation, or determining a dramatically increased rate of guanine nucleotide dissociation.[147,154] Although these studies documented a complex pattern of intrinsic biochemical properties, expression of these mutants in COS-7 cells promoted higher levels of phosphorylated MEK and ERK proteins, indicating hyperactivation of the MAPK cascade. Of note, available records indicate that germline *KRAS* mutations are rarely observed in human cancers. This finding is consistent with functional data indicating a reduced perturbing effect of these mutations on protein function compared with the cancer-associated ones.[147,154] This is also supported by the observation that malignancies have been rarely observed in subjects with NS or CFCS due to a KRAS mutation.[56,59,147,148]

More recently, two germline missense mutations (Thr50Ile and Gly60Glu amino acid changes) have been reported to account for a few NS cases.[155] A germline *KRAS* mutation affecting Gly60 (Gly60Arg) has been reported in CFCS,[56] and the somatic *NRAS* mutation Gly60Glu rarely has been observed to occur in malignancies (COSMIC database). Of note, mutations of the neighboring Gln61 are frequent oncogenic RAS alterations and confer impaired GTPase function. In contrast, no somatic or germline *RAS* gene mutation affecting Thr50 has been described so far. Thr50 is an exposed residue located in the β2-β3 loop connecting the two switch regions but is not predicted to be directly involved in GTP/GDP binding, GTP hydrolysis, or effector interaction. Functional characterization of these disease-associated mutations documented that, similarly to what was observed for *KRAS* and *HRAS* mutation, they confer enhanced stimulus-dependent MAPK activation but apparently through different molecular mechanisms.[155]

NF1 gene

The *NF1* gene (OMIM 16200) encodes neurofibromin, a cytoplasmic multifunction protein ubiquitously expressed with highest expression in neurons, Schwann cells, oligodendrocytes, astrocytes, and leukocytes.[156] The *NF1* coding sequence spans over 61 exons distributed over more than 300 Kb. A number of variant transcripts have been identified, which originate from alternative splicing, exon skipping, and short sequence insertions. While neurofibromin is a large protein, it has no clearly defined functional domains, except the SEC14 and RasGAP domains, the latter mediating the RAS-specific GAP function of the protein. Besides its role as a negative modulator of RAS signaling, neurofibromin is involved in the modulation of other signaling pathways, including the protein kinase A (PKA)-cAMP pathway, and in microtubule-associated processes.[157–160] Moreover, a possible function within the nucleus has been hypothesized.[161] For a general overview see Trovó-Marqui and Tajara.[156]

In the majority of affected individuals, disease-causing *NF1* mutations consist of intragenic frameshift, nonsense, splicing, or missense mutations arrayed along the entire length of the gene. A small proportion of affected individuals (<5%) carry a heterozygous deletion of approximately 1.5 Mb, which is the result of unequal homologous recombination and encompasses the entire *NF1* and a few adjacent genes.[162–164] There is poor correlation between classes of mutations and disease features, with the exception of patients with large deletions who tend to have more severe phenotype and higher numbers of neurofibromata. Phenotypic heterogeneity is high among affected individuals, even within families coinheriting the same *NF1* mutation.

NF1 is a tumor suppressor gene, and loss of the wild-type allele has been demonstrated to occur somatically as a second hit in malignancies associated with the trait.[165–167]

SPRED1 gene

The SPROUTY-related ENA/VASP homology-1 (EVH1) domain-containing protein 1 (SPRED1, OMIM 609291) is a member of the SPROUTY/SPRED family of membrane-associated proteins that function as modulators of growth factor-induced activation of the RAS-MAPK pathway. SPRED1 is characterized by an *N*-terminal EVH-1 domain, which binds to proline-rich sequences, a central KIT-binding domain (KBD) with unknown function, and a *C*-terminal SPRY domain implicated in protein translocation to the plasma membrane, RAF binding, and heterodimerization with other SPRED proteins.[168] SPRED1 negatively controls MAPK signaling and operates downstream of RAS by suppressing RAF phosphorylation and activation,[169] although a target upstream of RAS has also been suggested.[170] Heterozygous loss-of-function mutations in *SPRED1* cause Legius syndrome.[93] The mutational spectrum is wide, including nonsense, frameshift and missense mutations, as well as small in-frame deletions.[93–96] Most mutated alleles produce unstable proteins that are likely to be quickly degraded in the cell, but others encode nonfunctional proteins. It has been demonstrated that disease-causing mutations resulted in increased RAF1 kinase activity, enhanced phosphorylation levels of MEK and ERK proteins, and increased activity of ELK1, a transcription factor functioning downstream the MAPK cascade.[93] Of note, as documented in NF1, second hits in the gene have been documented in melanocytes from café-au-lait spots of affected subjects, indicating that this feature is caused by biallelic inactivation of *SPRED1*, as previously demonstrated for NF1.[171] Mutational screening performed on kidney and cell nervous system cancers do not suggest a major contribution of somatic *SPRED1* mutations to these cancers (COSMIC database).

RAF1 and BRAF genes

The RAF1 (OMIM 164760), BRAF (OMIM 164757), and ARAF (OMIM 311010) proteins are members of a small family of serine-threonine kinases that function as RAS effectors.[7,172,173] These kinases phosphorylate and activate the dual specificity kinases MEK1 and MEK2, which in turn promote the activation of the MAPKs, ERK1 and ERK2. The three members of the RAF family play different roles in the activation of the RAS-MAPK signaling cascade. BRAF has considerably higher MEK kinase activity compared to RAF1 and ARAF. These proteins also differ in their expression profiles as well as in the regulatory mechanisms controlling their function, and appear to have unique roles during development.[174–176] Consistent with these observations, somatic *BRAF* mutations frequently occur in malignant melanomas and in thyroid, colorectal, and ovarian cancers (COSMIC database), while *ARAF* and *RAF1* missense changes are observed rarely in malignancies. These kinases are characterized by three functional domains, known as conserved regions 1–3 (CR1–3) (Fig. 6). The *N*-terminal CR1 contains the region involved in GTP-RAS binding and a cysteine-rich sequence that mediates protein interaction with the cytosolic surface of the cell membrane. CR2 contains a negative regulatory domain controlling protein translocation to the membrane and its catalytic activation, while the *C*-terminal CR3 comprises the kinase domain of the protein.[172]

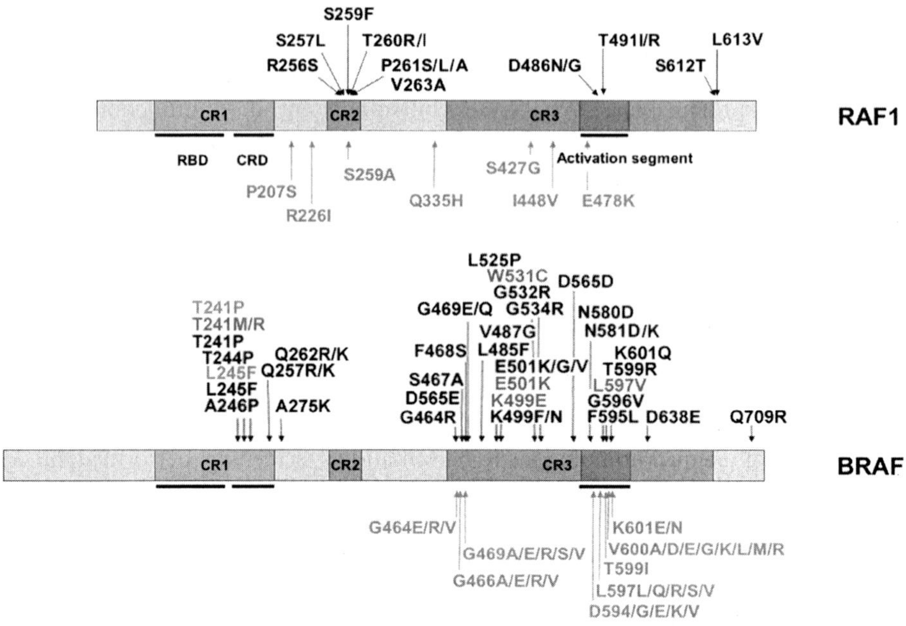

Figure 6. RAF1 and BRAF domain structure and location of affected residues in human disease. The domains of the RAF1 (above) and BRAF (below) proteins are indicated (CR, conserved region; RBD, RAS binding domain; CRD, cysteine-rich domain). Germline (RAF1: NS and LS; BRAF: NS, LS, and CFCS) and somatic (associated with cancer) mutations are reported above and below each cartoon, respectively. Only somatic BRAF mutations with prevalence $\geq$1.5%, according the COSMIC database (http://www.sanger.ac.uk/genetics/CGP/cosmic/) are reported. The BRAF Thr599Ile substitution, which rarely occurs in cancer and is homologous to the NS-causing RAF1 Thr491Ile change, is also reported. BRAF missense changes associated with a phenotype fitting NS or LS are colored blue and orange, respectively.

We and another group identified missense mutations in *RAF1* (OMIM 164760) in subjects with NS with a prevalence of mutations ranging between 5–15% (Fig. 2).[29,177] Mutations in the same gene were also identified in two of six subjects with LS without a mutation in *PTPN11*.[29] *RAF1* mutations affect residues clustering in three regions of the protein (Fig. 6). The first cluster (70% of total *RAF1* defects) involves the *N*-terminal consensus 14–3-3 recognition sequence or adjacent residues within the CR2 region. Of note, Arg256, Ser257, Ser259, and Pro261, which are the invariant residues within this motif, were all found to be mutated. The second group (15% of *RAF1* lesions) includes mutations affecting residues within the activation segment region of the kinase domain (Asp486 and Thr491). Of note, several *BRAF* missense mutations detected in solid tumors alter the activation segment, including some (Asp594Gly and Thr599Ile) homologous to those of *RAF1* identified in subjects with NS. Finally, the third cluster (15% of *RAF1* mutations) affect two adjacent residues (Ser612 and Leu613) located at the *C*-terminus. Phenotype analysis of the NS and LS subjects with *RAF1* mutations is notable for the observation that a large percentage of cases (75%) exhibit HCM, compared to HCM prevalence of 18% in the general NS population.[29,177] Moreover, this genotype–phenotype correlation seems to be allele specific as HCM appears to be associated with mutations affecting the *N*-terminal 14–3-3 consensus site or the *C*-terminus. In addition to the two cases with LS, multiple nevi, lentigines, and/or café-au-lait spots were detectable in one-third of NS patients with *RAF1* mutations, suggesting a predisposition to hyperpigmented cutaneous lesions.

In RAF1's inactive state, the *N*-terminal portion of the protein interacts with and inactivates the kinase domain at the *C*-terminus. This autoinhibited conformation is stabilized by 14–3-3 protein dimers that bind to phosphorylated Ser259 and Ser621.[7,172,173] Dephosphorylation of Ser259, which is possibly mediated by protein phosphatase 1C (PP1C) or protein phosphatase 2A (PP2A), is required for stable interaction with GTP-RAS, allowing protein translocation to the plasma membrane and phosphorylation of regulatory residues required for RAF1 catalytic activation. Functional characterization of a selected panel of RAF1 mutants supports the idea that mutations can differentially perturb protein function and intracellular signaling. In particular, mutations affecting the *N*-terminal 14–3-3 binding motif or the *C*-terminus of the protein, promote enhanced kinase activity and increased activation of the MAPK cascade compared to wild-type protein.[29,177] In contrast, Asp486Asn and Thr491Ile, representing the mutation cluster in the activation segment, were observed to be kinase impaired and differentially perturb signaling through the MAPK cascade. Pandit and coworkers provided evidence supporting the model that the increased activation promoted by the amino acid substitutions affecting the *N*-terminal 14–3-3 binding site results from a loss of 14–3-3-mediated inactivation,[29] while the mechanism through which mutations at the *C*-terminus activate RAF1's kinase activity remains to be explained.

Two groups independently discovered that *BRAF* is a major disease gene for CFCS (50–75% of cases) (Fig. 2).[56,57] More recently, germline *BRAF* mutations have been documented also in a small percentage of subjects with phenotypes fitting or suggestive of NS or LS (<2% of cases) (Fig. 2).[30,31,177,178] Of note, NS-associated mutations largely do not overlap with those occurring in CFCS, suggesting a genotype–phenotype correlation. The recurrence of the Trp531Cys amino acid substitution as well as the clustering of mutations at the Thr^{241} further supports the idea that the phenotype resulting from germline *BRAF* defects might be allele specific. Among subjects with NS, the available clinical data indicate an association with neonatal growth failure and feeding difficulties, mild-to-moderate cognitive deficits, and hypotonia, and a higher prevalence of multiple nevi and dark colored lentigines.[30] As adults, they display the full-blown phenotype of NS, which is more severe compared to those associated with *PTPN11* and *SOS1* mutations. In these individuals, however, polyhydramnios, HCM, and CFCS-related skin features are uncommon or absent,[52,179] and cardiac defects, neurological impairment, and feeding problems appear to be less severe compared to what is generally observed in CFCS.[30] While the occurrence of distinct *BRAF* gene mutations in subjects with a phenotype fitting NS would indicate that some features might be allele specific, the observation that a subgroup of *BRAF* mutation-positive subjects exhibits an "intermediate" phenotype suggests that a clinical continuum characterized by a differential combination and severity of features is associated with defects in the *BRAF* gene.[30] These findings emphasize the difficulty in identifying efficient clinical criteria to define CFCS, LS, or NS nosologically, and make evident the usefulness of a molecular-based definition of these clinically overlapping conditions that might direct clinicians toward the appropriate management of patients. Of note, the observation of a patient with LS and normal intelligence who was found to carry a novel sequence change in *BRAF*,[31] further illustrates that the phenotypic spectrum caused by *BRAF* mutations is broader than previously assumed and that mental retardation is not an invariably associated feature.

More than 40 germline *BRAF* mutations have been identified (Fig. 6). All are missense defects that are not randomly distributed, clustering in the cysteine-rich domain and in the amino-terminal portion and activation segment of the kinase domain. Most mutations are recurrent, with substitutions of residues Gln257 and Glu501 accounting for approximately 40% of all defects. Germline *BRAF* mutations are rarely observed as somatic mutations contributing to oncogenesis and NIH-3T3 cell colony focus formation assay data indicate that CFCS-associated BRAF mutants have a reduced transforming capability compared with the recurrent oncogenic Val600Glu BRAF protein, and that the NS- and LS-associated BRAF mutants do not confer enhanced transformation to cells.[30] While these results support the view that the mutations transmitted through the germline have less potency in deregulating BRAF's function, they also suggest that NS- and LS-associated mutations are less activating compared to those associated with CFCS, which parallels observations indicating that phenotypes arising from germline *BRAF* defects might be allele specific. Of note, *in vitro* functional studies indicate that germline and somatic amino acid changes can confer either increased or decreased activity upon the mutant protein,[30,57,180] suggesting that multiple alternative mechanisms are likely to be involved in the functional dysregulation of the kinase.

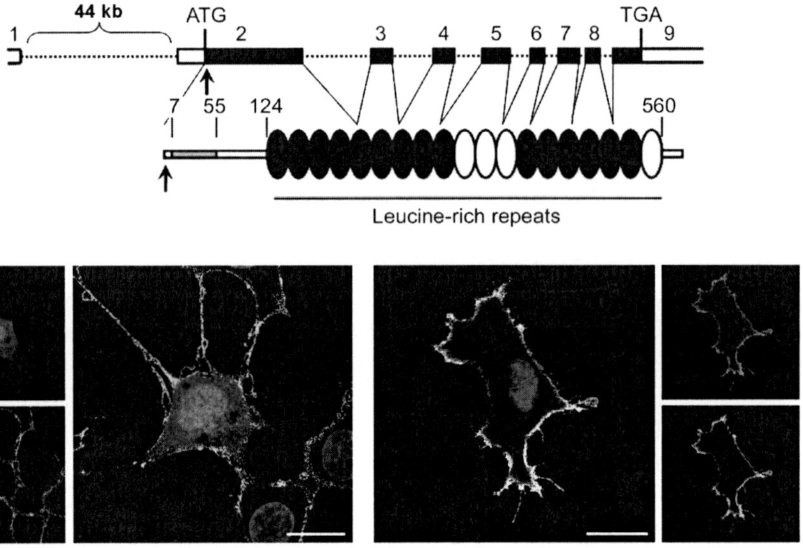

Figure 7. The disease-causing 4A > G (Ser2Gly) change in *SHOC2* promotes protein myristoylation and cell membrane targeting. (A) *SHOC2* genomic organization and protein structure. The coding exons are shown at the top as numbered filled boxes. Intronic regions are reported as dotted lines. SHOC2 motifs comprise an *N*-terminal lysine-rich region (gray colored) followed by 19 leucine-rich repeats. Numbers above the domain structure indicate the amino acid boundaries of those domains. (B) Confocal laser scanning microscopy analysis documents that $SHOC2^{wt}$ (red) is uniformly distributed in the cytoplasm and nucleus (left-hand side) of transiently transfected Cos-1 cells (DMEM supplemented with 10% heat-inactivated FBS), while $SHOC2^{S2G}$ (red) colocalizes with ganglioside M1 (green) to the cell membrane.[39] Nuclei are visualized by DAPI staining (blue). Bars indicate 20 μm.

SHOC2 gene

We recently discovered that the invariant 4A>G missense (Ser2Gly) change in the *SHOC2* gene (OMIM 602775), which encodes a cytoplasmic leucine-rich repeat-containing protein that positively modulates RAS-MAPK signal flow,[181–185] underlies NS/LAH.[39] This amino acid change was demonstrated to promote *N*-myristoylation of the protein, and causes aberrant targeting of SHOC2 to the plasma membrane and impaired translocation to the nucleus upon growth factor stimulation (Fig. 7).[39] *N*-myristoylation is an irreversible form of protein fatty acylation occurring during protein synthesis, in which myristate, a 14-carbon saturated fatty acid, is covalently added to an *N*-terminal glycine residue after excision of the initiator methionine residue by methionylaminopeptidase.[186–188] It is a relatively common lipid modification of many signal transducers, and contributes to direct protein anchoring to cellular membranes, which is required for their proper function.

SHOC2 is a widely expressed protein composed almost entirely by leucine-rich repeats (LRR) and has a lysine-rich sequence at the *N*-terminus (Fig. 7). Since LRRs can provide a structural framework for protein–protein interactions, SHOC2 is believed to function as a scaffold linking RAS to downstream signal transducers. It has recently been reported that SHOC2 functions as a regulatory subunit of the catalytic subunit of protein phosphatase 1 (PP1C).[184] By binding GTP-MRAS, SHOC2 promotes PP1C translocation to the membrane, allowing PP1C-mediated RAF1 dephosphorylation at residue Ser259, which is required for stable RAF1 translocation to the plasma membrane and catalytic activation. According to this model, constitutive membrane translocation of the disease-causing $SHOC2^{S2G}$ protein is expected to promote prolonged PP1C-mediated RAF1 dephosphorylation at Ser259 and, consequently, sustained RAF1-stimulated MAPK activation. Initial functional characterization of the mutant protein indicates that the expression of $SHOC2^{S2G}$ mutant enhances ERK activation in a cell type-specific fashion.[39] The observations that subcellular localization of SHOC2 is restricted to the nucleus following EGF stimulation and that $SHOC2^{S2G}$ ectopic expression in *C. elegans* engenders protruding vulva, a neomorphic phenotype previously associated with aberrant signaling, suggest that the disease-causing mutant might exert a wider perturbing effect on intracellular signaling.

MEK1 and MEK2 genes

MEK1 (OMIM 176872) and the functionally related MEK2 (OMIM 601263), belong to a family of dual specificity kinases that phosphorylate ERK proteins at tyrosine and serine/threonine residues.[189] The MEK proteins share a conserved structure including a negative regulatory domain at the *N*-terminus and a single protein kinase domain. MEK1 and MEK2 are both effectors of RAF proteins but appear to play nonredundant roles. In particular, genetic evidence from mouse models indicates that MEK1 function is required during embryonic development,[190] while MEK2 is dispensable.[191] Of note, even though MEK kinase activity is necessary for cell transformation via the MAPK cascade,[192] and constitutively active MEK mutants promote transformation,[193] mutations in these genes have not been reported in human cancers so far (COSMIC database). Missense mutations and in-frame deletions in *MEK1* and *MEK2* account for approximately 20% of CFCS.[57–60,178,194] Recently, one germline mutation in *MEK1*, predicting the Asp67Asn amino acid substitution, has been documented in two unrelated subjects exhibiting a phenotype fitting NS.[59] According to this study, *MEK1* gene mutations would account for approximately 3% of *PTPN11*- and *SOS1*-mutation negative NS cases. In NS and CFCS, *MEK* gene mutations are mostly missense changes affecting residues located in the regulatory region and the *N*-terminal portion of the catalytic domain. Functional characterization of a panel of CFCS-causing mutants documented that they are more active than the wild-type protein in stimulating ERK phosphorylation.[57,195] No datum on the effect of the Asp67Asn change on MEK1 function and MAPK signaling is currently available.

Concluding remarks

While dysregulation of the RAS-MAPK signaling pathway has been known for decades as a major event in oncogenesis, discoveries derived from a massive disease gene hunting effort have only recently established a novel scenario in which the upregulation of this signaling cascade underlies a group of clinically related developmental disorders. Based on the relatively high prevalence of some of these disorders, the dysregulation of this signaling pathway also represents one of the most common events affecting developmental processes. These discoveries have also provided us with novel molecular mechanisms causing enhanced signaling.

During the last years, screening efforts and genotype–phenotype correlation analyses have documented that mutations in identified disease genes are associated with distinct phenotypes, explaining, in part, the wide clinical diversity characterizing these disorders. It should be considered that the overlapping features observed among disorders of the NCFCS family, the wide breadth of phenotypes within each trait, and absence of clinical features with pathognomic value and consensus on specific and routinely used diagnostic criteria make diagnosis of these diseases challenging, particularly during early infancy and in adulthood. Such a failure has direct impact on patient management because of the prognostic relevance of diagnosis (prevalence and extent of cardiac disease, degree of cognitive and growth impairment, cancer predisposition). Molecular diagnosis now offers the opportunity to overcome the weakness of subjective clinical criteria and represents a highly informative prognostic tool, with direct impact on counseling and appropriate patient management.

Acknowledgments

The authors apologize to colleagues whose work was not cited due to limited space. Research in the authors' laboratories is supported in part by grants from Telethon-Italy (GGP10020), ERARE 2009 (NSEuroNet), and "*Associazione Italiana Sindromi di Costello e Cardiofaciocutanea*" to M.T., and from the National Institutes of Health (HL71207) and March of Dimes (06-1047) to B.D.G.

Conflicts of interest

The authors have royalties for PTPN11 mutation testing for Noonan syndrome that, in aggregate, exceed $10,000, received from the following entities: GeneDx, Correlegan, Baylor College of Medicine, Harvard Medical School/Partners, and Prevention Genetics.

References

1. Mitin, N., K.L. Rossman & C.J. Der. 2005. Signaling interplay in Ras superfamily function. *Curr. Biol.* **15:** R563–R574.

2. Karnoub, A.E. & R.A. Weinberg. 2008. Ras oncogenes: split personalities. *Nat. Rev. Mol. Cell Biol.* **9:** 517–531.
3. Sundaram, M.V. 2006. RTK/Ras/MAPK signaling. *WormBook* **11:** 1–19.
4. McKay, M.M. & D.K. Morrison. 2007. Integrating signals from RTKs to ERK/MAPK. *Oncogene* **26:** 3113–3121.
5. Bos, J.L. 1989. Ras oncogenes in human cancer: a review. *Cancer Res.* **49:** 4682–4689.
6. Rajalingam, K., R. Schreck, U.R. Rapp & S. Albert. 2007. Ras oncogenes and their downstream targets. *Bioch. Biophys. Acta* **1773:** 1177–1195.
7. Leicht, D.T., V. Balan, A. Kaplun, *et al.* 2007. Raf kinases: function, regulation and role in human cancer. *Biochim. Biophys. Acta* **1773:** 1196–1212.
8. Schubbert, S., K. Shannon & G. Bollag. 2007. Hyperactive Ras in developmental disorders and cancer. *Nat. Rev. Cancer* **7:** 295–308.
9. Rauen, K.A., L. Schoyer, F. McCormick, *et al.* 2010. Proceedings from the 2009 genetic syndromes of the Ras/MAPK pathway: from bedside to bench and back. *Am. J. Med. Genet. A* **152A:** 4–24.
10. Bentires-Alj, M., M.I. Kontaridis & B.G. Neel. 2006. Stops along the RAS pathway in human genetic disease. *Nat. Med.* **12:** 283–285.
11. Tidyman, W.E. & K.A. Rauen. 2009. The RASopathies: developmental syndromes of Ras/MAPK pathway dysregulation. *Curr. Opin. Genet. Dev.* **19:** 230–236.
12. Noonan, J.A. 1994. Noonan syndrome. An update and review for the primary pediatrician. *Clin. Pediatr. (Phila)* **33:** 548–555.
13. Allanson, J.E. 2007. Noonan syndrome. *Am. J. Med. Genet. C Semin. Med. Genet.* **145C:** 274–279.
14. Van Der Burgt, I. 2007. Noonan syndrome. *Orphanet J. Rare Dis.* **2:** 4.
15. Burch, M., M. Sharland, E. Shinebourne, *et al.* 1993. Cardiologic abnormalities in Noonan syndrome: phenotypic diagnosis and echocardiographic assessment of 118 patients. *J. Am. Coll. Cardiol.* **22:** 1189–1192.
16. Marino, B., M.C. Digilio, A. Toscano, *et al.* 1999. Congenital heart diseases in children with Noonan syndrome: an expanded cardiac spectrum with high prevalence of atrioventricular canal. *J. Pediatr.* **135:** 703–706.
17. Shah, N., M. Rodriguez, D.S. Louis, *et al.* 1999. Feeding difficulties and foregut dysmotility in Noonan's syndrome. *Arch. Dis. Child* **81:** 28–31.
18. Cesarini, L, P. Alfieri, F. Pantaleoni, *et al.* 2009. Cognitive profile of disorders associated with dysregulation of the RAS/MAPK signaling cascade. *Am. J. Med. Genet. A* **149A:** 140–146.
19. Pierpont, E.I., M.E. Pierpont, N.J. Mendelsohn, *et al.* 2009. Genotype differences in cognitive functioning in Noonan syndrome. *Genes Brain. Behav.* **8:** 275–282.
20. Bader-Meunier, B., G. Tchernia, F. Mielot, *et al.* 1997. Occurrence of myeloproliferative disorder in patients with Noonan syndrome. *J. Pediatr.* **130:** 885–889.
21. Fukuda, M., K. Horibe, Y. Miyajima, *et al.* 1997. Spontaneous remission of juvenile chronic myelomonocytic leukemia in an infant with Noonan syndrome. *J. Pediatr. Hematol. Oncol.* **19:** 177–179.
22. Choong, K., M.H. Freedman, D. Chitayat, *et al.* 1999. Juvenile myelomonocytic leukemia and Noonan syndrome. *J. Pediatr. Hematol. Oncol.* **21:** 523–527.
23. Tartaglia, M., C.M. Niemeyer & A. Fragale. 2003. Somatic mutations in PTPN11 in juvenile myelomonocytic leukemia, myelodysplastic syndromes and acute myeloid leukemia. *Nat. Genet.* **34:** 148–150.
24. Kratz, C.P., C.M. Niemeyer, R.P. Castleberry, *et al.* 2005. The mutational spectrum of PTPN11 in juvenile myelomonocytic leukemia and Noonan syndrome/myeloproliferative disease. *Blood* **106:** 2183–2185.
25. Nora, J.J., A.H. Nora, A.K. Sinha, *et al.* The Ullrich-Noonan syndrome (Turner phenotype). *Am. J. Dis. Child* **127:** 48–55.
26. Tartaglia, M., G. Zampino & B.D. Gelb. 2010. Noonan syndrome: clinical aspects and molecular pathogenesis. *Mol. Syndromol.* **1:** 2–26.
27. Digilio, M.C., E. Conti, A. Sarkozy, *et al.* 2002. Grouping of multiple-lentigines/LEOPARD and Noonan syndromes on the PTPN11 gene. *Am. J. Hum. Genet.* **71:** 389–394.
28. Legius, E., C. Schrander-Stumpel, E. Schollen, *et al.* 2002. PTPN11 mutations in LEOPARD syndrome. *J. Med. Genet.* **39:** 571–574.
29. Pandit, B., A. Sarkozy, L.A. Pennacchio, *et al.* 2007. Gain-of-function RAF1 mutations cause Noonan and LEOPARD syndromes with hypertrophic cardiomyopathy. *Nat. Genet.* **39:** 1007–1012.
30. Sarkozy, A., C. Carta, S. Moretti, *et al.* 2009. Germline BRAF mutations in Noonan, LEOPARD, and cardiofaciocutaneous syndromes: molecular diversity and associated phenotypic spectrum. *Hum. Mutat.* **30:** 695–702.
31. Koudova, M., E. Seemanova & M. Zenker. 2009. Novel BRAF mutation in a patient with LEOPARD syndrome and normal intelligence. *Eur. J. Med. Genet.* **52:** 337–340.
32. Gorlin, R.J., R.C. Anderson & J.H. Moller. 1971. The leopard (multiple lentigines) syndrome revisited. *Birth Defects Orig. Artic. Ser.* **7:** 110–115.
33. Digilio, M.C., A. Sarkozy, A. de Zorzi, *et al.* 2006. LEOPARD syndrome: clinical diagnosis in the first year of life. *Am. J. Med. Genet. A* **140A:** 740–674.
34. Voron, D.A., H.H. Hatfield & M.D. Kalkhoff. 1976. Multiple lentigines syndrome: case report and review of the literature. *Am. J. Med.* **60:** 447–456.
35. Gorlin, J.R., M.M. Cohen & L.S. Levn. 1990. Leopard syndrome. In *Syndromes of the Head and Neck*. Gorlin, J.R., M.M. Cohen & L.S. Levn, Eds.: 461–464. Oxford University Press. New York.
36. Somerville, J. & R.E. Bonham-Carter. 1972. The heart in lentiginosis. *Br. Heart J.* **34:** 58–66.
37. Sarkozy, A., M.C. Digilio, G. Zampino, *et al.* 2009. LEOPARD Syndromeclinical aspects and molecular pathogenesis. In *Noonan Syndrome and Related Disorders*. Vol. **17**, Zenker, M., Ed.: 55–65. Monogr. Hum. Genet. Karger, Basel.
38. Mazzanti, L., E. Cacciari, A. Cicognani, *et al.* 2003. Noonan-like syndrome with loose anagen hair: a new syndrome? *Am. J. Med. Genet. A* **118A:** 279–286.

39. Cordeddu, V., E. Di Schiavi, L.A. Pennacchio, *et al.* 2009. Mutation of SHOC2 promotes aberrant protein N-myristoylation and causes Noonan-like syndrome with loose anagen hair. *Nat. Genet.* **41:** 1022–1026.
40. Cantatore-Francis, J.L. & S.J. Orlow. 2009. Practical guidelines for evaluation of loose anagen hair syndrome. *Arch. Dermatol.* **145:** 1123–1128.
41. Tosti, A. & B.M. Piraccini. 2002. Loose anagen hair syndrome and loose anagen hair. *Arch. Dermatol.* **138:** 521–522.
42. Cohen, M.M. Jr. & R.J. Gorlin. 1991. Noonan-like/multiple giant cell lesion syndrome. *Am. J. Med. Genet.* **40:** 159–166.
43. Tartaglia, M., K. Kalidas, A. Shaw, *et al.* 2002. PTPN11 mutations in Noonan syndrome: molecular spectrum, genotype-phenotype correlation, and phenotypic heterogeneity. *Am. J. Hum. Genet.* **70:** 1555–1563.
44. Sarkozy, A., M.G. Obregon, E. Conti, *et al.* 2004. A novel PTPN11 gene mutation bridges Noonan syndrome, multiple lentigines/LEOPARD syndrome and Noonan-like/multiple giant cell lesion syndrome. *Eur. J. Hum. Genet.* **12:** 1069–1072.
45. Jafarov, T., N. Ferimazova & E. Reichenberger. 2005. Noonan-like syndrome mutations in PTPN11 in patients diagnosed with cherubism. *Clin. Genet.* **68:** 190–191.
46. Lee, J.C., M. Tartaglia, B.D. Gelb, *et al.* 2005. Phenotypic and genotypic characterization of Noonan-like/multiple giant cell lesion syndrome. *J. Med. Genet.* **42:** e11.
47. Mascheroni, E., M.C. Digilio, E. Cortis, *et al.* 2008. Pigmented villonodular synovitis in a patient with Noonan syndrome and SOS1 gene mutation. *Am. J. Med. Genet. A* **146A:** 2966–2967.
48. Beneteau, C., H. Cavé, A. Moncla, *et al.* 2009. SOS1 and PTPN11 mutations in five cases of Noonan syndrome with multiple giant cell lesions. *Eur. J. Hum. Genet.* **17:** 1216–1221.
49. Hanna, N., B. Parfait, I.M. Talaat, *et al.* 2009. SOS1: a new player in the Noonan-like/multiple giant cell lesion syndrome. *Clin. Genet.* **75:** 568–571.
50. Neumann, T.E., J. Allanson, I. Kavamura. 2009. Multiple giant cell lesions in patients with Noonan syndrome and cardio-facio-cutaneous syndrome. *Eur. J. Hum. Genet.* **17:** 420–425.
51. Reynolds, J.F., G. Neri, J.P. Herrmann, *et al.* 1986. New multiple congenital anomalies/mental retardation syndrome with cardio-facio-cutaneous involvement: the CFC syndrome. *Am. J. Med. Genet.* **25:** 413–427.
52. Roberts, A., J. Allanson, S.K. Jadico, *et al.* 2006. The cardiofaciocutaneous syndrome. *J. Med. Genet.* **43:** 833–842.
53. Fryer, A.E., P.J. Holt & H.E. Hughes. 1991. The cardio-facio-cutaneous (CFC) syndrome and Noonan syndrome: are they the same? *Am. J. Med. Genet.* **38:** 548–551.
54. Neri, G., M. Zollino & J.F. Reynolds. 1991. The Noonan-CFC controversy. *Am. J. Med. Genet.* **39:** 367–370.
55. Wieczorek, D., F. Majewski & G. Gillessen-Kaesbach. 1997. Cardio-facio-cutaneous (CFC) syndrome-a distinct entity? Report of three patients demonstrating the diagnostic difficulties in delineation of CFC syndrome. *Clin. Genet.* **52:** 37–46.
56. Niihori, T., Y. Aoki, Y. Narumi, *et al.* 2006. Germline KRAS and BRAF mutations in cardio-facio-cutaneous syndrome. *Nat. Genet.* **38:** 294–296.
57. Rodriguez-Viciana, P., O. Tetsu, W.E. Tidyman, *et al.* 2006. Germline mutations in genes within the MAPK pathway cause cardio-facio-cutaneous syndrome. *Science* **311:** 1287–1290.
58. Narumi, Y., Y. Aoki, T. Niihori, *et al.* 2006. Molecular and clinical characterization of cardio-facio-cutaneous (CFC) syndrome: overlapping clinical manifestations with Costello syndrome. *Am. J. Med. Genet. A* **143A:** 799–807.
59. Nava, C., N. Hanna, C. Michot, *et al.* 2007. Cardio-facio-cutaneous and Noonan syndromes due to mutations in the RAS/MAPK signalling pathway: genotype phenotype relationships and overlap with Costello syndrome. *J. Med. Genet.* **44:** 763–771.
60. Schulz, A.L., B. Albrecht, C. Arici, *et al.* 2008. Mutation and phenotypic spectrum in patients with cardio-facio-cutaneous and Costello syndrome. *Clin. Genet.* **73:** 62–70.
61. Costello, J.M. 1971. A new syndrome. *NZ Med. J.* **74:** 397.
62. Costello, J.M. 1977. A new syndrome: mental subnormality and nasal papillomata. *Aust. Paediatr. J.* **13:** 114–118.
63. Zampino, G., P. Mastroiacovo, R. Ricci, *et al.* 1993. Costello syndrome: further clinical delineation, natural history, genetic definition, and nosology. *Am. J. Med. Genet.* **47:** 176–183.
64. Hennekam, R.C. 2003. Costello syndrome: an overview. *Am. J. Med. Genet. C Semin. Med. Genet.* **117C:** 42–48.
65. Gripp, K.W. 2005. Tumor predisposition in Costello syndrome. *Am. J. Med. Genet. C Semin. Med. Genet.* **137C:** 72–77.
66. Aoki, Y., T. Niihori, H. Kawame, *et al.* 2005. Germline mutations in HRAS proto-oncogene cause Costello syndrome. *Nat. Genet.* **37:** 1038–1040.
67. Gripp, K.W., A.E. Lin, D.L. Stabley, *et al.* 2006. HRAS mutation analysis in Costello syndrome: genotype and phenotype correlation. *Am. J. Med. Genet. A* **140A:** 1–7.
68. Estep, A.L., W.E. Tidyman, M.A. Teitell, *et al.* 2006. HRAS mutations in Costello syndrome: detection of constitutional activating mutations in codon 12 and 13 and loss of wild-type allele in malignancy. *Am. J. Med. Genet. A* **140A:** 8–16.
69. Kerr, B., M.A. Delrue, S. Sigaudy, *et al.* 2006. Genotype-phenotype correlation in Costello syndrome: HRAS mutation analysis in 43 cases. *J. Med. Genet.* **43:** 401–405.
70. Zampino, G., F. Pantaleoni, C. Carta, *et al.* 2007. Diversity, parental germline origin, and phenotypic spectrum of de novo HRAS missense changes in Costello syndrome. *Hum. Mutat.* **28:** 265–272.
71. Friedman, J.M. & P.H. Birch. 1997. Type 1 neurofibromatosis: a descriptive analysis of the disorder in 1,728 patients. *Am. J. Med. Genet.* **70:** 138–143.
72. Williams, V.C., J. Lucas, M.A. Babcock, *et al.* 2009. Neurofibromatosis type 1 revisited. *Pediatrics* **123:** 124–33.
73. National Institutes of Health Consensus Development Conference Statement. 1988. Neurofibromatosis. *Arch. Neurol.* **45:** 575–578.
74. Shen, M.H., P.S. Harper & M. Upadhyaya. 1996. Molecular genetics of neurofibromatosis type 1 (NF1). *J. Med. Genet.* **33:** 2–17.

75. Fernel, R.E. 2007. Neurofibromatosis 1. *Eur. J. Hum. Genet.* **15:** 131–138.
76. Lee, M.-J. & D.A. Stephenson. 2007. Recent developments in neurofibromatosis type 1. *Curr. Opin. Neurol.* **20:** 135–141.
77. Allanson, J.E., J.G. Hall & M.I. Van Allen. 1985. Noonan phenotype associated with neurofibromatosis. *Am. J. Med. Genet.* **21:** 457–462.
78. Opitz, J.M. & D.D. Weaver. 1985. The neurofibromatosis-Noonan syndrome. *Am. J. Med. Genet.* **21:** 477–490.
79. Carey, J.C. 1998. Neurofibromatosis-Noonan syndrome. *Am. J. Med. Genet.* **75:** 263–264.
80. Baralle, D., C. Mattocks, K. Kalidas, *et al.* 2003. Different mutations in the NF1 gene are associated with neurofibromatosis-Noonan syndrome (NFNS). *Am. J. Med. Genet.* **119A:** 1–8.
81. De Luca, A., I. Bottillo, A. Sarkozy, *et al.* 2005. NF1 gene mutations represent the major molecular event underlying neurofibromatosis-Noonan syndrome. *Am. J. Hum. Genet.* **77:** 1092–1101.
82. Hüffmeier, U., M. Zenker, J. Hoyer, *et al.* 2006. A variable combination of features of Noonan syndrome and neurofibromatosis type I are caused by mutations in the NF1 gene. *Am. J. Med. Genet. A* **140A:** 2749–2756.
83. Bertola, D.R., A.C. Pereira, F. Passetti, *et al.* 2005. Neurofibromatosis-Noonan syndrome: molecular evidence of the concurrence of both disorders in a patient. *Am. J. Med. Genet. A* **136:** 242–245.
84. Thiel, C., M. Wilken, M. Zenker, *et al.* 2009. Independent NF1 and PTPN11 mutations in a family with neurofibromatosis-Noonan syndrome. *Am. J. Med. Genet. A* **149A:** 1263–1267.
85. Nyström, A.M., S. Ekvall, J. Allanson, *et al.* 2009. Noonan syndrome and neurofibromatosis type I in a family with a novel mutation in NF1. *Clin. Genet.* **76:** 524–534.
86. Watson, G.H. 1967. Pulmonary stenosis, cafe-au-lait spots, and dull intelligence. *Arch. Dis. Child.* **42:** 303–307.
87. Partington, M.W., G.W. Burggraf, J.E. Fay & E. Frontini. 1985. Pulmonary stenosis, cafe au lait spots and dull intelligence: the Watson syndrome revisited. *Proc. Greenwood Genet. Center* **4:** 105.
88. Allanson, J.E., M. Upadhyaya, G.H. Watson, *et al.* 1989. Molecular linkage analysis of Watson syndrome. *Am. J. Hum. Genet.* **45**(suppl.): A38.
89. Allanson J.E., M. Upadhyaya, G.H. Watson, *et al.* 1991. Watson syndrome: is it a subtype of type 1 neurofibromatosis? *J. Med. Genet.* **28:** 752–756.
90. Upadhyaya, M., M. Sarfarazi, S. Huson, *et al.* 1990. Linkage of Watson's syndrome to chromosome 17 markers. *J. Med. Genet.* **27:** 209.
91. Upadhyaya, M., M. Shen, A. Cherryson, *et al.* 1992. Analysis of mutations at the neurofibromatosis 1 (NF1) locus. *Hum. Mol. Genet.* **1:** 735–740.
92. Tassabehji, M., T. Strachan, M. Sharland, *et al.* 1993. Tandem duplication within a neurofibromatosis type I (NF1) gene exon in a family with features of Watson syndrome and Noonan syndrome. *Am. J. Hum. Genet.* **53:** 90–95.
93. Brems, H., M. Chmara, M. Sahbatou, *et al.* 2007. Germline loss-of-function mutations in SPRED1 cause a neurofibromatosis 1-like phenotype. *Nat. Genet.* **39:** 1120–1126.
94. Pasmant, E., A. Sabbagh, N. Hanna, *et al.* 2009. SPRED1 germline mutations causes a neurofibromatosis type 1 overlapping phenotype. *J. Med. Genet.* **46:** 425–430.
95. Spurlock, G., E. Bennett, N. Chuzhanova, *et al.* 2009. SPRED1 mutations (Legius syndrome): another clinically useful genotype for dissecting the neurofibromatosis type 1 phenotype. *J. Med. Genet.* **46:** 431–437.
96. Messiaen, L., S. Yao, H. Brems, *et al.* 2009. Clinical and mutational spectrum of neurofibromatosis type 1-like syndrome. *JAMA* **302:** 2111–2118.
97. Neel, B.G., H. Gu & L. Pao. 2003. The 'Shp'ing news: SH2 domain-containing tyrosine phosphatases in cell signaling. *Trends Biochem. Sci.* **28:** 284–293.
98. Tartaglia, M., C.M. Niemeyer, K. M. Shannon & M.L. Loh. 2004. SHP-2 and myeloid malignancies. *Curr. Opin. Hematol.* **11:** 44–50.
99. Dance, M., A. Montagner, J.P. Salles, *et al.* 2008. The molecular functions of Shp2 in the Ras/Mitogen-activated protein kinase (ERK1/2) pathway. *Cell Signal.* **20:** 453–459.
100. Xu, D. & C.K. Qu. 2008. Protein tyrosine phosphatases in the JAK/STAT pathway. *Front. Biosci.* **13:** 4925–4932.
101. Hof, P., S. Pluskey, S. Dhe-Paganon, *et al.* 1998. Crystal structure of the tyrosine phosphatase shp-2. *Cell* **92:** 441–450.
102. Tartaglia, M., E.L. Mehler, R. Goldberg, *et al.* 2001. Mutations in PTPN11, encoding the protein tyrosine phosphatase SHP-2, cause Noonan syndrome. *Nat. Genet.* **29:** 465–468.
103. Zenker, M., G. Buheitel, R. Rauch, *et al.* 2004. Genotype-phenotype correlations in Noonan syndrome. *J. Pediatr.* **144:** 368–374.
104. Keilhack, H., F.S. David, M. McGregor, *et al.* 2005. Diverse biochemical properties of shp2 mutants. Implications for disease phenotypes. *J. Biol. Chem.* **280:** 30984–30993.
105. Tartaglia, M., S. Martinelli, L. Stella, *et al.* 2006. Diversity and functional consequences of germline and somatic PTPN11 mutations in human disease. *Am. J. Hum. Genet.* **78:** 279–290.
106. Martinelli, S., P. Torreri, M. Tinti, *et al.* 2008. Diverse driving forces underlie the invariant occurrence of the T42A, E139D, I282V and T468M SHP2 amino acid substitutions causing Noonan and LEOPARD syndromes. *Hum. Mol. Genet.* **17:** 2018–2029.
107. Hanna, N., A. Montagner, W.H. Lee, *et al.* Reduced phosphatase activity of SHP-2 in LEOPARD syndrome: consequences for PI3K binding on Gab1. *FEBS Lett.* **580:** 2477–2482.
108. Kontaridis, M.I., K.D. Swanson, F.S. David, *et al.* Ptpn11 (shp2) mutations in LEOPARD syndrome have dominant negative, not activating, effects. *J. Biol. Chem.* **281:** 6785–6792.
109. Oishi, K., H. Zhang, W.J. Gault, *et al.* 2009. Phosphatase-defective LEOPARD syndrome mutations in PTPN11 gene have gain-of-function effects during Drosophila development. *Hum. Mol. Genet.* **18:** 193–201.
110. Edouard, T., J.P. Combier, A. Nédélec, *et al.* 2010. Functional effects of PTPN11 (SHP2) mutations causing

LEOPARD syndrome on epidermal growth factor-induced phosphoinositide 3-kinase/AKT/glycogen synthase kinase 3beta signaling. *Mol. Cell. Biol.* **30:** 2498–2507.
111. Niemeyer, C.M. & C.P. Kratz. 2008. Paediatric myelodysplastic syndromes and juvenile myelomonocytic leukaemia: molecular classification and treatment options. *Br. J. Haematol.* **140:** 610–624.
112. Loh, M.L., S. Vattikuti, S. Schubbert, *et al.* 2004. Mutations in PTPN11 implicate the SHP-2 phosphatase in leukemogenesis. *Blood* **103:** 2325–2331.
113. Tartaglia, M., S. Martinelli, G. Cazzaniga, *et al.* 2004. Genetic evidence for lineage- and differentiation stage-related contribution of somatic PTPN11 mutations to leukemogenesis in childhood acute leukemia. *Blood* **104:** 307–313.
114. Tartaglia, M., S. Martinelli, I. Iavarone, *et al.* 2005. Somatic PTPN11 mutations in childhood acute myeloid leukaemia. *Br. J. Haematol.* **129:** 333–339.
115. Loh, M.L., M.G. Reynolds, S. Vattikuti, *et al.* 2004. PTPN11 mutations in pediatric patients with acute myeloid leukemia: results from the children's cancer group. *Leukemia* **18:** 1831–1834.
116. Chan, R.J., M.B. Leedy, V. Munugalavadla, *et al.* 2005. Human somatic PTPN11 mutations induce hematopoietic-cell hypersensitivity to granulocyte-macrophage colony-stimulating factor. *Blood* **105:** 3737–3742.
117. Schubbert, S., K. Lieuw, S.L. Rowe, *et al.* 2005. Functional analysis of leukemia-associated PTPN11 mutations in primary hematopoietic cells. *Blood* **106:** 311–317.
118. Shchelochkov, O.A., A. Patel, G.M. Weissenberger, *et al.* 2008. Duplication of chromosome band 12q24.11q24.23 results in apparent Noonan syndrome. *Am. J. Med. Genet. A* **146A:** 1042–1048.
119. Graham, J.M., Jr N. Kramer, B.A. Bejjani, *et al.* 2009. Genomic duplication of PTPN11 is an uncommon cause of Noonan syndrome. *Am. J. Med. Genet. A* **149A:** 2122–2128.
120. Quilliam, L.A., R. Khosravi-Far, S.Y. Huff & C.J. Der. 1995. Guanine nucleotide exchange factors: activators of the Ras superfamily of proteins. *Bioessays* **17:** 395–404.
121. Quilliam, L.A., J.F. Rebhun & A.F. Castro. A growing family of guanine nucleotide exchange factors is responsible for activation of Ras-family GTPases. *Prog. Nucleic Acid Res. Mol. Biol.* **71:** 391–444.
122. Nimnual, A., D. Bar-Sagi. 2002. The two hats of SOS. *Sci. STKE* **145:** PE36.
123. Roberts, A.E., T. Araki, K.D. Swanson, *et al.* 2007. Germline gain-of-function mutations in SOS1 cause Noonan syndrome. *Nat. Genet.* **39:** 70–74.
124. Tartaglia, M., L.A. Pennacchio, C. Zhao, *et al.* 2007. Gain-of-function SOS1 mutations cause a distinctive form of Noonan syndrome. *Nat. Genet.* **39:** 75–79.
125. Chardin, P., J.H. Camonis, N.W. Gale, *et al.* 1993. Human Sos1: a guanine nucleotide exchange factor for Ras that binds to GRB2. *Science* **260:** 1338–1343.
126. Nimnual, A.S., B.A. Yatsula & D. Bar-Sagi. 1998. Coupling of Ras and Rac guanosine triphosphatases through the Ras exchanger Sos. *Science* **279:** 560–563.
127. Chen, R.H., S. Corbalan-Garcia & D. Bar-Sagi. 1997. The role of the PH domain in the signal-dependent membrane targeting of Sos. *EMBO J.* **16:** 1351–1359.
128. Sondermann, H., B. Nagar, D. Bar-Sagi & J. Kuriyan. 2005. Computational docking and solution x-ray scattering predict a membrane-interacting role for the histone domain of the Ras activator Son of sevenless. *Proc. Natl. Acad. Sci. USA* **102:**16632–16637.
129. Zenker, M., D. Horn, D. Wieczorek, *et al.* 2007. SOS1 is the second most common Noonan gene but plays no major role in cardio-facio-cutaneous syndrome. *J. Med. Genet.* **44:** 651–656.
130. Narumi, Y., Y. Aoki, T. Niihori, *et al.* 2007. Molecular and clinical characterization of cardio-facio-cutaneous (CFC) syndrome: overlapping clinical manifestations with Costello syndrome. *Am. J. Med. Genet. A* **143A:** 799–807.
131. Nyström, A.M., S. Ekvall, E. Berglund, *et al.* 2008. Noonan and cardio-facio-cutaneous syndromes: two clinically and genetically overlapping disorders. *J. Med. Genet.* **45:** 500–506.
132. Margarit, S.M., H. Sondermann, B.E. Hall, *et al.* 2003. Structural evidence for feedback activation by Ras.GTP of the Ras-specific nucleotide exchange factor SOS. *Cell* **112:** 685–695.
133. Sondermann, H., S.M. Soisson, S. Boykevisch, *et al.* 2004. Structural analysis of autoinhibition in the Ras activator Son of sevenless. *Cell* **119:** 393–405.
134. Kratz, C.P., C.M. Niemeyer, C. Thomas, *et al.* 2007. Mutation analysis of Son of Sevenless in juvenile myelomonocytic leukemia. *Leukemia* **21:** 1108–1109.
135. Swanson, K.D., J.M. Winter, M. Reis, *et al.* 2008. SOS1 mutations are rare in human malignancies: implications for Noonan Syndrome patients. *Genes Chrom. Cancer* **47:** 253–259.
136. Wennerberg, K., K.L. Rossman & C.J. Der. 2005. The Ras superfamily at a glance. *J. Cell Sci.* **118:** 843–846.
137. Barbacid, M. 1987. Ras genes. *Annu. Rev. Biochem.* **56:** 779–827.
138. Adjei, A.A. 2001. Blocking oncogenic Ras signaling for cancer therapy. *J. Natl. Cancer Inst.* **93:** 1062–1074.
139. Clanton, D.J., S. Hattori & T.Y. Shih. 1986. Mutations of the Ras gene product p21 that abolish guanine nucleotide binding. *Proc. Natl. Acad. Sci. USA* **83:** 5076–5080.
140. Der, C.J., B.T. Pan & G.M. Cooper. 1986. RasH mutants deficient in GTP binding. *Mol. Cell. Biol.* **6:** 3291–3294.
141. Walter, M., S.G. Clark & A.D. Levinson. 1986. The oncogenic activation of human p21ras by a novel mechanism. *Science* **233:** 649–652.
142. Denayer, E., A. Parret, M. Chmara, *et al.* 2008. Mutation analysis in Costello syndrome: functional and structural characterization of the HRAS p.Lys117Arg mutation. *Hum. Mutat.* **29:** 232–239.
143. Gremer, L., A. De Luca, T. Merbitz-Zahradnik, *et al.* 2010. Duplication of Glu37 in the switch I region of HRAS impairs effector/GAP binding and underlies Costello syndrome by promoting enhanced growth factor-dependent MAPK and AKT activation. *Hum. Mol. Genet.* **19:** 790–802.
144. Seeburg, P.H., W.W. Colby, D.J. Capon, *et al.* 1984. Biological properties of human c-Ha-ras1 genes mutated at codon 12. *Nature* **312:** 71–75.

145. Van Der Burgt, I., W. Kupsky, S. Stassou, *et al.* 2007. Myopathy caused by HRAS germline mutations: implications for disturbed myogenic differentiation in the presence of constitutive HRas activation. *J. Med. Genet.* **44:** 459–462.
146. Carta, C., F. Pantaleoni, G. Bocchinfuso, *et al.* 2006. Germline missense mutations affecting KRAS isoform b are associated with a severe noonan syndrome phenotype. *Am. J. Hum. Genet.* **79:** 129–135.
147. Schubbert, S., M. Zenker, S.L. Rowe, *et al.* 2006. Germline KRAS mutations cause Noonan syndrome. *Nat. Genet.* **38:** 331–336.
148. Zenker, M., K. Lehmann, A.L. Schulz, *et al.* 2007. Expansion of the genotypic and phenotypic spectrum in patients with KRAS germline mutations. *J. Med. Genet.* **44:** 131–135.
149. Kratz, C.P., G. Zampino, M. Kriek, *et al.* 2009. Craniosynostosis in patients with Noonan syndrome caused by germline KRAS mutations. *Am. J. Med. Genet.* **149A:** 1036–1040.
150. Friday, B.B. & A.A. Adjei. 2005. K-ras as a target for cancer therapy. *Biochim. Biophys. Acta* **1756:** 127–144.
151. Johnson, L., D. Greenbaum, K. Cichowski, *et al.* 1997. K-ras is an essential gene in the mouse with partial functional overlap with N-ras. *Genes Dev.* **11:** 2468–2481.
152. Koera, K., K. Nakamura, K. Nakao, *et al.* 1997. K-ras is essential for the development of the mouse embryo. *Oncogene* **15:** 1151–1159.
153. Plowman, S.J., D.J. Williamson, M.J. O'Sullivan, *et al.* 2003. While K-ras is essential for mouse development, expression of the K-ras 4a splice variant is dispensable. *Mol. Cell. Biol.* **23:** 9245–9250.
154. Schubbert, S., G. Bollag, N. Lyubynska, *et al.* 2007. Biochemical and functional characterization of germ line KRAS mutations. *Mol. Cell. Biol.* **27:** 7765–7770.
155. Cirstea, I.C., K. Kutsche, R. Dvorsky, *et al.* 2010. A restricted spectrum of NRAS mutations causes Noonan syndrome. *Nat. Genet.* **42:**27–29.
156. Trovó-Marqui, A.B. & E.H. Tajara. 2006. Neurofibromin: a general outlook. *Clin. Genet.* **70:** 1–13.
157. The, I., G.E. Hannigan, G.S. Cowley, *et al.* 1997. Rescue of a Drosophila NF1 mutant phenotype by protein kinase A. *Science* **276:** 791–794.
158. Xu, H. & D.H. Gutmann. 1997. Mutations in the GAP-related domain impair the ability of neurofibromin to associate with microtubules. *Brain Res.* **759:** 149–152.
159. Tong J, Hannan F, Zhu Y, *et al.* 2002. Neurofibromin regulates G protein-stimulated adenylyl cyclase activity. *Nat. Neurosci.* **5:** 95–96.
160. Dasgupta, B., L.L. Dugan & D.H. Gutmann. 2003. The neurofibromatosis 1 gene product neurofibromin regulates pituitary adenylate cyclase-activating polypeptide-mediated signaling in astrocytes. *J. Neurosci.* **23:** 8949–8954.
161. Vandenbroucke, I., P. Van Oostveldt, E. Coene, *et al.* 2004. Neurofibromin is actively transported to the nucleus. *FEBS Lett.* **560:** 98–102.
162. Dorschner, M.O., V.P. Sybert, M. Weaver, *et al.* 2000. NF1 microdeletion breakpoints are clustered at flanking repetitive sequences. *Hum. Mol. Genet.* **9:** 35–46.
163. Riva, P., L. Corrado, F. Natacci, *et al.* 2000. NF1 microdeletion syndrome: refined FISH characterization of sporadic and familial deletions with locus-specific probes. *Am. J. Hum. Genet.* **66:** 100–109.
164. Jenne, D.E., S. Tinschert, H. Reimann, *et al.* 2001. Molecular characterization and gene content of breakpoint boundaries in patients with neurofibromatosis type 1 with 17q11.2 microdeletions. *Am. J. Hum. Genet.* **69:** 516–527.
165. Sawada, S., S. Florell, S.M. Purandare, *et al.* 1996. Identification of NF1 mutations in both alleles of a dermal neurofibroma. *Nat. Genet.* **14:** 110–112.
166. Serra, E., S. Puig, D. Otero, *et al.* 1997. Confirmation of a double-hit model for the NF1 gene in benign neurofibromas. *Am. J. Hum. Genet.* **61:** 512–519.
167. Side, L., B. Taylor, M. Cayouette, *et al.* 1997. Homozygous inactivation of the NF1 gene in bone marrow cells from children with neurofibromatosis type 1 and malignant myeloid disorders. *New Eng. J. Med.* **336:** 1713–1720.
168. Bundschu, K., U. Walter & K. Schuh. 2007. Getting a first clue about SPRED functions. *Bioessays* **29:** 897–907.
169. Wakioka, T., A. Sasaki, R. Kato, *et al.* 2001. Spred is a Sprouty-related suppressor of Ras signalling. *Nature* **412:** 647–651.
170. King, J.A., A.F. Straffon, G.M. D'Abaco, *et al.* 2005. Distinct requirements for the Sprouty domain for functional activity of Spred proteins. *Biochem. J.* **388:** 445–454.
171. De Schepper, S., O. Maertens, T. Callens, *et al.* 2008. Somatic mutation analysis in NF1 café au lait spots reveals two NF1 hits in the melanocytes. *J. Invest. Dermatol.* **128:** 1050–1053.
172. Wellbrock, C., M. Karasarides & R. Marais. 2004. The RAF proteins take centre stage. *Nat. Rev. Mol. Cell Biol.* **5:** 875–885.
173. Schreck, R. & U.R. Rapp. 2006. Raf kinases: oncogenesis and drug discovery. *In. J. Cancer* **119:** 2261–2271.
174. Pritchard, C.A., L. Bolin, R. Slattery, *et al.* 1996. Post-natal lethality and neurological and gastrointestinal defects in mice with targeted disruption of the A-Raf protein kinase gene. *Curr. Biol.* **6:** 614–617.
175. Wojnowski, L., A.M. Zimmer, T.W. Beck, *et al.* 1997. Endothelial apoptosis in Braf-deficient mice. *Nat. Genet.* **16:** 293–297.
176. Mikula, M., M. Schreiber, Z. Husak, *et al.* 2001. Embryonic lethality and fetal liver apoptosis in mice lacking the c-raf-1 gene. *EMBO J.* **20:** 1952–1962.
177. Razzaque, M.A, T. Nishizawa, Y. Komoike, *et al.* 2007. Germline gain-of-function mutations in RAF1 cause Noonan syndrome. *Nat. Genet.* **39:** 1013–1017.
178. Nyström, A.M., S. Ekvall, E. Berglund, *et al.* 2008. Noonan and cardio-facio-cutaneous syndromes: two clinically and genetically overlapping disorders. *J. Med. Genet.* **45:** 500–506.
179. Armour, C.M. & J.E. Allanson. 2008. Further delineation of cardio-facio-cutaneous syndrome: clinical features of 38 individuals with proven mutations. *J. Med. Genet.* **45:** 249–254.
180. Dhomen, N. & R. Marais. 2007. New insight into BRAF mutations in cancer. *Curr. Opin. Genet. Dev.* **17:** 31–39.
181. Selfors, L.M., J.L. Schutzman, C.Z. Borland & M.J. Stern. 1998. SOC-2 encodes a leucine-rich repeat protein

implicated in fibroblast growth factor receptor signaling. *Proc. Natl. Acad. Sci. USA* **95:** 6903–6908.

182. Sieburth, D.S., Q. Sun & M. Han. 1998. SUR-8, a conserved Ras-binding protein with leucine-rich repeats, positively regulates Ras-mediated signaling in C. elegans. *Cell* **94:**119–130.
183. Li, W., M. Han & K.L. Guan. 2000. The leucine-rich repeat protein SUR-8 enhances MAP kinase activation and forms a complex with Ras and Raf. *Genes Dev.* **14:** 895–900.
184. Rodriguez-Viciana, P., J. Oses-Prieto, A. Burlinqame, & F. McCormick. 2006. A phosphatase holoenzyme comprised of Shoc2/Sur8 and the catalytic subunit of PP1 functions as an M-Ras effector to modulate Raf activity. *Mol. Cell* **22:** 217–30.
185. Matsunaga-Udagawa R, Y. Fujita, S. Yoshiki, *et al.* 2010. The scaffold protein Shoc2/SUR-8 accelerates the interaction of Ras and Raf. *J. Biol. Chem.* **285:** 7818–7826.
186. Boutin, J.A. 1997. Myristoylation. *Cell Signal.* **9:** 15–35.
187. Farazi, T.A., G. Waksman & J.I. Gordon. 2001. The biology and enzymology of protein N-myristoylation. *J. Biol. Chem.* **276:** 39501–39504.
188. Resh, M.D. 2006. Trafficking and signaling by fatty-acylated and prenylated proteins. *Nat. Chem. Biol.* **2:** 584–590.
189. Zheng, C.F. & K.L. Guan. 1993. Properties of MEKs: the kinases that phosphorylate and activate the extracellular signal-regulated kinases. *J. Biol. Chem.* **268:** 23933–23939.
190. Giroux, S., M. Tremblay, D. Bernard, *et al.* 1999. Embryonic death of Mek1-deficient mice reveals a role for this kinase in angiogenesis in the labyrinthine region of the placenta. *Curr. Biol.* **9:** 369–372.
191. Bélanger, L.F., S. Roy, M. Tremblay, *et al.* 2003. Mek2 is dispensable for mouse growth and development. *Mol. Cell. Biol.* **23:** 4778–4787.
192. Cowley, S., H. Paterson, P. Kemp & C.J. Marshall. 1994. Activation of MAP kinase kinase is necessary and sufficient for PC12 differentiation and for transformation of NIH 3T3 cells. *Cell* **77:** 841–852.
193. Mansour, S.J., W.T. Matten, A.S. Hermann, *et al.* 1994. Transformation of mammalian cells by constitutively active MAP kinase kinase. *Science* **265:** 966–970.
194. Dentici, M.L., A. Sarkozy, F. Pantaleoni, *et al.* 2009. Spectrum of MEK1 and MEK2 gene mutations in cardio-facio-cutaneous syndrome and genotype-phenotype correlations. *Eur. J. Hum. Genet.* **17:** 733–740.
195. Rauen, K.A., W.E. Tidyman, A.L. Estep, *et al.* 2010. Molecular and functional analysis of a novel MEK2 mutation in cardio-facio-cutaneous syndrome: transmission through four generations. *Am. J. Med. Genet. A* **152A:** 807–814.

Ann. N.Y. Acad. Sci. ISSN 0077-8923

ANNALS OF THE NEW YORK ACADEMY OF SCIENCES
Issue: *The Year in Human and Medical Genetics*

Impact of Mendelian inheritance in cardiovascular disease

Kim L. McBride[1,2] and Vidu Garg[1,3]

[1]Departments of Pediatrics, The Ohio State University College of Medicine, Columbus, Ohio. [2]Center for Molecular and Human Genetics, Research Institute at Nationwide Children's Hospital, and [3]The Heart Center and Center for Cardiovascular and Pulmonary Research, Research Institute at Nationwide Children's Hospital, Columbus, Ohio

Addresses for correspondence: Vidu Garg, M.D., The Heart Center and Center for Cardiovascular and Pulmonary Research, Nationwide Children's Hospital, 700 Children's Drive W302, Columbus, OH 43205. vidu.garg@nationwidechildrens.org; and Kim L. McBride, M.D., Center for Molecular and Human Genetics, Research Institute at Nationwide Children's Hospital, 700 Children's Drive W410, Columbus, OH 43205. kim.mcbride@nationwidechildrens.org

Cardiovascular disease is a leading cause of mortality worldwide. While the etiology for the majority of cardiovascular disease is presumed to be a combination of genetic and environmental factors, developments in understanding the basic biology of cardiac disorders have been greatly advanced through discoveries made studying heart diseases that exhibit Mendelian forms of inheritance. Most of these diseases primarily affect children and young adults and include cardiomyopathies, arrhythmias, aortic aneurysms, and congenital heart defects. The discovery of the genetic etiologies for these diseases have had significant impact on our understanding of more complex forms of cardiovascular disease and in some cases have led to novel diagnostic and treatment modalities. In this review, we will summarize these seminal genetic discoveries, highlighting a few that have resulted in significant impact on human disease, and discuss the potential utility of studying Mendelian-inherited heart disease with the development of new genetic technologies and our increased understanding of the human genome.

Keywords: genetics; congenital cardiovascular malformations; inherited cardiomyopathy; congenital arrhythmias; aortic aneurysm

Introduction

Diseases of the cardiovascular system are a leading cause of death in both children[1] and adults.[2] Adult-onset cardiovascular disease—hypertension, atherosclerotic heart disease, valvar disease—remain significant health problems and leading causes of death in developed countries,[2] and the prevalence is increasing in developing countries.[3] Inherited cardiomyopathies and arrhythmias are among the most common human genetic disorders. In addition, congenital malformations of the heart and great vessels are the most common type of birth defect and are the leading cause of death in infants less than one year of age, despite recent advancements in medical and surgical therapy.[4] Improved treatments have led to a greater number of adult survivors, such that there are now more adults than children with a congenital cardiovascular malformation (CVM).[5] Survivors of these defects encounter significant life-long morbidity and frequently have a shortened life span.[6] These cardiovascular diseases that affect children and young adults have a strong genetic contribution and accordingly attracted interest from researchers to identify disease-causing genes.

A significant amount of effort has been spent on identifying the genetic contributors to cardiovascular disease. A number of genome-wide association studies for complex genetic diseases that affect adults such as stroke, hypertension, and atherosclerosis have been performed, with limited success in identifying the underlying genetic susceptibility.[7] In contrast, many classes of cardiovascular disease that affect children and young adults that are due to single gene disorders (cardiomyopathies, arrhythmias) are now well worked out, while those occupying a continuum from single gene through oligogenic to complex genetic etiologies (CVMs, aortopathies) are just now being elucidated. With the knowledge

doi: 10.1111/j.1749-6632.2010.05791.x

of individual causative genes for these disorders exhibiting Mendelian inheritance, much insight has been gained in disease pathophysiology and resulted in novel treatments.

In this review, we will demonstrate the importance of single gene defects in cardiovascular disease, focusing on cardiomyopathies, aortopathies, arrhythmias, and CVMs. Recent advances in clinical diagnostic testing and novel therapies will be highlighted. While cardiovascular diseases in both children and adults have a strong genetic contribution, our review will focus primarily on diseases first presenting in children and young adults where the majority of significant genetic discoveries have occurred.

Cardiomyopathy

In this section, we have focused on hypertrophic cardiomyopathy (HCM) and dilated cardiomyopathy (DCM), as major advances in understanding the genetic basis of these diseases have been made and have impacted clinical care.

Hypertrophic cardiomyopathy

Familial HCM is defined as left ventricular thickening, which occurs in the absence of a syndrome or other cardiovascular conditions that increase afterload (i.e., aortic stenosis or hypertension) and possess the characteristic histology of myocyte disarray.[8] Manifestations include dyspnea, chest pain, syncope, palpitations, or sudden death. Symptoms may present from infancy to old age but are most apparent in adolescence. HCM is among the most common genetic disorders, with a prevalence of 1/500.[9] It remains the most frequent cause of sudden cardiac death (SCD) in young competitive athletes in the United States, accounting for one in three SCDs.[10]

Although first described as a clinical entity in the 1950s,[11] the familial nature of the disease was not elucidated until the 1980s. This was confirmed in 1990 with the description of the first gene implicated in HCM, *MYH7*.[12] Over 20 genes have now been characterized as causing familial HCM (Table 1A), and in the process demonstrated that HCM is a disease of the sarcomere.[13] Four of these genes, *MYH7*, *MYBPC3*, *TNNT2*, and *TNNI3*, account for roughly 80% of all HCM for which a specific gene mutation can be identified.[14] Over 900 different mutations have been found, with many mutations unique to individual families. The most common mutation, *MYBPC3* p.Arg502Trp, occurs in only 2.4% of individuals of European ancestry with HCM.[15]

With this increased knowledge of the genetics of HCM, it was thought that genotype–phenotype correlations could be identified. What was discovered was wide variability in presentation between families with the same mutation and even within a family with the same mutation.[16] Only a few specific genotype–phenotype correlations exist, and only broad generalizations can be made. *MYBPC3* mutations tend to result in a later onset of symptoms.[17] Mutations in *MYH7* tend to have a somewhat earlier age of onset, and the p.Arg403Gln and p.Arg453Cys mutations appear to confer more severe disease.[18] However, these mutations have proven to be very rare (<1%) among those with HCM.[19] *TNNT2* mutations cause less hypertrophy but impart a higher risk of SCD.[20–22] Presence of a mutation in any of the known genes that cause HCM is associated with more severe disease compared to those with HCM in whom a mutation is not identified.[23] Inheritance of mutations in two or more different genes known to cause HCM has been seen in a number of cases, and results in an earlier onset and more severe disease.[14,24–27]

Clinical gene testing for HCM has been available for over 5 years. Despite a reasonable rate of mutation identification (40% for isolated to 70% for familial cases), testing has not entered routine clinical practice. Many reasons exist for this, including high cost and limited utility for the individual patient as testing will have little impact on the prognosis or treatment. Most patients and their families are interested in testing once genetic counseling is provided.[28] The greatest benefit exists for relatives of the patient, as once the mutation in the family is identified, clinically unaffected at-risk relatives may be tested and those without the mutation may safely discontinue recommended clinical screening. While identification of a mutation in an unaffected individual may have negative psychological consequences, studies have demonstrated that the majority of individuals identified as carriers of disease-causing mutations positively value genetic counseling and predictive testing.[29,30]

Dilated cardiomyopathy

DCM is a much more clinically heterogeneous disorder. Individuals with DCM present with

Table 1. Disease genes for cardiomyopathy

Gene	Gene MIM	Chromosome location	Disease
(A) Disease genes for hypertrophic cardiomyopathy (HCM)			
ACTC1	102540	15q14	HCM
CSRP3	600824	11p15.1	HCM
MYBPC3	600958	11p11.2	HCM
MYH6	160710	14q12	HCM
MYH7	160760	14q12	HCM
MYL2	160781	12q23-q24	HCM
MYL3	160790	3p21.3-p21.2	HCM
PRKAG2	602743	7q36	HCM
TNNC1	191040	3p21.3-p14.3	HCM
TNNI3	191044	19q13.4	HCM
TNNT2	191045	1q32	HCM
TPM1	191010	15q22.1	HCM
TTN	188840	2q31	HCM
VCL	193065	10q22.1-q23	HCM
(B) Disease genes for dilated cardiomyopathy (DCM)			
ABCC9	601439	12p12.1	DCM
ACTC	102540	15q14	DCM
ACTN2	102573	1q42-q43	DCM
CSRP3	600824	11p15.1	DCM
DES	125660	2q35	DCM
DSG2	125671	18q12.1-q12.2	DCM
DSP	125647	6p24	DCM
EYA4	603550	6q23-q24	DCM
FCMD	607440	9q31	DCM
LAMP2	309060	Xq24	DCM
LDB3	605906	10q22-q23	DCM ± LVNC
LMNA	150330	1q21.2	DCM ± CD
MYBPC3	600958		DCM
MYH6	160710	14q12	DCM
MYH7	160760	14q12	DCM
NEXN	613121	1p32-p31	DCM
PLN	172405	6q22.1	DCM
PSEN1	104311	14q24.3	DCM
PSEN2	600759	1q31-q42	DCM
RBM20	613171	10q25.2	DCM
SCN5A	600163	3p	DCM
SGCD	601411	5q33	DCM
TCAP	604488	17q12	DCM
TMPO	188380	12q22	DCM
TNN	188840	2q31	DCM
TNNC1	191040	3p21.3-p14.3	DCM
TNNI3	191044	19q13.4	DCM
TNNT2	191045	1q32	DCM
TPM1	191010	15q22.1	DCM
VCL	193065	10q22-q23	DCM
(C) Disease genes for left ventricular noncompaction (LVNC) and arrhythmogenic right ventricular dysplasia (ARVD)			
DTNA	601239	18q12.1	LVNC
TAZ	300394	Xq28	LVNC
DSC2	125645	18q12.1	ARVD
DSG2	125671	18q12.1-q12	ARVD
DSP	125647	6p24	ARVD
JUP	173325	17q21	ARVD
PKP2	602861	12p11	ARVD
RYR2	180902	1q42-q43	ARVD
TGFB3	190230	14q23	ARVD
TMEM43	612048	3p23	ARVD

CD, conduction defect.

symptoms of heart failure, arrhythmias or conduction disturbance, and thromboembolic disease (from an intracardiac thrombus). DCM is defined by left ventricular enlargement and evidence of systolic dysfunction.[31] A wide variety of acquired causes are known for DCM, including ischemic heart disease, valve disease, congenital heart disease, toxins, chemotherapy, infectious myocarditis, radiation, and many others.[32] Additionally, many specific genetic diseases present with DCM as a component of their manifestations, including muscular dystrophies and inborn errors of metabolism (IEM), making idiopathic DCM a diagnosis of exclusion. The incidence of idiopathic DCM for children is estimated at 0.57/100,000 while in adults it is 1/2,500.[33,34] Most experts believe this is an underestimate as individuals with DCM may be asymptomatic for a prolonged time before heart failure becomes evident.

Familial DCM is defined as the occurrence of idiopathic DCM in two or more close relatives. Scattered case reports and small studies in the 1980s suggested DCM was a familial disease in only a small percentage of cases.[35] Full recognition that DCM could be a familial disease did not occur until a prospective echocardiography study on relatives of individuals with DCM in 1992,[36] which reported familial DCM in 20% of cases. It is now estimated that from 30% to 48% of individuals with DCM have an affected relative.[37–39] This genetic knowledge has sparked investigation of seemingly sporadic DCM. Two recent studies demonstrated that half of pregnancy or peripartum DCM cases have a family history of DCM, and mutations can be identified in 5%.[40,41]

In contrast to HCM, the genetic etiology of DCM is varied and involves many structures or functions of the cardiomyocyte, that when perturbed, results in a loss of contractile force generation or transmission.[42] Most genes implicated in familial DCM encode for cytoskeletal (dystrophin-sarcoglycan complex) or sarcomere proteins but also include ion channels and transcription factors. Over 40 loci and more than 25 genes have been described (Table 1B), many of which are implicated in syndromic forms of DCM or also cause HCM. It is likely that many more genes may be involved in DCM, as the currently known genes account for <40% of familial DCM.

The most commonly involved genes in DCM are *LMNA*, *MYH7*, and *TNNT2*, which account for 5%, 4%, and 3% of all familial DCM, respectively, while other implicated genes are found in <1%.[43,44] Combined, mutations in sarcomere genes account for up to 10% of mutations in DCM.[45]

Unlike HCM, the role for genetic testing in DCM is limited due to the extensive genetic heterogeneity and low yield (currently 20%). The yield increases in families with DCM and conduction defects, where mutations in *LMNA* may be found in 40%.[44,46]

Other cardiomyopathies

Left ventricular noncompaction (LVNC) accounts for ~10% of all cardiomyopathies, and is characterized by spongy myocardial appearance with increased trabeculations.[47] Arrythmogenic right ventricular cardiomyopathy/dysplasia (ARVC/D) consists of fibrofatty infiltration and dilation of the right ventricle.[48] Multiple disease loci and genes have been identified for LVNC and ARVC/D (Table 1C). Many inherited disorders have secondary hypertrophic or DCM, including specific syndromes (Noonan and other Ras pathway syndromes), mitochondrial disorders,[49] IEM,[50] and muscular dystrophies[51] and are beyond the scope of this review.

Novel technologies for clinical cardiomyopathy testing

Traditionally, gene testing is performed using Sanger sequencing technology. Analyzing multiple genes can quickly result in expensive test panels. Novel sequencing technologies have been used to lower cost, improve turn-around time, and perform more comprehensive testing by including more genes in the test panel. Oligonucleotide hybridization-based resequencing DNA microarrays use 25 bp oligonucleotides that vary at a single central position.[52] Four probes (A, C, G, and T) are constructed for each basepair in the gene. Fluorescently labeled patient DNA is added to the array, and the dye of maximal intensity identifies the base at the position interrogated. Resequencing arrays are easily scalable and amenable to quickly adding new genes to the test panel. They are currently used for HCM[53,54] and DCM[55] in clinical testing.

Next generation sequencing has just entered the clinical realm.[56] Various strategies have been devised to reduce the cost by sequencing specific targets through capture by microarray[57,58] or fluid methods,[59] rather than the whole genome. "Bait" sequence is used to capture the "target" region of interest in the patient's DNA, which may then be sequenced on a variety of platforms. Clinical testing using this method is already available for HCM and DCM.

Thoracic aortic aneurysms

The thoracic aorta is a complex structure with elastic and contractile properties that allow it to receive blood, transmit the pulse wave, and reduce workload of the ventricle. Thoracic aortic aneurysms (TAAs) occur due to a weakening of the arterial wall, and predispose to dissection. TAAs, particularly of the ascending aorta, represent an entity distinct from abdominal aortic aneurysms.[60] The incidence of TAA is 3.5/100,000 with rates increasing with advancing age, but frequent misdiagnosis suggests this is an underestimate.[61] TAAs may be sporadic, familial, or syndromic.

Marfan syndrome

Marfan syndrome (MFS) is a well-known multisystem connective tissue disorder with a defining feature of aortic root enlargement at the sinuses of Valsalva that progresses to aortic dissection by 20–40 years of age.[62] MFS accounts for only 5% of all TAAs.

It is still diagnosed clinically using family history and the clinical features (Ghent criteria) to identify major involvement of at least two systems (skeletal, ophthalmologic, cardiovascular, dura) and involvement of a third (the preceding and pulmonary, dermatologic).[63] Three groups converged on *FBN1* as the causative gene in 1991.[64–66] Individuals meeting Ghent criteria will have a mutation in *FBN1* in 70–93% of cases.[67] Some genotype–phenotype correlation exists, as missense mutations eliminating or producing a cysteine are more likely to cause ectopia lentis, and mutations in exons 24–32 have a more severe phenotype, including neonatal MFS.[68] Originally thought to be a structural disorder, recent work has elegantly shown that the primary pathophysiology is due to alterations in TGFB signaling.[69] Loss of fibrillin-1 sequestration of TGFB in the extracellular matrix leads to an increase in active TGFB signaling that results in aortic root dilation, lung bullae formation, and impaired muscle regeneration.

Loeys-Dietz syndrome

Following the identification of *FBN1* as the causative gene for MFS, there were observations of individuals with a clinical overlap but without *FBN1* mutations. Several of these families were investigated by linkage analysis, and mutations in *TGFBR1* and *TGFBR2* were found.[70] Individuals with Loeys-Dietz syndrome have aortopathy (arterial tortuosity, aneurysms, and dissections), skeletal abnormalities (pectus deformities, arachnodactyly, talipes equinovarus), translucent skin, and craniofacial abnormalities (bifid uvula, ocular hypertelorism). Compared to MFS, aneurysms may occur in any medium-sized artery, and the arterial disease is much more aggressive with earlier dissection and age of death.[71] Mutations in either gene cause a similar phenotype.

Ehlers-Danlos syndromes

The Ehlers-Danlos syndromes (EDS) are a collection of connective tissue disorders characterized by variably severe joint hypermobility and increased skin elasticity.[72] With the exception of the kyphoscoliotic type, all are autosomal dominant in inheritance. EDS, vascular type (formerly EDS IV), is caused by mutations in *COL3A1*. It is characterized by joint hypermobility, increased skin translucency, and rupture of medium-sized arteries, bowel, and uterus. Although TAAs may occur, it is not the most common vessel involved. EDS, kyphoscoliotic form (formerly type VI), is due to mutations in *PLOD1*. It also may have arterial dissection of the medium-sized vessels or aorta. It is distinguished by severe neonatal hypotonia, early and progressive scoliosis, and rupture of the globe. EDS, classic type (types I and II) are due to mutations in *COL5A1* or *COL5A2*, and EDS, hypermobility type (type III or benign familial hypermobility) is likely genetically heterogeneous. Echocardiography observations indicate mild aortic root dilation may be present in up to 1/3 of individuals with the classic or hypermobility types.[73,74] In the absence of severe dilation, there does not appear to be a significant risk of aortic dissection; however there are currently no clinical guidelines on long-term management. Longitudinal studies are required to determine if the observed mild dilation is progressive.

Familial TAA

The genetic contribution to nonsyndromic TAA has only recently been defined. The first familial case of nonsyndromic TAA was described 25 years ago,[75] but not until the late 1990s was the inheritance and segregation studied in detail.[76–78] Family studies indicate roughly 20% of individuals with a TAA have a similarly affected relative, and in those families, individuals with TAA presented at an earlier age.[77,79] TAA families also have higher rates of aneurysms anywhere in the arterial tree. Aortic dilation is common among individuals with a bicuspid aortic valve (BAV),[80,81] with multiple reports now demonstrating cosegregation of TAA and BAV in families.[82] Families with TAA may be divided into two phenotypic groups by location, using the ligamentum arteriosum. TAA above the ligamentum is highly familial and not associated with atherosclerosis, while TAA below is associated with atherosclerosis.[60] This likely reflects the embryonic origin of the smooth muscle cells in the aorta, as cardiac neural crest cells only contribute to the aortic arch above the ligamentum arteriosum.[83]

To date, a total of seven loci have been described and four genes identified for familial TAA (Table 2).[84–87] The four identified genes appear to explain roughly 20% of all familial TAA. In addition, families with BAV and aortic dilation or TAA segregating with *NOTCH1* mutations have been identified.[88] Little genotype–phenotype correlation exists, but individuals with mutations in *ACTA2* may also be predisposed to stroke and exhibit moya–moya phenomenon on cerebral angiography.[89]

Table 2. Disease genes for thoracic aortic aneurysms

Gene	Gene MIM	Chromosome location	Disease
ACTA2	102620	10q22	TAA
MYH11	160745	16p13	TAA & PDA
TGFBR1	190181	9q22	TAA
TGFBR2	190182	3p22	TAA

PDA, patent ductus arteriosus.

Novel therapies for TAA

Much excitement has been generated by the demonstration of nearly complete cessation of aortic root dilation in a mouse model of MFS with the use of losartan.[90] Losartan is an angiotensin II receptor blocker (ARB), but as demonstrated in the mouse model, it also acts to suppress TGFB signaling. Preliminary results on the use of losartan in patients with MFS appear promising. A small group of pediatric patients who were switched to losartan after treatment failure with beta-blockers demonstrated significant slowing of aortic root growth.[91] Several trials are in progress comparing standard beta-blocker therapy to losartan in the United States and Europe.[92–94] The role ARBs might play in other syndromic, familial or sporadic TAA is unknown, as is the role of the TGFB pathway in these other diseases. The greatest promise may exist for LDS, which is also due to TGFB pathway dysregulation.

Ventricular arrhythmias

Diseases of the cardiac conduction system are very prevalent in the world and affect an estimated millions of individuals in the United States alone.[95] The majority of these diseases are associated with adult cardiovascular disease, but the elucidation of the genetic causes of inherited arrhythmias, including long QT syndrome (LQTS), Brugada syndrome, and catecholaminergic polymorphic ventricular tachycardia (CPVT), which exhibit Mendelian inheritance patterns, have resulted in an increased understanding of conduction system disease. While conduction system disease encompasses a spectrum of arrhythmias, those of ventricular origin have a propensity to be lethal and present in SCD. Here, we highlight the discovery of the genetic etiologies of LQTS, Brugada syndrome, and CPVT and how these findings have affected clinical management and care.

Long QT syndrome

LQTS is an inherited cardiac disease, which presents with syncope and SCD, due to the ventricular tachyarrythmias (VT), torsade de pointes. It is characterized by prolongation of the QT interval on electrocardiogram. The diagnosis is made using clinical criteria[96] and the disease has an estimated prevalence of 1 in 2,000–5,000 individuals.[97,98] Four clinical subtypes of LQTS have been defined and include the more common Romano–Ward syndrome,[99,100] Jervell–Nielsen syndrome (with congenital deafness),[101] Andersen syndrome (with periodic paralysis and malformations),[102,103] and Timothy syndrome (with cardiac and other birth defects, autism).[104]

The etiology for LQTS had remained unknown until 1995, when two simultaneous publications identified mutations in *HERG* (LQTS1) and *SCN5A* (LQTS3) through linkage analysis of large kindreds.[105,106] These discoveries led to the concept of ion channelopathies as the cause of inherited arrhythmias. Since then, 12 LQTS subtypes have been identified (Table 3A), with the first three subtypes comprising the majority of clinical cases.[107,108] For these three most common subtypes (LQTS1, LQTS2, and LQTS3), numerous features that tend to occur with each subtype have been identified. Each subtype has unique triggers: exercise for LQTS1, startling due to noises for LQTS2, and episodes of bradycardia (i.e., sleep) for LQTS3.[109] In addition, each of these three subtypes has distinct electrocardiographic findings when examining T wave morphology and are often used to direct genetic testing.[110] The subtypes also have variable responses in regards to drug therapy, as LQTS1 is more responsive to beta-blocker therapy when compared to LQTS3 where mexiletene (a sodium channel blocker) is often used, and have different prognostic implications with LQTS3 being highly lethal as compared to LQTS1.[111] This increased knowledge has led to the use of commercially available genetic testing, and allowed for screening of family members and improved clinical management. Although much knowledge has been gained, clear genotype–phenotype associations regarding SCD risk prediction and treatment efficacy remain elusive.

Table 3. Disease genes for cardiac conduction system diseases

Type	Gene	Gene MIM
(A) Disease genes for long QT syndrome		
LQT1	KCNQ1	607542
LQT2	KCNH2	152427
LQT3	SCN5A	600163
LQT4	ANK2	106410
LQT5	KCNE1	176261
LQT6	KCNE2	603796
LQT7	KCNJ2	600681
LQT8	CACNA1C	114205
LQT9	CAV3	601253
LQT10	SCN4B	608256
LQT11	AKAP9	604001
LQT12	SNTA1	601017
BrS type	**Gene name**	**Gene MIM**
(B) Disease genes for Brugada syndrome		
BrS1	SCN5A	600163
BrS2	GPD1L	611778
BrS3	CACNA1C	114205
BrS4	CACNB2	600003
BrS5	SCN1B	600235
BrS6	KCNE3	604433
BrS7	SCN3B	608214
Type	**Gene**	**Gene MIM**
(C) Disease genes for catecholaminergic polymorphic ventricular tachycardia		
Autosomal dominant	RYR2	180902
Autosomal recessive	CASQ2	114251

Brugada syndrome

With the discovery of ion channel mutations as a cause of VT, investigators began investigating other familial cases of idiopathic VT. The best described of these is Brugada syndrome, which is characterized by ST-segment elevation in the right precordial leads (V1-V3) on electrocardiogram and an increased risk of SCD due to polymorphic VT.[112,113] The disease was known to have an autosomal dominant pattern of inheritance with incomplete penetrance, but the genetic cause was unclear until 1998. Using a candidate gene approach and studying several well-phenotyped small families with autosomal dominant disease, investigators identified mutations in *SCN5A* as the cause of Brugada syndrome.[114] Since then additional molecular genetic studies have identified seven molecular subtypes of Brugada syndrome but *SCN5A* mutations account for the majority of cases (Table 3B).[115] Similar to LQTS, these discoveries have resulted in commercially available genetic testing and improved genetic counseling but novel therapeutics and treatments are still under development.

Catecholaminergic polymorphic ventricular tachycardia

CPVT is a relatively rare but important cause of SCD in young children.[116] The disease, which is characterized by adrenergically mediated VT has been described for over three decades.[117] Individuals typically develop symptoms of syncope, VT, SCD with exercise and suffer from significant mortality prior to 40 years of age.[118] It is a genetically inherited condition and presents in both an autosomal dominant and recessive condition (Table 3C). The autosomal dominant form was mapped to a region on chromosome 1q42–43[119] and subsequently mutations in the human cardiac ryanodine receptor (*RYR2*) were identified as the genetic etiology.[120,121] The autosomal recessive form had been mapped to a region on chromosome 1p13-p21 and the identification of *Ryr2* as the cause of the autosomal dominant form allowed investigators to focus on other genes critical for calcium handling on the sarcoplasmic reticulum. Calsequestrin 2 (*CASQ2*) was located in this 8 Mb interval and mutations in this gene were found to segregate with the disease in families. These genetic discoveries led to a greater understanding of the biochemical effects of the mutations to cause sarcoplasmic reticulum leak of calcium[122,123] and the development of mouse models for CPVT.[124,125] These molecular insights and investigations have allowed for the recent discovery that flecainide, an FDA-approved antiarrhythmic drug, was efficacious in suppressing VT in mouse models of CPVT and humans with CPVT.[126] Larger clinical trials are currently being planned to determine if this is a novel therapy for this highly lethal disease.

These seminal discoveries have increased our understanding of SCD and opened a field of channelopathies. While these genetic discoveries have impacted patient care, a significant amount of research still needs to be done. For example, mutations in *SC5NA* cause both LQTS and Brugada syndrome and the reason for the phenotypic differences remain unclear. Of note, these discoveries have allowed us to expand into conduction system disease not related to familial cases of SCD. For example, genetic polymorphisms in the LQTS genes likely result in drug-induced long QT[127] and a nucleotide variation in *SC5NA* results in increased arrhythmias in African American population.[128] These genetic discoveries have also assisted in the identification of the cause of SCD as approximately 25–35% of sudden unexplained deaths in children and young adults have no autopsy findings. It is estimated that up to 10% of sudden infant death syndrome (SIDS) cases can be attributable to a mutation in an LQTS or CPVT susceptibility gene.[129] An increased understanding of these molecular pathways regulating ion channels will help clarify the disease phenotype differences.

Cardiovascular malformations

CVMs constitute the most common group of birth defects, accounting for ~25% of all birth defects. The birth prevalence for all CVMs is 6–8 per 1,000 live births, with rates increasing with advances in imaging technology that led to discovery of milder defects.[130–132] Improved medical treatment and surgical palliation has led to a burgeoning population of adult survivors of CVMs, but deaths from CVMs still account for 40% of all deaths from birth defects.

Both birth defect registry studies and population-based case-control studies have been performed for CVMs to define the epidemiology and identify specific etiologies. The Baltimore Washington Infant Study (BWIS) was instrumental in defining population-based risk factors for CVMs classified into specific mechanistic groups.[133] This study identified many important features of CVMs at a population level, such as sex and race differences for specific groups of defects, recurrence of specific CVMs within families, and the increased risk for CVMs in maternal diabetes. While a number of studies, including the BWIS, have searched for environmental factors at the population level, the resulting data has been less conclusive.[134] Only a limited number of prenatal exposures of the fetus are known environmental risk factors and they include alcohol, retinoic acid, and anticonvulsants, and maternal diseases such as rubella, influenza, diabetes, and phenylketonuria.

Multiple studies show CVMs possess high heritability and increased recurrence risk above the general population suggesting a role for genetic contributors.[135] Identified genetic causes include chromosomal microdeletions or microduplications (22q11), and aneuploidy (trisomies 13, 18, and 21, and monosomy X) (Table 4) and were uncovered by standard cytogenetic methodology. Although for the vast majority of cases, the underlying cause is unknown, it is clear that the complex process of heart development is guided by a combination of a carefully orchestrated genetic program and hemodynamic forces. This intricate process of cardiac morphogenesis with its multiple cellular contributors is controlled by a set of highly conserved molecular pathways. Studies using diverse species, from flies to mice, have led to the identification of many genes critical for normal cardiac development.[136,137] This knowledge of cardiac developmental biology has assisted investigators in the elucidation of the genetic contributors of human CVM. In this section, we discuss the genetic causes of syndromic and nonsyndromic CVM that were discovered using Mendelian cases of disease and highlight how the knowledge of cardiac development genes assisted in gene discovery.

Single gene defects associated with syndromic and nonsyndromic CVM

With advances in genetic technology and the completion of the Human Genome Project, single gene defects leading to syndromes associated with congenital heart disease have been elucidated and they are summarized in Table 5A. Some of the earliest work was the discovery that mutation of *Fibrillin 1* (*FBN1*) was the cause of MFS (discussed earlier). Since then the genetic basis of numerous syndromes have been identified and each is characterized by a unique constellation of birth defects. One of the first was Holt-Oram syndrome, which is characterized by atrial and ventricular septal defects, progressive atrioventricular conduction system disease, and radial limb and thumb anomalies. Classic positional approaches were utilized to first identify the genetic locus on chromosome 12q2[138] and subsequently, disease-causing mutations were discovered in the transcription factor, TBX5.[139] Another common syndrome, which frequently exhibits CVMs, is Noonan syndrome. The phenotype consists of cardiac defects, typically pulmonary valve stenosis and

Table 4. Common syndromes resulting from anueploidy and microdeletions

Syndrome	Cardiac anomalies	% with CHD
Trisomy 13	ASD, VSD, PDA, HLHS	80
Trisomy 18	ASD, VSD, PDA, TOF, DORV, CoA, BAV	90–100
Trisomy 21 (Down syndrome)	ASD, VSD, AVSD, TOF	40–50
Monosomy X (Turner syndrome)	CoA, BAV, AS, HLHS	25–35
1p36 deletion	VSD, AVS, TOF, CoA, LVNC	70
7q11.23 deletion (Williams-Beuren syndrome)	Supravalvar AS, PPS	50–85
11q23 deletion (Jacobsen syndrome)	VSD, AVS, CoA, HLHS	50–60
22q11.2 deletion (DiGeorge Syndrome)	IAA Type B, aortic arch anomalies, truncus arteriosus, TOF	75

CHD, congenital heart defect; ASD, atrial septal defect; VSD, ventricular septal defect; PDA, patent ductus arteriosus; HLHS, hypoplastic left heart syndrome; TOF, tetralogy of Fallot; DORV, double outlet right ventricle; CoA, coarctation of aorta; BAV, bicuspid aortic valve; AVSD, atrioventricular septal defect; IAA, interrupted aortic arch; AS, aortic stenosis; PPS, peripheral pulmonic stenosis; LVNC, left ventricular noncompaction.

HCM, as well as cognitive disability, characteristic facies, and bleeding disorders. Using small families with autosomal dominantly inherited disease, investigators used a candidate gene-based approach to identify mutations in *PTPN11*, a gene involved in Ras signaling and implicated by murine studies to be critical for semilunar valvulogenesis, to be the cause of 50% of cases.[140,141] Subsequent studies have found that mutations of other genes involved in the Ras signaling pathway including *RAF1*, *SOS1*, and *KRAS* were also associated with a similar spectrum of disease. In addition, LEOPARD and Costello syndromes, which exhibit a similar phenotype as Noonan syndrome, are the result of mutations in Ras signaling pathway members.[142,143] Alagille syndrome is characterized by intrahepatic bile duct paucity and CVMs, including peripheral pulmonic stenosis, pulmonary valve stenosis, and tetralogy of Fallot. Disease-segregating mutations in *JAG1*, a gene encoding a ligand in the Notch signaling pathway, were identified by studying several small families with autosomal dominant disease.[144,145] Consistent with this, mutations in a NOTCH receptor, *NOTCH2*, have also been identified in subjects with Alagille syndrome using a candidate gene approach.[146] A rare genetic syndrome that is characterized by dysmorphic facies and digit anomalies along with congenital heart disease (specifically patent ductus arteriosus) was found to be caused by mutations in the transcription factor, *TFAP2b* using traditional approaches after the identification of large kindreds.[147] Finally, heterotaxy syndrome, which is randomization of cardiac, pulmonary, and gastrointestinal situs, is frequently associated with CVMs, specifically atrioventricular septal defects, and transposed great arteries. Disease-causing mutations in *ZIC3* were identified in an X-linked form of heterotaxy.[148] After this discovery, *ZIC3* was shown to be critical in left-right development *in vivo*[149] and subsequently mutations in *CFC1*, *ACVR2B*, and *LEFTYA*, genes that regulate left-right asymmetry in the developing embryo were found to be associated with heterotaxy.[150]

Over the past decade, single gene defects associated with isolated or nonsyndromic congenital heart disease have been discovered (Table 5B). The first gene identified to be associated with nonsyndromic CVM was *NKX2.5*, which was identified by linkage analysis in families with autosomal dominant inherited atrial septal defects with atrioventricular conduction delay.[151] Mutations in *GATA4*, a zinc finger transcription factor known to interact with *NKX2.5*, have been linked to isolated atrial septal defects without conduction system abnormalities using large multigeneration families. Interestingly, a mutation in Gata4 specifically disrupted an interaction with *TBX5*[152] and mutations in a common downstream target of GATA4 and TBX5, myosin heavy chain 6 (*MYH6*), have been identified as another cause of atrial septal defects.[153] More recently, investigators have linked another *GATA* family member, *GATA6*, to truncus arteriosus using a candidate gene approach.[154] Similarly another T box family member, *TBX20*, was

Table 5. Disease genes for cardiovascular malformations

Syndrome	Cardiac anomalies	Causative gene(s)	Gene MIM
(A) Disease genes for syndromic cardiovascular malformations			
Alagille syndrome	PS, TOF, ASD, peripheral pulmonary stenosis	JAG1, NOTCH2	601920, 600275
Char syndrome	PDA	*TFAP2β*	601601
CHARGE syndrome	ASD, VSD, valve defects	*CHD7, SEMA3E*	608892, 608116
Costello syndrome	PS, HCM, cardiac conduction abnormalities	HRAS	190020
Ellis van Creveld	ASD	*EVC, EVC2*	604831, 607261
Heterotaxy syndrome	DILV, DORV, d-TGA, AVSD	*ZIC3, CFC1*	300265, 605194
Holt-Oram syndrome	ASD, VSD, AVSD, progressive AV conduction system disease	*TBX5*	601620
LEOPARD syndrome	PS and cardiac conduction abnormalities	PTPN11, RAF1	176876, 164760
Noonan syndrome	PS with dysplastic pulmonary valve, AVSD, HCM, CoA	PTPN11, KRAS, RAF1, SOS1	176876, 190070, 164760, 182530
Rubinstein Taybi	ASD, VSD	*CREBBP, EP300*	600140, 602700
Smith Lemli Opitz	VSD, ASD, AVSD	*DHCR7*	602858

Cardiac anomalies	Gene	Gene MIM
(B) Disease genes for nonsyndromic cardiovascular malformations		
ASD, VSD	*GATA4*	600576
Truncus arteriosus	*GATA6*	601656
ASD, hypertrophic cardiomyopathy	*MYH6*	160710
ASD, atrioventricular conduction delay, TOF, tricuspid valve abnormalities	*NKX2.5*	600584
BAV, early valve calcification	*NOTCH1*	190198
ASD, VSD, cardiomyopathy, cardiac valve anomalies	*TBX20*	606061

PS, pulmonic valve stenosis; AVSD, atrioventricular septal defect; HCM, hypertrophic cardiomyopathy; CoA, coarctation of aorta; TOF, tetralogy of Fallot; ASD, atrial septal defect; VSD, ventricular septal defect; AV, atrioventricular; DILV, double inlet left ventricle; DORV, double outlet right ventricle; TGA, transposition of the great arteries; PDA, patent ductus arteriosus; BAV, bicuspid aortic valve; PHTN, pulmonary hypertension.

identified to cause CVM, predominantly cardiac septal and valve defects, in several small families with autosomal dominant disease.[155] Additionally, mutations in *NOTCH1* have been identified as a cause of aortic valve malformations, including bicuspid aortic valve and early aortic valve calcification, via genome-wide linkage analysis of an affected family.[156] Subsequent studies have expanded the phenotype to a spectrum of malformations that affect left-sided cardiac structures.[157] These discoveries that syndromic and nonsyndromic CVM are caused by single gene defects were facilitated by studies in families with Mendelian-inherited disease and even in those cases where candidate gene approaches were utilized, the availability of small pedigrees allowed for increased confidence that disease-segregating mutations were truly causal. Ultimately, these initial discoveries will reveal more about the molecular pathways that are critical for cardiac morphogenesis and relevant to human CVMs.

Unlike for the other cardiovascular diseases described earlier, the translation of the genetic discoveries for CVMs to the clinic is still in its early stages. The knowledge of genetic etiologies for syndromic cases has been utilized for genetic testing in only a subset of cases that are difficult to define. Often, determining the diagnosis in syndromic cases has significant prognostic implications, which are useful for the clinician especially in regards to neurodevelopmental outcomes. At this time, the genetic knowledge that has been gained from the discovery of genetic causes of nonsyndromic CVM has focused on development of transgenic mouse models, which have allowed for an increased understanding of the molecular pathways regulating heart development. The information has been less useful clinically as the identified genetic causes have ultimately been found to represent only a small percentage of cases of CVM[158–161] and no clear genotype–phenotype correlations have been identified. At this time, the etiology for most isolated cases of nonsyndromic CVM remains unknown, and it is likely that rare functional sequence variations in genes critical for cardiac morphogenesis are present in affected individuals. Future studies utilizing new diagnostic methods, such as next-generation sequencing, may ultimately provide a more accurate genetic predisposition risk for CVM. Ultimately, the discovery of CVM-causing genes and associated animal models may allow us to test environmental influences on genetic predispositions and possibly devise novel preventive measures that reduce the incidence of congenital heart defects.

Summary

From chromosomal aneuploidy to single gene defects causing cardiovascular disease, genetics has a much greater influence on the development of various types of heart disease than previously appreciated.[135] This review summarizes many genetic causes for cardiovascular diseases, which were identified by positional cloning approaches utilizing families with Mendelian-inherited forms of disease. Several new methodologies hold great promise to identify many more disease-associated genetics variants. We are just beginning to understand the significance of copy number variations in cardiac disease.[162–164] As the cost of DNA sequencing decreases, sequencing of entire genomes will lead to an increased rate of gene discovery in cardiovascular disease likely utilizing even smaller families with Mendelian disease.[56,165] Armed with these new cardiac disease-causing genes and the development of international collaborative databases with phenotypic information will allow for the study of clinical outcomes and prognosis based on genetic predispositions. This new genetic knowledge of cardiovascular diseases will lead to improved genetic counseling and the development of novel therapeutics to benefit future generations.

Addendum

At the time of press, revised Ghent criteria for Marfan syndrome were published to improve diagnosis and reduce misdiagnosis (Loeys *et al. J. Med Genet*. 2010 **47:** 476–485.

Acknowledgments

The authors would like to thank S. Koenig for assistance in preparation of the tables. V.G. is supported by grants from the NIH/NHLBI (R01HL088965) and the Children's Heart Foundation. K.L.M. is supported by NIH/NHLBI (R01HL090506) and The Research Institute at Nationwide Children's Hospital.

Conflicts of Interest

The authors declare no conflicts of interest.

References

1. Heron, M. *et al.* 2010. Annual summary of vital statistics: 2007. *Pediatrics* **125:** 4–15.
2. Lloyd-Jones, D. *et al.* 2010. Executive summary: heart disease and stroke statistics–2010 update: a report from the American Heart Association. *Circulation* **121:** 948–954.
3. Gaziano, T.A. *et al.* 2010. Growing epidemic of coronary heart disease in low- and middle-income countries. *Curr. Probl. Cardiol.* **35:** 72–115.
4. Centers for Disease Control and Prevention. 2007. Hospital stays, hospital charges, and in-hospital deaths among infants with selected birth defects–United States, 2003. *MMWR Morb. Mortal. Wkly. Rep.* **56:** 25–29.
5. Marelli, A.J. *et al.* 2007. Congenital heart disease in the general population: changing prevalence and age distribution. *Circulation* **115:** 163–172.
6. Pillutla, P., K.D. Shetty & E. Foster. 2009. Mortality associated with adult congenital heart disease: trends in the US population from 1979 to 2005. *Am. Heart J.* **158:** 874–879.
7. Arking, D.E. & A. Chakravarti. 2009. Understanding cardiovascular disease through the lens of genome-wide association studies. *Trends Genet.* **25:** 387–394.

8. Maron, B.J. *et al.* 2003. American College of Cardiology/European Society of Cardiology clinical expert consensus document on hypertrophic cardiomyopathy. A report of the American College of Cardiology Foundation Task Force on Clinical Expert Consensus Documents and the European Society of Cardiology Committee for Practice Guidelines. *J. Am. Coll. Cardiol.* **42:** 1687–1713.
9. Maron, B.J. *et al.* 1995. Prevalence of hypertrophic cardiomyopathy in a general population of young adults. Echocardiographic analysis of 4111 subjects in the CARDIA Study. Coronary Artery Risk Development in (Young) Adults. *Circulation* **92:** 785–789.
10. Maron, B.J. *et al.* 2009. Sudden deaths in young competitive athletes: analysis of 1866 deaths in the United States, 1980–2006. *Circulation* **119:** 1085–1092.
11. Teare, D. 1958. Asymmetrical hypertrophy of the heart in young adults. *Br. Heart J.* **20:** 1–8.
12. Geisterfer-Lowrance, A.A. *et al.* 1990. A molecular basis for familial hypertrophic cardiomyopathy: a beta cardiac myosin heavy chain gene missense mutation. *Cell* **62:** 999–1006.
13. Bowles, N.E., K.R. Bowles & J.A. Towbin. 2000. The "final common pathway" hypothesis and inherited cardiovascular disease. The role of cytoskeletal proteins in dilated cardiomyopathy. *Herz* **25:** 168–175.
14. Morita, H. *et al.* 2008. Shared genetic causes of cardiac hypertrophy in children and adults. *N. Engl. J. Med.* **358:** 1899–1908.
15. Saltzman, A.J. *et al.* 2010. Short communication: the cardiac myosin binding protein C Arg502Trp mutation: a common cause of hypertrophic cardiomyopathy. *Circ. Res.* **106:** 1549–1552.
16. Bos, J.M., J.A. Towbin & M.J. Ackerman. 2009. Diagnostic, prognostic, and therapeutic implications of genetic testing for hypertrophic cardiomyopathy. *J. Am. Coll. Cardiol.* **54:** 201–211.
17. Niimura, H. *et al.* 2002. Sarcomere protein gene mutations in hypertrophic cardiomyopathy of the elderly. *Circulation* **105:** 446–451.
18. Watkins, H. *et al.* 1992. Characteristics and prognostic implications of myosin missense mutations in familial hypertrophic cardiomyopathy. *N. Engl. J. Med.* **326:** 1108–1114.
19. Ackerman, M.J. *et al.* 2002. Prevalence and age-dependence of malignant mutations in the beta-myosin heavy chain and troponin T genes in hypertrophic cardiomyopathy: a comprehensive outpatient perspective. *J. Am. Coll. Cardiol.* **39:** 2042–2048.
20. Moolman, J.C. *et al.* 1997. Sudden death due to troponin T mutations. *J. Am. Coll. Cardiol.* **29:** 549–555.
21. Varnava, A.M. *et al.* 2001. Hypertrophic cardiomyopathy: histopathological features of sudden death in cardiac troponin T disease. *Circulation* **104:** 1380–1384.
22. Watkins, H. *et al.* 1995. Mutations in the genes for cardiac troponin T and alpha-tropomyosin in hypertrophic cardiomyopathy. *N. Engl. J. Med.* **332:** 1058–1064.
23. Olivotto, I. *et al.* 2008. Myofilament protein gene mutation screening and outcome of patients with hypertrophic cardiomyopathy. *Mayo Clin. Proc.* **83:** 630–638.
24. Ingles, J. *et al.* 2005. Compound and double mutations in patients with hypertrophic cardiomyopathy: implications for genetic testing and counselling. *J. Med. Genet.* **42:** e59.
25. Girolami, F. *et al.* 2010. Clinical features and outcome of hypertrophic cardiomyopathy associated with triple sarcomere protein gene mutations. *J. Am. Coll. Cardiol.* **55:** 1444–1453.
26. Richard, P. *et al.* 2003. Hypertrophic cardiomyopathy: distribution of disease genes, spectrum of mutations, and implications for a molecular diagnosis strategy. *Circulation* **107:** 2227–2232.
27. Van Driest, S.L. *et al.* 2004. Myosin binding protein C mutations and compound heterozygosity in hypertrophic cardiomyopathy. *J. Am. Coll. Cardiol.* **44:** 1903–1910.
28. Fitzgerald-Butt, S.M. *et al.* 2009. Parental knowledge and attitudes toward hypertrophic cardiomyopathy genetic testing. *Pediatr. Cardiol.* **31:** 195–202.
29. Christiaans, I. *et al.* 2009. Quality of life and psychological distress in hypertrophic cardiomyopathy mutation carriers: a cross-sectional cohort study. *Am. J. Med. Genet. A.* **149A:** 602–612.
30. Ingles, J. *et al.* 2008. Psychosocial impact of specialized cardiac genetic clinics for hypertrophic cardiomyopathy. *Genet. Med.* **10:** 117–120.
31. Mestroni, L. *et al.* 1999. Guidelines for the study of familial dilated cardiomyopathies. Collaborative Research Group of the European Human and Capital Mobility Project on Familial Dilated Cardiomyopathy. *Eur. Heart J.* **20:** 93–102.
32. Jefferies, J.L. & J.A. Towbin. 2010. Dilated cardiomyopathy. *Lancet* **375:** 752–762.
33. Codd, M.B. *et al.* 1989. Epidemiology of idiopathic dilated and hypertrophic cardiomyopathy. A population-based study in Olmsted County, Minnesota, 1975–1984. *Circulation* **80:** 564–572.
34. Towbin, J.A. *et al.* 2006. Incidence, causes, and outcomes of dilated cardiomyopathy in children. *JAMA* **296:** 1867–1876.
35. Burkett, E.L. & R.E. Hershberger. 2005. Clinical and genetic issues in familial dilated cardiomyopathy. *J. Am. Coll. Cardiol.* **45:** 969–981.
36. Michels, V.V. *et al.* 1992. The frequency of familial dilated cardiomyopathy in a series of patients with idiopathic dilated cardiomyopathy. *N. Engl. J. Med.* **326:** 77–82.
37. Baig, M.K. *et al.* 1998. Familial dilated cardiomyopathy: cardiac abnormalities are common in asymptomatic relatives and may represent early disease. *J. Am. Coll. Cardiol.* **31:** 195–201.
38. Grunig, E. *et al.* 1998. Frequency and phenotypes of familial dilated cardiomyopathy. *J. Am. Coll. Cardiol.* **31:** 186–194.
39. Keeling, P.J. *et al.* 1995. Familial dilated cardiomyopathy in the United Kingdom. *Br. Heart J.* **73:** 417–421.
40. Morales, A. *et al.* 2010. Rare variant mutations in pregnancy-associated or peripartum cardiomyopathy. *Circulation* **121:** 2176–2182.
41. van Spaendonck-Zwarts, K.Y. *et al.* 2010. Peripartum cardiomyopathy as a part of familial dilated cardiomyopathy. *Circulation* **121:** 2169–2175.
42. Towbin, J.A. & N.E. Bowles. 2002. The failing heart. *Nature* **415:** 227–233.

43. Hershberger, R.E. *et al.* 2008. Coding sequence mutations identified in MYH7, TNNT2, SCN5A, CSRP3, LBD3, and TCAP from 313 patients with familial or idiopathic dilated cardiomyopathy. *Clin. Transl. Sci.* **1:** 21–26.
44. Parks, S.B. *et al.* 2008. Lamin A/C mutation analysis in a cohort of 324 unrelated patients with idiopathic or familial dilated cardiomyopathy. *Am. Heart J.* **156:** 161–169.
45. Kamisago, M. *et al.* 2000. Mutations in sarcomere protein genes as a cause of dilated cardiomyopathy. *N. Engl. J. Med.* **343:** 1688–1696.
46. van Tintelen, J.P. *et al.* 2007. High yield of LMNA mutations in patients with dilated cardiomyopathy and/or conduction disease referred to cardiogenetics outpatient clinics. *Am. Heart J.* **154:** 1130–1139.
47. Maron, B.J. *et al.* 2006. Contemporary definitions and classification of the cardiomyopathies: an American Heart Association Scientific Statement from the Council on Clinical Cardiology, Heart Failure and Transplantation Committee; Quality of Care and Outcomes Research and Functional Genomics and Translational Biology Interdisciplinary Working Groups; and Council on Epidemiology and Prevention. *Circulation* **113:** 1807–1816.
48. Basso, C. *et al.* 2009. Arrhythmogenic right ventricular cardiomyopathy. *Lancet* **373:** 1289–1300.
49. Scaglia, F. *et al.* 2004. Clinical spectrum, morbidity, and mortality in 113 pediatric patients with mitochondrial disease. *Pediatrics* **114:** 925–931.
50. Wilcken, D.E. 2003. Overview of inherited metabolic disorders causing cardiovascular disease. *J. Inherit. Metab. Dis.* **26:** 245–257.
51. Finsterer, J. & C. Stollberger. 2008. Primary myopathies and the heart. *Scand. Cardiovasc. J.* **42:** 9–24.
52. Beaudet, A.L. & J.W. Belmont. 2008. Array-based DNA diagnostics: let the revolution begin. *Annu. Rev. Med.* **59:** 113–129.
53. Fokstuen, S. *et al.* 2008. A DNA resequencing array for pathogenic mutation detection in hypertrophic cardiomyopathy. *Hum. Mutat.* **29:** 879–885.
54. Waldmuller, S. *et al.* 2008. Array-based resequencing assay for mutations causing hypertrophic cardiomyopathy. *Clin. Chem.* **54:** 682–687.
55. Zimmerman, R.S. *et al.* 2010. A novel custom resequencing array for dilated cardiomyopathy. *Genet. Med.* **12:** 268–278.
56. Lupski, J.R. *et al.* 2010. Whole-genome sequencing in a patient with Charcot-Marie-Tooth neuropathy. *N. Engl. J. Med.* **362:** 1181–1191.
57. Okou, D.T. *et al.* 2007. Microarray-based genomic selection for high-throughput resequencing. *Nat. Methods* **4:** 907–909.
58. Albert, T.J. *et al.* 2007. Direct selection of human genomic loci by microarray hybridization. *Nat. Methods* **4:** 903–905.
59. Gnirke, A. *et al.* 2009. Solution hybrid selection with ultra-long oligonucleotides for massively parallel targeted sequencing. *Nat. Biotechnol.* **27:** 182–189.
60. Elefteriades, J.A. & E.A. Farkas. 2010. Thoracic aortic aneurysm clinically pertinent controversies and uncertainties. *J. Am. Coll. Cardiol.* **55:** 841–857.
61. Clouse, W.D. *et al.* 2004. Acute aortic dissection: population-based incidence compared with degenerative aortic aneurysm rupture. *Mayo Clin. Proc.* **79:** 176–180.
62. Keane, M.G. & R.E. Pyeritz. 2008. Medical management of Marfan syndrome. *Circulation* **117:** 2802–2813.
63. De Paepe, A. *et al.* 1996. Revised diagnostic criteria for the Marfan syndrome. *Am J. Med. Genet.* **62:** 417–426.
64. Dietz, H.C. *et al.* 1991. Marfan syndrome caused by a recurrent de novo missense mutation in the fibrillin gene. *Nature* **352:** 337–339.
65. Maslen, C.L. *et al.* 1991. Partial sequence of a candidate gene for the Marfan syndrome. *Nature* **352:** 334–337.
66. Lee, B. *et al.* 1991. Linkage of Marfan syndrome and a phenotypically related disorder to two different fibrillin genes. *Nature* **352:** 330–334.
67. Stheneur, C. *et al.* 2009. Identification of the minimal combination of clinical features in probands for efficient mutation detection in the FBN1 gene. *Eur. J. Hum. Genet.* **17:** 1121–1128.
68. Faivre, L. *et al.* 2007. Effect of mutation type and location on clinical outcome in 1,013 probands with Marfan syndrome or related phenotypes and FBN1 mutations: an international study. *Am. J. Hum. Genet.* **81:** 454–466.
69. Neptune, E.R. *et al.* 2003. Dysregulation of TGF-beta activation contributes to pathogenesis in Marfan syndrome. *Nat. Genet.* **33:** 407–411.
70. Loeys, B.L. *et al.* 2005. A syndrome of altered cardiovascular, craniofacial, neurocognitive and skeletal development caused by mutations in TGFBR1 or TGFBR2. *Nat. Genet.* **37:** 275–281.
71. Loeys, B.L. *et al.* 2006. Aneurysm syndromes caused by mutations in the TGF-beta receptor. *N. Engl. J. Med.* **355:** 788–798.
72. Callewaert, B. *et al.* 2008. Ehlers-Danlos syndromes and Marfan syndrome. *Best Pract. Res. Clin. Rheumatol.* **22:** 165–189.
73. McDonnell, N.B. *et al.* 2006. Echocardiographic findings in classical and hypermobile Ehlers-Danlos syndromes. *Am. J. Med. Genet. A.* **140:** 129–136.
74. Wenstrup, R.J. *et al.* 2002. Prevalence of aortic root dilation in the Ehlers-Danlos syndrome. *Genet. Med.* **4:** 112–117.
75. Nicod, P. *et al.* 1989. Familial aortic dissecting aneurysm. *J. Am. Coll. Cardiol.* **13:** 811–819.
76. Biddinger, A. *et al.* 1997. Familial thoracic aortic dilatations and dissections: a case control study. *J. Vasc. Surg.* **25:** 506–511.
77. Coady, M.A. *et al.* 1999. Familial patterns of thoracic aortic aneurysms. *Arch. Surg.* **134:** 361–367.
78. Milewicz, D.M. *et al.* 1998. Reduced penetrance and variable expressivity of familial thoracic aortic aneurysms/dissections. *Am. J. Cardiol.* **82:** 474–479.
79. Albornoz, G. *et al.* 2006. Familial thoracic aortic aneurysms and dissections–incidence, modes of inheritance, and phenotypic patterns. *Ann. Thorac. Surg.* **82:** 1400–1405.
80. Davies, R.R. *et al.* 2007. Natural history of ascending aortic aneurysms in the setting of an unreplaced bicuspid aortic valve. *Ann. Thorac. Surg.* **83:** 1338–1344.
81. Keane, M.G. *et al.* 2000. Bicuspid aortic valves are associated with aortic dilatation out of proportion to coexistent valvular lesions. *Circulation* **102:** III35–III39.

82. Biner, S. *et al.* 2009. Aortopathy is prevalent in relatives of bicuspid aortic valve patients. *J. Am. Coll. Cardiol.* **53:** 2288–2295.
83. Jiang, X. *et al.* 2000. Fate of the mammalian cardiac neural crest. *Development* **127:** 1607–1616.
84. Tran-Fadulu, V. *et al.* 2009. Analysis of multigenerational families with thoracic aortic aneurysms and dissections due to TGFBR1 or TGFBR2 mutations. *J. Med. Genet.* **46:** 607–613.
85. Guo, D.C. *et al.* 2007. Mutations in smooth muscle alpha-actin (ACTA2) lead to thoracic aortic aneurysms and dissections. *Nat. Genet.* **39:** 1488–1493.
86. Pannu, H. *et al.* 2005. Mutations in transforming growth factor-beta receptor type II cause familial thoracic aortic aneurysms and dissections. *Circulation* **112:** 513–520.
87. Zhu, L. *et al.* 2006. Mutations in myosin heavy chain 11 cause a syndrome associating thoracic aortic aneurysm/aortic dissection and patent ductus arteriosus. *Nat. Genet.* **38:** 343–349.
88. McKellar, S.H. *et al.* 2007. Novel NOTCH1 mutations in patients with bicuspid aortic valve disease and thoracic aortic aneurysms. *J. Thorac. Cardiovasc. Surg.* **134:** 290–296.
89. Guo, D.C. *et al.* 2009. Mutations in smooth muscle alpha-actin (ACTA2) cause coronary artery disease, stroke, and Moyamoya disease, along with thoracic aortic disease. *Am. J. Hum. Genet.* **84:** 617–627.
90. Habashi, J.P. *et al.* 2006. Losartan, an AT1 antagonist, prevents aortic aneurysm in a mouse model of Marfan syndrome. *Science* **312:** 117–121.
91. Brooke, B.S. *et al.* 2008. Angiotensin II blockade and aortic-root dilation in Marfan's syndrome. *N. Engl. J. Med.* **358:** 2787–2795.
92. Gambarin, F.I. *et al.* 2009. Rationale and design of a trial evaluating the effects of losartan vs. nebivolol vs. the association of both on the progression of aortic root dilation in Marfan syndrome with FBN1 gene mutations. *J. Cardiovasc. Med. (Hagerstown)* **10:** 354–362.
93. Lacro, R.V. *et al.* 2007. Rationale and design of a randomized clinical trial of beta-blocker therapy (atenolol) versus angiotensin II receptor blocker therapy (losartan) in individuals with Marfan syndrome. *Am. Heart J.* **154:** 624–631.
94. Radonic, T. *et al.* 2010. Losartan therapy in adults with Marfan syndrome: study protocol of the multi-center randomized controlled COMPARE trial. *Trials* **11:** 3.
95. Keating, M.T. & M.C. Sanguinetti. 2001. Molecular and cellular mechanisms of cardiac arrhythmias. *Cell* **104:** 569–580.
96. Schwartz, P.J. *et al.* 1993. Diagnostic criteria for the long QT syndrome. An update. *Circulation* **88:** 782–784.
97. Crotti, L. *et al.* 2008. Congenital long QT syndrome. *Orphanet. J. Rare Dis.* **3:** 18.
98. Goldenberg, I., W. Zareba & A.J. Moss. 2008. Long QT syndrome. *Curr. Probl. Cardiol.* **33:** 629–694.
99. Romano, C., G. Gemme & R. Pongiglione. 1963. Rare cardiac arrythmias of the pediatric age. II. Syncopal attacks due to paroxysmal ventricular fibrillation. (Presentation of 1st case in Italian Pediatric Literature). *Clin. Pediatr. (Bologna)* **45:** 656–683.
100. Ward, O.C. 1964. A new familial cardiac syndrome in children. *J. Ir. Med. Assoc.* **54:** 103–106.
101. Jervell, A. & F. Lange-Nielsen. 1957. Congenital deaf-mutism, functional heart disease with prolongation of the Q-T interval and sudden death. *Am. Heart J.* **54:** 59–68.
102. Tawil, R. *et al.* 1994. Andersen's syndrome: potassium-sensitive periodic paralysis, ventricular ectopy, and dysmorphic features. *Ann. Neurol.* **35:** 326–330.
103. Tristani-Firouzi, M. *et al.* 2002. Functional and clinical characterization of KCNJ2 mutations associated with LQT7 (Andersen syndrome). *J. Clin. Invest.* **110:** 381–388.
104. Splawski, I. *et al.* 2004. Ca(V)1.2 calcium channel dysfunction causes a multisystem disorder including arrhythmia and autism. *Cell* **119:** 19–31.
105. Wang, Q. *et al.* 1995. SCN5A mutations associated with an inherited cardiac arrhythmia, long QT syndrome. *Cell* **80:** 805–811.
106. Curran, M.E. *et al.* 1995. A molecular basis for cardiac arrhythmia: HERG mutations cause long QT syndrome. *Cell* **80:** 795–803.
107. Hedley, P.L. *et al.* 2009. The genetic basis of long QT and short QT syndromes: a mutation update. *Hum. Mutat.* **30:** 1486–1511.
108. Lehnart, S.E. *et al.* 2007. Inherited arrhythmias: a National Heart, Lung, and Blood Institute and Office of Rare Diseases workshop consensus report about the diagnosis, phenotyping, molecular mechanisms, and therapeutic approaches for primary cardiomyopathies of gene mutations affecting ion channel function. *Circulation* **116:** 2325–2345.
109. Schwartz, P.J. *et al.* 2001. Genotype-phenotype correlation in the long-QT syndrome: gene-specific triggers for life-threatening arrhythmias. *Circulation* **103:** 89–95.
110. Moss, A.J. *et al.* 1995. ECG T-wave patterns in genetically distinct forms of the hereditary long QT syndrome. *Circulation* **92:** 2929–2934.
111. Moss, A.J. *et al.* 2000. Effectiveness and limitations of beta-blocker therapy in congenital long-QT syndrome. *Circulation* **101:** 616–623.
112. Antzelevitch, C. *et al.* 2003. Brugada syndrome: 1992–2002: a historical perspective. *J. Am. Coll. Cardiol.* **41:** 1665–1671.
113. Brugada, P. & J. Brugada. 1992. Right bundle branch block, persistent ST segment elevation and sudden cardiac death: a distinct clinical and electrocardiographic syndrome. A multicenter report. *J. Am. Coll. Cardiol.* **20:** 1391–1396.
114. Chen, Q. *et al.* 1998. Genetic basis and molecular mechanism for idiopathic ventricular fibrillation. *Nature* **392:** 293–296.
115. Hedley, P.L. *et al.* 2009. The genetic basis of Brugada syndrome: a mutation update. *Hum. Mutat.* **30:** 1256–1266.
116. Leenhardt, A. *et al.* 1995. Catecholaminergic polymorphic ventricular tachycardia in children. A 7-year follow-up of 21 patients. *Circulation* **91:** 1512–1519.
117. Reid, D.S. *et al.* 1975. Bidirectional tachycardia in a child. A study using His bundle electrography. *Br. Heart J.* **37:** 339–344.
118. Mohamed, U., C. Napolitano & S.G. Priori. 2007. Molecular and electrophysiological bases of catecholaminergic

polymorphic ventricular tachycardia. *J. Cardiovasc. Electrophysiol.* **18:** 791–797.
119. Swan, H. *et al.* 1999. Arrhythmic disorder mapped to chromosome 1q42-q43 causes malignant polymorphic ventricular tachycardia in structurally normal hearts. *J. Am. Coll. Cardiol.* **34:** 2035–2042.
120. Laitinen, P.J. *et al.* 2001. Mutations of the cardiac ryanodine receptor (RyR2) gene in familial polymorphic ventricular tachycardia. *Circulation* **103:** 485–490.
121. Priori, S.G. *et al.* 2001. Mutations in the cardiac ryanodine receptor gene (hRyR2) underlie catecholaminergic polymorphic ventricular tachycardia. *Circulation* **103:** 196–200.
122. Wehrens, X.H. *et al.* 2003. FKBP12.6 deficiency and defective calcium release channel (ryanodine receptor) function linked to exercise-induced sudden cardiac death. *Cell* **113:** 829–840.
123. Wehrens, X.H., S.E. Lehnart & A.R. Marks. 2005. Ryanodine receptor-targeted anti-arrhythmic therapy. *Ann. N.Y. Acad. Sci.* **1047:** 366–375.
124. Cerrone, M. *et al.* 2007. Arrhythmogenic mechanisms in a mouse model of catecholaminergic polymorphic ventricular tachycardia. *Circ. Res.* **101:** 1039–1048.
125. Knollmann, B.C. *et al.* 2006. Casq2 deletion causes sarcoplasmic reticulum volume increase, premature Ca2+ release, and catecholaminergic polymorphic ventricular tachycardia. *J. Clin. Invest.* **116:** 2510–2520.
126. Watanabe, H. *et al.* 2009. Flecainide prevents catecholaminergic polymorphic ventricular tachycardia in mice and humans. *Nat. Med.* **15:** 380–383.
127. Kannankeril, P.J. & D.M. Roden. 2007. Drug-induced long QT and torsade de pointes: recent advances. *Curr. Opin. Cardiol.* **22:** 39–43.
128. Splawski, I. *et al.* 2002. Variant of SCN5A sodium channel implicated in risk of cardiac arrhythmia. *Science* **297:** 1333–1336.
129. Tester, D.J. & M.J. Ackerman. 2009. Cardiomyopathic and channelopathic causes of sudden unexplained death in infants and children. *Annu. Rev. Med.* **60:** 69–84.
130. Bruneau, B.G. 2008. The developmental genetics of congenital heart disease. *Nature* **451:** 943–948.
131. Hoffman, J.I. & S. Kaplan. 2002. The incidence of congenital heart disease. *J. Am. Coll. Cardiol.* **39:** 1890–1900.
132. Oyen, N. *et al.* 2009. National time trends in congenital heart defects, Denmark, 1977–2005. *Am. Heart J.* **157:** 467–473.
133. Ferencz, C. 1993. *Epidemiology of Congenital Heart Disease: The Baltimore-Washington Infant Study 1981–1989.* Futura Publishing. Mount Kisco, NY.
134. Jenkins, K.J. *et al.* 2007. Noninherited risk factors and congenital cardiovascular defects: current knowledge: a scientific statement from the American Heart Association Council on Cardiovascular Disease in the Young: endorsed by the American Academy of Pediatrics. *Circulation* **115:** 2995–3014.
135. Pierpont, M.E. *et al.* 2007. Genetic basis for congenital heart defects: current knowledge: a scientific statement from the American Heart Association Congenital Cardiac Defects Committee, Council on Cardiovascular Disease in the Young: endorsed by the American Academy of Pediatrics. *Circulation* **115:** 3015–3038.
136. Olson, E.N. 2006. Gene regulatory networks in the evolution and development of the heart. *Science.* **313:** 1922–1927.
137. Srivastava, D. 2006. Making or breaking the heart: from lineage determination to morphogenesis. *Cell* **126:** 1037–1048.
138. Basson, C.T. *et al.* 1994. The clinical and genetic spectrum of the Holt-Oram syndrome (heart-hand syndrome). *N. Engl. J. Med.* **330:** 885–891.
139. Basson, C.T. *et al.* 1997. Mutations in human TBX5 [corrected] cause limb and cardiac malformation in Holt-Oram syndrome. *Nat. Genet.* **15:** 30–35.
140. Chen, B. *et al.* 2000. Mice mutant for Egfr and Shp2 have defective cardiac semilunar valvulogenesis. *Nat. Genet.* **24:** 296–299.
141. Tartaglia, M. *et al.* 2001. Mutations in PTPN11, encoding the protein tyrosine phosphatase SHP-2, cause Noonan syndrome. *Nat. Genet.* **29:** 465–468.
142. Aoki, Y. *et al.* 2008. The RAS/MAPK syndromes: novel roles of the RAS pathway in human genetic disorders. *Hum. Mutat.* **29:** 992–1006.
143. Schubbert, S., G. Bollag & K. Shannon. 2007. Deregulated Ras signaling in developmental disorders: new tricks for an old dog. *Curr. Opin. Genet. Dev.* **17:** 15–22.
144. Li, L. *et al.* 1997. Alagille syndrome is caused by mutations in human Jagged1, which encodes a ligand for Notch1. *Nat. Genet.* **16:** 243–251.
145. Oda, T. *et al.* 1997. Mutations in the human Jagged1 gene are responsible for Alagille syndrome. *Nat. Genet.* **16:** 235–242.
146. McDaniell, R. *et al.* 2006. NOTCH2 mutations cause Alagille syndrome, a heterogeneous disorder of the notch signaling pathway. *Am. J. Hum. Genet.* **79:** 169–173.
147. Satoda, M. *et al.* 2000. Mutations in TFAP2B cause Char syndrome, a familial form of patent ductus arteriosus. *Nat. Genet.* **25:** 42–46.
148. Gebbia, M. *et al.* 1997. X-linked situs abnormalities result from mutations in ZIC3. *Nat. Genet.* **17:** 305–308.
149. Purandare, S.M. *et al.* 2002. A complex syndrome of left-right axis, central nervous system and axial skeleton defects in Zic3 mutant mice. *Development* **129:** 2293–2302.
150. Zhu, L., J.W. Belmont & S.M. Ware. 2006. Genetics of human heterotaxias. *Eur. J. Hum. Genet.* **14:** 17–25.
151. Schott, J.J. *et al.* 1998. Congenital heart disease caused by mutations in the transcription factor NKX2–5. *Science* **281:** 108–111.
152. Garg, V. *et al.* 2003. GATA4 mutations cause human congenital heart defects and reveal an interaction with TBX5. *Nature* **424:** 443–447.
153. Ching, Y.H. *et al.* 2005. Mutation in myosin heavy chain 6 causes atrial septal defect. *Nat. Genet.* **37:** 423–428.
154. Kodo, K. *et al.* 2009. GATA6 mutations cause human cardiac outflow tract defects by disrupting semaphorin-plexin signaling. *Proc. Natl. Acad. Sci. USA* **106:** 13933–13938.
155. Kirk, E.P. *et al.* 2007. Mutations in cardiac T-box factor gene TBX20 are associated with diverse cardiac

pathologies, including defects of septation and valvulogenesis and cardiomyopathy. *Am. J. Hum. Genet.* **81:** 280–291.
156. Garg, V. *et al.* 2005. Mutations in NOTCH1 cause aortic valve disease. *Nature* **437:** 270–274.
157. McBride, K.L. *et al.* 2008. NOTCH1 mutations in individuals with left ventricular outflow tract malformations reduce ligand-induced signaling. *Hum. Mol. Genet.* **17:** 2886–2893.
158. Goldmuntz, E., E. Geiger & D.W. Benson. 2001. NKX2.5 mutations in patients with tetralogy of fallot. *Circulation* **104:** 2565–2568.
159. Rajagopal, S.K. *et al.* 2007. Spectrum of heart disease associated with murine and human GATA4 mutation. *J. Mol. Cell Cardiol.* **43:** 677–685.
160. Schluterman, M.K. *et al.* 2007. Screening and biochemical analysis of GATA4 sequence variations identified in patients with congenital heart disease. *Am. J. Med. Genet. A.* **143A:** 817–823.
161. Tomita-Mitchell, A. *et al.* 2007. GATA4 sequence variants in patients with congenital heart disease. *J. Med. Genet.* **44:** 779–783.
162. Greenway, S.C. *et al.* 2009. De novo copy number variants identify new genes and loci in isolated sporadic tetralogy of Fallot. *Nat. Genet.* **41:** 931–935.
163. Redon, R. *et al.* 2006. Global variation in copy number in the human genome. *Nature* **444:** 444–454.
164. Richards, A.A. *et al.* 2008. Cryptic chromosomal abnormalities identified in children with congenital heart disease. *Pediatr. Res.* **64:** 358–363.
165. Roach, J.C. *et al.* 2010. Analysis of genetic inheritance in a family quartet by whole-genome sequencing. *Science* **328:** 636–639.

Ann. N.Y. Acad. Sci. ISSN 0077-8923

ANNALS OF THE NEW YORK ACADEMY OF SCIENCES
Issue: *The Year in Human and Medical Genetics*

Clues to immune tolerance: the monogenic autoimmune syndromes

Michael Waterfield[1,2] and Mark S. Anderson[1]

[1]Diabetes Center, University of California, San Francisco, California. [2]Department of Pediatrics, University of California, San Francisco, California

Address for correspondence: Mark S. Anderson, M.D., Ph.D., UCSF Diabetes Center, 513 Parnassus Avenue, Box 0540, San Francisco, CA 94143. manderson@diabetes.ucsf.edu

Autoimmune disease affects a significant proportion of the population. The etiology of most autoimmune diseases is largely unknown, but it is thought to be multifactorial with both environmental and genetic influences. Rare monogenic autoimmune diseases, however, offer an invaluable window into potential disease mechanisms. In this review, we will discuss the autoimmune polyglandular syndrome (APS1), the immunedysregulation, polyendocrinopathy, enteropathy, X-linked syndrome (IPEX), and autoimmune lymphoproliferative syndrome (ALPS). Significantly, the information gained from the study of these diseases has provided new insights into more common autoimmune disease and have yielded new diagnostics and therapeutic opportunities.

Keywords: IPEX; ALPS; APS1; autoimmunity; immune tolerance; regulatory T cells

Introduction

Autoimmune disease results from a breakdown in the mechanisms that maintain tolerance to self-tissues. The exact cause of most autoimmune diseases is unknown and is thought to be due to a combination of genetic and environmental factors. The multifactorial nature of autoimmune diseases has made it difficult to determine their underlying pathogenesis. Genome-wide association studies have yielded correlations between multiple small nucleotide polymorphisms (SNPs) and autoimmune diseases. However, for most of the associated loci, the signals are not strong enough to directly implicate a specific gene, and the role of the candidate genes in a particular locus can only be speculated.[1] The loci that have been potentially linked with autoimmune diseases involve genes that regulate many aspects of the immune system including lymphocyte activation, microbial recognition, cytokines and cytokines receptors. Moreover, many of these gene variants are associated with multiple autoimmune diseases.[2] For example, similar major histocompatibility locus (*MHC*) haplotypes have been linked to multiple autoimmune disorders and SNPs in the protein tyrosine phosphatase inhibitor gene, *PTPN22*, have been connected to rheumatoid arthritis, type I diabetes, and Crohn's disease. Although polymorphisms in the *MHC* and *PTPN22* have been linked to autoimmune disease a direct genotype to phenotype correlation is lacking.[1,2] In contrast, studies with rare monogenic autoimmune diseases have yielded invaluable insights into the underlying mechanisms of immune tolerance because of their strong penetrance.

Here we highlight a handful of disorders including the autoimmune polyglandular syndrome (APS) type I (OMIM 240300) also know as autoimmune polyendocrinopathy candidiasis-ectodermal dystrophy (APECED); the immunedysregulation, polyendocrinopathy, enteropathy, X-linked (IPEX) syndrome (OMIM 304790); and autoimmune lymphoproliferative syndrome (ALPS) (OMIM 601859) (Table 1). Studies of these disorders have advanced our knowledge of the underlying mechanisms of autoimmunity, and how the breakdown of tolerance can lead to autoimmune disease. In this review, we will cover the clinical manifestations, genetics, diagnosis/treatment, and in-depth pathophysiology of these disorders.

doi: 10.1111/j.1749-6632.2010.05818.x

 Ann. N.Y. Acad. Sci. 1214 (2010) 138–155

Table 1. Major characteristics of monogenic autoimmune diseases

	APS1	IPEX	ALPS
Defective gene(s)	*AIRE*	*FOXP3*	*FAS* *FasL* *Caspase 10* *NRAS*
Inheritance	Autosomal recessive	X-linked	Autosomal dominant
major symptoms	Hypoparathyroidism Mucocutaneous candidiasis Addison's disease	Diarrhea Type I diabetes Thyroiditis Eczema	Lymphadenopathy Splenomegaly Autoimmune Cytopenias Malignancy
Treatment	Hormone replacement antifungals	Bone marrow transplant	Immunosuppression

Autoimmune polyglandular syndrome type I

APS1 clinical features

APS1 is a rare disease that usually manifests in childhood. The clinical diagnosis of APS1 is based on the presence of at least two of three major features: mucocutaneous candidiasis (CMC), adrenal insufficiency, or hypoparathyroidism, with the most common presenting symptom being mucocutaneous candidiasis in infancy (Table 1).[3]

In a Finnish cohort of patients with APS1, 50% of patients had developed CMC by age 5, 70% by age 10, and 94% by age 20.[4] The infection usually begins in the mouth but may spread to other regions of the GI tract including the esophagus, vagina, stomach, and intestines. The severity of the mucositis caused by CMC infection varies greatly with some patients having only minor buccal redness and ulcerations to other patients developing esophageal strictures. Patients with CMC are at increased risk of squamous cell carcinoma of the esophagus and require periodic screening.[5]

Hypoparathyroidism is the second most common manifestation of APS1. In a Finnish cohort the prevalence was 33% by age 5, 66% at age 10 and 85% at age 30.[4] Interestingly, if patients develop adrenal insufficiency prior to hypoparathyroidism they are somewhat protected from subsequently developing hypoparathyroidism.[6] Clinically, hypoparathyroidism has a wide range of symptoms from muscle cramps and paresthesia to frank seizures secondary to very low serum calcium levels. Adrenal insufficiency is the third most common manifestation with 40% of Finnish patients showing symptoms by age 10 and 78% of patients symptomatic at age 30.[4] Symptoms of hypoadrenalism include fatigue, salt craving, hypotension, weight loss, and increased skin pigmentation and arise due to loss of adrenocortical dysfunction.

APS1 can have multiple other associated autoimmune manifestations including autoimmune hepatitis, type I diabetes, keratitis, autoimmune gonadal failure, pernicious anemia, interstitial lung disease, or celiac disease to name a few.[3,7] These manifestations can vary widely in APS1 patients and overall show low penetrance. Additionally, patients with APS1 may have enamel dysplasia or other signs of ectodermal dystrophy.[7,8] Among patients with APS1 there is great variability in the timing of the disease. In the Finnish cohort, the youngest patient presented at 2 months and the oldest at 18 years, while in a Norwegian cohort, the youngest patient presented at a few months of age and the oldest at 43.[4,9] To further complicate the diagnosis, some patients will suffer from the less common disease manifestations before presenting with the major disease-defining manifestations.[4,7] There are reports of APS1 patients manifesting symptoms such as chronic diarrhea, fevers, periodic rash, or autoimmune hepatitis for several years before displaying clinical features of the diagnostic triad.

The heterogeneous nature of disease manifestations and the significant associated morbidity and mortality, make it imperative that APS1 patients are followed closely by experienced providers. Treatment of APS1 depends largely on the organ-specific manifestations. CMC in APS1 is treated with

chronic mucocutaneous candidiasis, hypoparathyroidism, or adrenal insufficiency, (2) one major component and a sibling with a definitive diagnosis, or (3) *AIRE* mutations in both genes. APS1 patients also develop autoantibodies which can both aid in diagnosis and also give insights into disease pathogenesis. For example, multiple independent groups found autoantibodies to type I interferons in the sera of APS1 patients with the correlation between APS1 and autoantibodies to type I interferons close to 100%.[49,50] The specific interferon subtypes that APS1 patients have reactivity to include interferon alpha and/or interferon omega. Patients with the G228W mutation also produce autoantibodies to type I interferons, potentially making them a useful diagnostic tool for all forms of the disease.[49] The major *in vivo* function of type I interferons is thought to be in protection from viral infections, however, APS1 patients do not show increased susceptibility to viral infections. Thus, the clinical relevance of type I interferon autoantibodies, other than as a diagnostic maker, is unclear.

In addition to autoantibodies against type I interferons, a recent set of studies have discovered autoantibodies that may explain the susceptibility to candida infection in patients with APS1. A consortium of researchers has discovered autoantibodies to T helper 17 (Th17) cytokines specifically IL-17A, IL-17F, and IL-22 in sera from APS1 patients.[51,52] The Th17 subset of T cells is thought to be important for host protection against fungi at epithelial surfaces. Thus, these studies have proposed that the CMC seen in APS1 patients is secondary to autoantibodies targeting these cytokines. While the data shows a significant correlation between autoantibodies to Th17 cytokines and CMC, further study needs to be performed to directly link a Th17 defect to CMC.

Another study, using sera from APS1 patients with hypoparathroidism identified the parathyroid specific protein NACHT leucine-rich repeat protein 5 (NALP5) as a putative parathyroid autoantigen.[53] Autoantibodies to NALP5 are specifically found in APS1 patients with hypoparathyroidism making the autoantibody response to NALP5 a potential diagnostic marker for hypoparathyroidism in these patients. With regard to adrenal insufficiency, APS1 patients also have autoantibodies to P450-containing enzymes that are important in steroid biosynthesis. Similar to patients with isolated autoimmune adrenal insufficiency (Addison's disease), APS1 patients have autoantibodies to 21-hydroxylase (a key enzyme in adrenal steroid biogenesis).[3] In addition, APS1 patients with adrenal sufficiency can also have autoantibodies to other enzymes involved in steroid biogenesis such as side-chain cleavage enzyme and aromatic L-amino acid decarboxylase.[54,55]

Research is currently underway to identify novel autoantibodies that could be used to help diagnose specific APS1 clinical manifestations or potentially other autoimmune diseases. Studies in *Aire* deficient mice have shown that autoantibodies result directly from failed antigen presentation in the thymus.[25–27] For example, *Aire* deficient mice make autoantibodies to lung epithelium and have an inflammatory infiltrate within the lung. The autoantibody target is the lung-specific protein vomeromodulin and the autoantibody arises secondary to failed thymic presentation of vomeromodulin. Similar to *Aire* deficient mice, there are also APS1 patients with interstitial lung disease (ILD) that shows nearly identical pathology to the disease in mice and interestingly, these patients also make autoantibodies to LPLUNC1 a vomeromodulin homologue.[27]

Thymic antigen expression in other autoimmune diseases

The importance of thymic antigen presentation may extend beyond APS1 and be relevant to other autoimmune diseases. For example, in type 1 diabetes thymic expression of insulin has been linked to disease. Population studies have shown that individuals with polymorphisms in the variable number of tandem repeat (VNTR) region upstream of the insulin promoter are more likely to develop disease secondary to lower thymic expression of insulin.[56,57] A similar situation occurs with myasthenia gravis and expression of the TRA *CHRNA1*(αchain of the acetylcholine receptor).[58] A single nucleotide polymorphism in the promoter of the *CHRNA1* gene has been associated with early-onset myasthenia in two populations. Interestingly, this polymorphism is associated with decreased CHRNA1 antigen expression in *ex vivo* human mTECs. Furthermore, a correlation seems to exist between thymic CHRNA1 expression and AIRE expression, indicating that AIRE may play a role in CHRNA1 expression in the human thymus. Finally, it is interesting to note that the known predisposition to autoimmunity in patients with thymic tumors may be acquiring autoimmunity through a defect in appropriate AIRE expression.[59]

Immunodysregulation polyendocrinopathy enteropathy X-linked syndrome

IPEX clinical features

IPEX is a rare, X-linked, syndrome that results in a severe systemic autoimmune disorder. Most patients with IPEX have the clinical triad of enteritis, endocrinopathy, and dermatitis (Table 1).[60,61] The presenting symptom is often intractable diarrhea with failure to thrive (FTT) requiring parenteral nutrition.[62] Histologically, the bowel shows villous atrophy and extensive lymphocytic infiltrate within the mucosa. Insulin-dependent type I diabetes and hypothyrodism are the most common endocrinopathies seen in IPEX. Similar to the GI disease, patients with endocrinopathies have inflammatory infiltrates within the pancreas and thyroid and will often have autoantibodies to pancreatic islet cells.[60,61] The dermatitis that affects IPEX patients is most commonly an eczematous rash that in some patients can be very severe affecting a large percentage of their body surface area. In addition to the diagnostic triad, IPEX patients can have multiple other autoimmune manifestations including cytopenias, renal disease, and hepatitis.[60,61] The symptoms in IPEX can fluctuate with exacerbations precipitated by infections, dietary allergans, or vaccinations. Laboratory studies often show eosinophilia and hypergammaglobulinemia. IPEX patients often display increased susceptibility to a wide range of infections including bacterial, viral, and fungal, a phenomenom which is ascribed to decreased barrier function secondary to severe gastritis and dermatitis.[61,62] Treatment of IPEX is with either immunosuppressive agents or bone marrow transplantation.[63,64] Aggressive treatment is warranted as this disease can be lethal if left untreated.

IPEX genetics

The discovery of the genetic mutation in IPEX was made possible secondary to observation of a mouse mutant called *scurfy* which has severe autoimmunity and lymphoproliferation similar to IPEX. The mutated gene in *scurfy* mice was found to be the *Foxp3* gene located on Xp11.3-q13.3.[65] After *Foxp3* was identified in mice, two groups sequenced the *FOXP3* gene in families with IPEX and discovered mutations within the human homologue.[66,67] In these families, only males are affected and female carriers with one functional copy of FOXP3 are asymptomatic. A wide range of mutations have been found in the *FOXP3* gene of IPEX patients including point mutations, deletions, and splicing mutations.[60,61] Foxp3 is a zinc-finger transcription factor required for the development and function of regulatory T cells (T_{reg} cells). T_{reg} cells are $CD4^+$, IL-2Rα ($CD25^+$), and $Foxp3^+$ T cells that are functionally anergic and upon activation produce cytokines and cell surface molecules that are important in diminishing immune responses in a variety of settings.[68,69] These cells make up approximately 10–15% of $CD4^+$ T cells and have proven to be critical mediators of peripheral tolerance.

IPEX pathogenesis

Prior to the discovery of Foxp3, the identity of T_{reg} cells was suspected when a seminal study showed lymphocyte suspensions depleted of $CD4^+CD25^+$ cells induced autoimmunity when injected into lymphopenic mice.[70] Interestingly, the autoimmunity could be prevented with immediate reconstitution of the mice with $CD4^+CD25^+$ T cells and the effects of the $CD4^+CD25^+$ T cells were dose dependent. This important study was the first to suggest a population of suppressor cells within the $CD4^+CD25^+$ T cell lineage. After the discovery of Foxp3, several groups determined that Foxp3 functions specifically in $CD4^+CD25^+$ T cells and is critical for the prevention of autoimmunity.[71,72] Bone marrow chimeras using bone marrow from *Foxp3* deficient mice into wild-type hosts induced an autoimmune syndrome similar to *scurfy* mice indicating that Foxp3 functions in hematopoietic cells.[71] Furthermore, using retroviral vectors that express Foxp3, it was shown that $CD4^+CD25^-$ T cells could be transformed into T_{reg} cells by expression of Foxp3 and these cells are able to suppress T effecter cells both *in vitro* and *in vivo*.[72] T_{reg} cell lineage specification was further defined using a $Foxp3^{GFP}$ reporter mouse that expresses functional Foxp3.[73] This mouse showed that Foxp3 is exclusively expressed in $CD4^+CD25^+$ T_{reg} cells and in no other tissues or hematopoietic cells. In addition, the importance of T_{reg} cell function throughout life is illustrated with the $Foxp3^{DTR}$ mouse and a study utilizing cre-mediated ablation of *Foxp3* in mature T_{reg} cells.[74,75] In both mouse models, selective ablation of T_{reg} cells in adult mice leads to death in three weeks secondary to autoimmunity. The ability of a specialized set of cells to enforce tolerance is defined as dominant tolerance, as this cell type is able to suppress self-reactivity by acting in a dominant fashion.[68,69] The model of dominant tolerance explains why female carriers of a mutated *Foxp3* are not affected. Despite random X-chromosome inactivation and only one-half of T cells harboring a functional *Foxp3* gene, carrier females are normal because the T_{reg} cell population carrying the wild-type allele is able to control autoimmunity in a dominant fashion.[76]

Significant research has been performed to determine the exact mechanism by which T_{reg} cells are able to dampen the immune system. It is likely that multiple mechanisms are involved including both contact dependent mechanisms (direct killing of activated T cells/antigen presenting cells (APCs)) and soluble factors (production of inhibitory cytokines such as IL-10 or TGF-β).[68,69] Two well characterized mediators of T_{reg} cell suppression are the cytotoxic T lymphocyte antigen (CTLA-4) and the cytokine IL-10.[77,78] CTLA-4 is a surface molecule on T cells which is in the same family as the co-stimulatory molecule CD28. Selective ablation of CTLA-4 on T_{reg} cells in mice results in severe autoimmunity. CTLA-4 is hypothesized to function on T_{reg} cells by sending an inhibitory signal to APCs and reducing their production of proinflammatory cytokines.[69,78] Additionally, loss of IL-10 production by T_{reg} cells results in gut and lung inflammation suggesting a role for specific cytokines in T_{reg} cell function.[77]

T_{reg} cell differentiation and Foxp3 expression

The extracellular signals required for T_{reg} cell differentiation in the thymus and Foxp3 expression is an area of intense research. It has been hypothesized that T_{reg} cells express self-reactive TCRs and that TCR specificity for self may guide T_{reg} cell differentiation. Several studies have shown that T_{reg} cell TCRs are skewed toward self-peptides and that TCR signaling is likely important for Foxp3 expression.[79,80] Several transcription factors downstream of the TCR have been implicated in Foxp3 expression including NFAT, NF-κB, CREB, AP1, and ATF.[81–84] IL-2 has also been implicated in T_{reg} cell differentiation because mature T_{reg} cells express high levels of CD25. Experiments utilizing a $Foxp3^{GFP}$ reporter mouse crossed onto the IL-2 and IL-2 receptor knockout background showed that IL-2 is not required for Foxp3 induction in the thymus, but it is required for the peripheral homeostasis of T_{reg} cells and the continued expression of Foxp3 in the periphery.[85] In support of this finding, mice lacking IL-2 or IL-2Rα have a severe lymphoproliferative syndrome.[86] Notably, recent case reports have described patients that have mutations in both copies of the IL-2 receptor gene and these patients present with severe autoimmunity similar to IPEX.[87,88] Additional signals implicated in T_{reg} cell differentiation include TCR co-stimulatory molecules such as CD28 and the cytokine TGF-β.[89,90]

To add a layer of complexity, T_{reg} cell differentiation can be further divided into natural and adaptive regulatory T cells.[91] Natural T_{reg} cells originate in the thymus while adaptive T_{reg} cells are induced in the periphery from a population of $CD4^+CD25^-$ T cells. Adaptive T_{reg} cells are mainly located in lymph nodes and chronically inflamed tissue.[92,93] The exact signals required for adaptive T_{reg} cell development are controversial; however, there is thought to be considerable overlap with the signals required for natural T_{reg} cell development including TCR signaling, costimulation with CTLA-4, TGF-β, and IL-2.[91] Additionally, retinoic acid produced by $CD103^+$ dendritic cells in the lamina propria of the gut has been implicated in the development of adaptive T_{reg} cells.[94,95] Furthermore, it has been hypothesized that adaptive T_{reg} cells may be particularly important in the gut to dampen the immune response to commensal bacteria and food antigens.

Molecular mechanism of Foxp3 function

Regulatory T cells (T_{reg} cells) represent a unique population of $CD4^+$ T cells and the transcription factor Foxp3 serves as a lineage specification marker. The Foxp3 protein contains many functional domains: an N terminal proline rich region that binds the nuclear factor of activated T cells (NFAT), a central domain that contains a C2H2 zinc finger and a leucine zipper that is thought to be involved in protein-protein interactions, and a C terminal region that contains the forkhead DNA binding domain and the nuclear targeting sequences (Fig. 3).[61] NFAT is a transcription factor required for activation of T cells and IL-2 production.

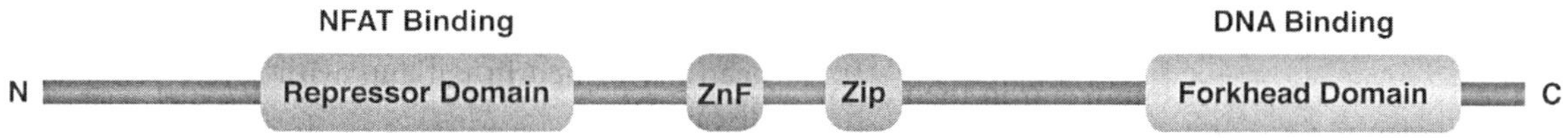

Figure 3. Functional Domains of Foxp3 protein. The N terminus contains a repressor domain important for binding the nuclear factor of activated T cells (NFAT) and inhibiting its function. The middle of the Foxp3 protein contains a zinc finger domain and a leucine zipper. The C terminus contains the forkhead DNA binding domain and a nuclear localization sequence.

Interestingly, the Foxp3/NFAT complex acts as a corepressor of cytokine promoters such as IL-2 and changes T cell phenotype from one of activation to tolerance.[96] In addition to NFAT, Foxp3 also binds to several other transcription factors including the Runt domain transcription factor Runx1, IRF4, STAT3, and the orphan nuclear receptors RORγ and RORα.[97–100]

Recently, two groups performed chromatin immunoprecipitation (ChIP) experiments to analyze Foxp3 DNA binding.[101,102] Previous microarray profiling of Foxp3$^+$ T cells compared to Foxp3$^-$ T cells revealed that Foxp3 induced the expression of many genes important for its function including CTLA-4, CD25, IL-10, and the TNF family member GITR.[73,103,104] Interestingly, ChIP analysis showed that Foxp3 only directly regulated approximately 10% of these genes and induced a group of transcription factors and micro RNAs that likely play an important role in T_{reg} cell biology.[68,101,102]

Human FOXP3$^+$ regulatory T cells

The isolation of T_{reg} cells from peripheral blood in humans has proven to be much more complex than the isolation of T_{reg} cells in mice. The main reason for this complexity is that upon T cell stimulation naïve human CD4 T cells will transiently express low-levels of FOXP3 making it difficult to differentiate T_{reg} cells from effecter T cells.[105,106] The exact function of the low-level expression of FOXP3 in activated T cells is unknown, however, these cells secrete inflammatory cytokines and are unable to suppress *in vitro*.[107,108] Due to the heterogeneous expression of FOXP3 in human CD4$^+$ T cells, it cannot be used alone as a marker of T_{reg} cells.[109,110] To circumvent this problem, many groups have attempted to use the level of FOXP3 protein and the methylation status of the FOXP3 promoter to identify T_{reg} cells.[109,111,112] Human T_{reg} cells are defined by high levels of FOXP3 protein associated with a completely demethylated FOXP3 promoter while activated T cells express lower levels of FOXP3 protein and the FOXP3 promoter is incompletely demethylated.

One difficulty in using FOXP3 protein to define T_{reg} cells is that FOXP3 is a nuclear protein and T cells must be fixed prior to analyzing protein expression. Therefore, T_{reg} cells isolated using FOXP3 cannot be used for further applications. In mice, CD25 can be used as a surrogate marker as most CD25$^+$ cells will express high levels of FOXP3 and will have a suppressive phenotype.[70] In contrast, activated human CD4$^+$ T cells that transiently express FOXP3 also express high levels of CD25 making it difficult to solely use this cell-surface marker for the isolation of human T_{reg} cells. For this reason, many groups are using additional markers such as CD127, CD49d, and CD45 isoforms for the isolation of human T_{reg} cells to be used in further functional assays.[109,113,114]

One marker that has been used successfully is the IL-7 receptor, CD127, which is found on both effecter and memory T cells. In contrast, FOXP3 positive human T regulatory cells express low levels of CD127. Using a combination of high surface expression of CD4 and CD25 along with low levels of CD127 has allowed several groups to isolate a pure population of human T_{reg} cells that express high levels of FOXP3 protein that retain suppressive function.[114,115] Another marker that has recently been used to aid in the purification of human T_{reg} cells is CD49d which is the α-chain of the integrin VLA-4($\alpha_4\beta_1$) and is found on the majority of proinflammatory effecter cells, but is absent on FOXP3 positive T_{reg} cells. By depleting CD49d$^+$ cells, one group has been able to enrich for T_{reg} suppressor cells within the CD25$^+$/CD127^{lo} subset of cells.[113] The ability to isolate a pure population of human T_{reg} cells is critical because T_{reg} cells may be important in the pathogenesis of several autoimmune diseases and adoptive transfer of T_{reg} cells has been proposed as a potential therapy for these diseases.

T_{reg} cells and autoimmune disease

While T_{reg} cells are clearly important in the pathogenesis of IPEX, a number of studies have recently attempted to address whether subtle changes in T_{reg} cell function or numbers are important in other polygenic autoimmune disease. In both rheumatoid arthritis (RA) and type 1 diabetes, there have been conflicting results as to whether T_{reg} cells are deficient. For example, in RA two groups found no difference in the ability of T_{reg} cells from RA patients to suppress effecter T cell proliferation, however, another group found a striking defect.[116–118] Similarly, for type 1 diabetes one group found a reduction in T_{reg} cell percentage of $CD4^+$ T cells, while several other groups have found no difference in T_{reg} cell frequency.[119–122] Further research needs to be performed to better understand the role of T_{reg} cells in more common autoimmune disease.

Autoimmune lymphoproliferative syndrome

ALPS clinical features

The autoimmune lymphoproliferative syndrome (ALPS) is characterized by lymphadenopathy, splenomegaly, hepatomegaly, autoimmune manifestations, and an increased risk of malignancy (Table 1).[123–125] The lymphadenopathy in ALPS is chronic and to be diagnosed a patient must have lymphadenopathy for >6 months.[124,125] The most common autoimmune manifestation of ALPS are cytopenias secondary to autoimmune destruction of blood cells (red blood cells, platelets, or neutrophils) affecting 70% of patients.[124,125] Destruction of hematopoietic cells is often autoantibody mediated and 92% of ALPS patients develop at least one autoantibody: anticardiolipin, direct antigen test (DAT), antiplatelet, antineutrophil, antinuclear, or rheumatoid factor.[125] Autoimmunity, however, can affect almost any organ system including the kidney (nephritis), liver (hepatitis), eye (uveitis), joints (arthritis), gut (colitis), and skin (urticaria). The exact risk of malignancy in ALPS patients is unknown, although it is postulated to be 10-20%.[126] The most common malignancy is lymphoma, and there are case reports of leukemias as well as several solid tumors.[124–126] The treatment of ALPS is based on the clinical manifestations with immunosuppressive agents used to treat autoimmune manifestations and chemotherapy for malignancy.[124–126] Bone-marrow transplants have been performed in patients with severe-refractory disease.[127]

ALPS genetics

Similar to APS1 and IPEX, the pathogenesis of ALPS has been delineated utilizing a mouse model that phenotypically resembled patients with ALPS. Mice homozygous for the *lpr* or *gld* mutations were found to have multiple autoimmune manifestations, autoantibody production, and an increase in nonmalignant TCRαβ $CD4^-CD8^-$ double-negative (DN) T cells.[128,129] The *lpr* and *gld* mutations were determined to be in Fas or its binding partner Fas ligand (FasL) which are both critical for the apoptosis of activated lymphocytes. Similar to *lpr* or *gld* mice, ALPS patients have increased numbers of DN T cells and were found to have mutations in *FAS* (ALPS-FAS), and *FASLG* (ALPS-FASLG), demonstrating that defective apoptosis is the cause of the disease.[130–132] A large number of *FAS* mutations (>70) have been discovered with two-thirds in the intracellular domain (ICD) and one-third in the extracellular domain (ECD). Clinical penetration is higher with ICD mutations (70–90% penetration) than ECD mutations (30% penetration).[125,133]

The first ALPS patients described had mutations in one *FAS* gene and the syndrome was inherited in an autosomal dominant fashion in which there was a family history of ALPS. Subsequent studies revealed patients with homozygous mutations in *FAS*.[131,134,135] Predictably, patients with homozygous mutations have severely diminished FAS expression and a more severe ALPS phenotype requiring early bone-marrow transplantation. Additionally, one family has been described in which two siblings are compound heterozygote's with distinct missense mutations in each *FAS* allele.[136] Interestingly, despite mutations in both *FAS* genes these patients have a milder ALPS phenotype. Also, unlike other patients with homozygous *FAS* mutations these patients did not have a family history of ALPS despite both parents being heterozygous for *FAS*. This may be due to the specific *FAS* mutation with missense mutations causing a milder phenotype.

Further work identified other mutations responsible for an ALPS or ALPS-like phenotype including caspase 10 (ALPS-CASP10), caspase 8 (CEDS), and the NRAS oncogene (RALD).[137–140] Caspase 10 is a

Table 2. Diagnostic classification of ALPS and ALPS-like disorders[140]

Previous nomenclature	Revised nomenclature	Gene
ALPS Type 0	ALPS-FAS	Germline homozygous *FAS* mutations
ALPS Type Ia	ALPS-FAS	Germline heterozygous *FAS* mutations
ALPS Type Im	ALPS-sFAS	Somatic *FAS* mutations
ALPS Type Ib	ALPS-FASLG	Germline *FAS ligand* mutations
ALPS Type IIa	ALPS-CASP10	Germline caspase 10 mutations
ALPS Type III	ALPS-U	Unknown mutation
ALPS Type IIb	CEDS	Germline caspase 8 mutations
ALPS Type IV	RALD	Somatic *NRAS* mutations
DALD	DALD	Unknown mutation
XLP1	XLP1	Germline *SH2D1A* mutations

protein downstream of Fas in the apoptotic pathway. Similar to ALPS patients with *FAS* or *FASLG* mutations, *CASPASE 10* mutations are usually inherited in an autosomal dominant fashion. Analogous to caspase 10, caspase 8 is dowstream of Fas in the apoptotic pathway, however, the inheritance of caspase 8 mutations and the phenotype of families with caspase 8 mutations differs from typical ALPS patients. For this reason, caspase 8 deficiency is given a unique acronym, CEDS, standing for Caspase-Eight Deficiency State.[141] Families with caspase 8 mutations display autosomal recessive inheritance and also have defects in T, B, and NK cell activation leading to susceptibility to bacterial and viral infections. NRAS is a small GTPase and activating mutations in NRAS cause decreased activity of the proapoptotic protein BIM.[142] This disease entity is now classified as RALD, for RAS-associated autoimmune leukoproliferative disease.[141]

To further complicate ALPS classification, patients with somatic mutations in Fas (not germline) have recently been discovered and have been named ALPS-sFAS.[143,144] Patients who have the ALPS phenotype but who do not have an identifiable mutation are called ALPS-U. Recently, an NIH consensus conference created a new ALPS classification scheme (Table 2) and diagnostic criteria (Table 3).[124,141,145,146]

Two other ALPS-like disorders are Dianzani Autoimmune Lyphoproliferative disease (DALD) and X-linked lymphoproliferative disease (XLP1).[147,148] The mutation for DALD is unknown and the mutation in XLP1 is in the *SH2D1A* gene. Patients with DALD present with lymphadenopathy, autoimmunity, and defective Fas-mediated apoptosis, however, these patients have normal numbers of DN T cells. Due to the normal number of DN T cells, DALD is given its own acronym because it is thought to be a unique genetic disorder.

ALPS pathogenesis

As discussed above, ALPS patients have a variety of defects in the apoptotic pathway. The exact mechanism, however, by which defects in apoptosis cause the autoimmunity seen in ALPS is unknown. One hypothesis is that failure of the immune system to contract after antigen stimulation may lead to the escape of autoreactive clones and autoimmunity. Upon antigen stimulation, lymphocytes express Fas and FasL on their surface in the process of activation induced cell death (AICD). In order for Fas or FasL to induce apoptosis they each must form a trimolecular complex.[149] The requirement of the trimolecular complex explains the autosomal dominant inheritance seen in ALPS-FAS, ALPS-FASL, and ALPS-sFAS.[130–132] In heterozygous patients the mutated allele functions as a dominant-negative inhibiting the function of the wild-type allele and disrupting the formation of a functional complex. In most cases, the mutated Fas protein still forms a trimolecular complex with wild-type Fas, however, the mutated Fas allele disrupts death domain signaling or binding of FasL.[125] Similarly, patients with ALPS-CASP10 also show autosomal dominant inheritance with the mutated Caspase 10 functioning in a dominate-negative fashion.

One important unanswered question regarding the pathogenesis of ALPS pertains to the

Table 3. Diagnostic criteria for ALPS

(a) Original Diagnostic Criteria[126]
1. Chronic nonmalignant lymphoproliferation
2. Elevated peripheral blood DNT's
3. Defective *in vitro* Fas-mediated apoptosis
Supporting: genetic mutations; autoimmunity
Diagnosis: Must meet all three criteria
(b) Revised diagnostic criteria[140]
Required
1. Chronic (>6 months), nonmalignant, noninfectious lymphadenopathy and/or splenomegaly
2. Elevated $CD3^+$ $TCR\alpha\beta^+$ $CD4^-CD8^-$ DNT cells (equal to or greater than 1.5% of total lymphocytes or 2.5% of $CD3^+$ lymphocytes) in the setting of normal or elevated lymphocyte counts.
Accessory
Primary
1. Defective lymphocyte apoptosis (in two separate assays);
2. Somatic or germline pathogenic mutation in *FAS*, *FASLG*, or *CASP10*;
Secondary
1. Elevated plasma sFASL levels (>200 pg/mL) OR elevated plasma IL-10 levels (>20 pg/mL) OR elevated serum or plasma Vitamin B12 levels (>1500 ng/L) or elevated plasma IL-18 levels above 500 pg/mL;
2. Typical immuno-histological findings as reviewed by an experienced hematopathologist;
3. Autoimmune cytopenias (hemolytic anemia, thrombocytopenia, or neutropenia) AND elevated IgG levels (polyclonal hypergammaglobulinemia);
4. Family history of a nonmalignant/noninfectious lymphoproliferation with or without autoimmunity.

Definitive Diagnosis: Both required criteria plus one primary accessory criterion.
Probable Diagnosis: Both required criteria plus one secondary accessory criterion.

unexplained heterogeneity in clinical manifestations among family members with the same ALPS genotype. Some family members will be asymptomatic while others will require significant immunosuppression. It has been proposed that the genetic background of the individual will determine the severity of the disease and that additional mutations in genes involved in tolerance may tip the balance toward autoimmunity.[150] Thus, similar to cancer autoimmunity may involve a multistep process in which additional mutations in rapidly dividing immune cells promotes the growth of autoimmune clones that cause disease. Support for this hypothesis comes from ALPS patients with somatic mutations in Fas.[143,144] These patients have normal germline FAS alleles; however, at some point in lymphocyte maturation a somatic mutation occurred in one FAS allele breaking tolerance. The finding that somatic mutations are able to break tolerance may explain the heterogeneity seen in ALPS patients as patients with more severe disease may have accumulated additional mutations promoting autoimmunity. Future research will aim to determine the other genes and environmental factors potentially modulating the ALPS phenotype and to further understand how defects in apoptosis lead to autoimmunity.

Summary

The study of monogenic autoimmune diseases has provided invaluable insights into the underlying mechanisms of immune tolerance. Mouse models of these diseases have allowed detailed studies of the pathogenesis of these unique syndromes and have shed light on the molecular mechanisms of autoimmunity. By further dissecting the molecular pathways involved in these diseases we may gain a greater understanding not only of the pathogenesis of monogenic autoimmune syndromes, but of more common autoimmune syndromes. The downstream genes or associated pathways may provide us with novel therapies or aid in the diagnosis of autoimmune disease. Therapies are already being developed from the lessons learned in these diseases. One new therapy that has emerged is the adoptive transfer of T_{reg} cells for the treatment of type 1 diabetes based on studies of a mouse model of diabetes showing that the transfer of a small number of antigen-specific T_{reg} cells could reverse diabetes.[151] A number of groups are in the process of moving toward using the adoptive transfer of T_{reg} cells for diverse autoimmune diseases such as type 1 diabetes and graft versus host disease.[152–154] Similarly, the novel autoantibodies discovered from the study in APS1 may aid in the diagnosis of more common autoimmune disease such as interstitial lung disease or autoimmune gonadal failure. In summary, the exciting research advances from the study of monogenic autoimmune diseases are making vital contributions to the diagnosis and treatment of more common autoimmune diseases.

Acknowledgments

We would like to thank Una Fan for her help with the figures and Mickie Cheng for critical review of the manuscript. This work was supported by grants from the NIH and the Burroughs Wellcome Fund to M.S.A. and from the Arthritis Foundation to M.W.

Conflicts of interest

The authors declare no conflicts of interest.

References

1. Zenewicz, L., C. Abraham, R. Flavell & J. Cho. 2010. Unraveling the genetics of autoimmunity. *Cell.* **140:** 791–797.
2. Gregersen, P. & L. Olsson. 2009. Recent advances in the genetics of autoimmune disease. *Annu. Rev. Immunol.* **27:** 363–391.
3. Husebye, E., J. Perheentupa, R. Rautemaa & O. Kämpe. 2009. Clinical manifestations and management of patients with autoimmune polyendocrine syndrome type I. *J. Intern. Med.* **265:** 514–529.
4. Perheentupa, J. 2006. Autoimmune polyendocrinopathy-candidiasis-ectodermal dystrophy. *J. Clin. Endocrinol. Metab.* **91:** 2843–2850.
5. Rautemaa, R., J. Hietanen, S. Niissalo, *et al.* 2007. Oral and oesophageal squamous cell carcinoma-a complication or component of autoimmune polyendocrinopathy-candidiasis-ectodermal dystrophy (APECED, APS-I). *Oral Oncol.* **43:** 607–613.
6. Gylling, M. *et al.* 2003. The hypoparathyroidism of autoimmune polyendocrinopathy-candidiasis-ectodermal dystrophy protective effect of male sex. *J. Clin. Endocrinol. Metab.* **88:** 4602–4608.
7. Ahonen, P., S. Myllarniemi I. Sipila, *et al.* 1990. Clinical variation of autoimmune polyendocrinopathy-candidiasis-ectodermal dystrophy (APECED) in a series of 68 patients. *N Engl. J. Med.* **322:** 1829–1836.
8. Myllärniemi, S. & J. Perheentupa. 1978. Oral findings in the autoimmune polyendocrinopathy-candidosis syndrome (APECS) and other forms of hypoparathyroidism. *Oral Surg. Oral Med. Oral Pathol.* **45:** 721–729.
9. Wolff, A. *et al.* 2007. Autoimmune polyendocrine syndrome type 1 in Norway: phenotypic variation, autoantibodies, and novel mutations in the autoimmune regulator gene. *J. Clin. Endocrinol. Metab.* **92:** 595–603.
10. Rosatelli, M. *et al.* 1998. A common mutation in Sardinian autoimmune polyendocrinopathy-candidiasis-ectodermal dystrophy patients. *Hum. Genet.* **103:** 428–434.
11. Zlotogora, J. & M. Shapiro. 1992. Polyglandular autoimmune syndrome type I among Iranian Jews. *J. Med. Genet.* **29:** 824–826.
12. Consortium, F.-G.A. 1997. An autoimmune disease, APECED, caused by mutations in a novel gene featuring two PHD-type zinc-finger domains. *Nat. Genet.* **17:** 399–403.
13. Nagamine, K. *et al.* 1997. Positional cloning of the APECED gene. *Nat. Genet.* **17:** 393–398.
14. Mathis, D., Benoist & C. Aire. 2009. *Annu. Rev. Immunol.* **27:** 287–312.
15. Scott, H. *et al.* 1998. Common mutations in autoimmune polyendocrinopathy-candidiasis-ectodermal dystrophy patients of different origins. *Mol. Endocrinol.* **12:** 1112–1119.
16. Gylling, M. *et al.* 2000. ss-Cell autoantibodies, human leukocyte antigen II alleles, and type 1 diabetes in autoimmune polyendocrinopathy-candidiasis-ectodermal dystrophy. *J. Clin. Endocrinol. Metab.* **85:** 4434–4440.
17. Halonen, M. *et al.* 2002. AIRE mutations and human leukocyte antigen genotypes as determinants of the autoimmune polyendocrinopathy-candidiasis-ectodermal dystrophy phenotype. *J. Clin. Endocrinol. Metab.* **87:** 2568–2574.
18. Cetani, F. *et al.* 2001. A novel mutation of the autoimmune regulator gene in an Italian kindred with autoimmune polyendocrinopathy-candidiasis-ectodermal dystrophy, acting in a dominant fashion and strongly cosegregating with hypothyroid autoimmune thyroiditis. *J. Clin. Endocrinol. Metab.* **86:** 4747–4752.
19. Cervato, S. *et al.* 2009. Evaluation of the AIRE gene mutations in a cohort of Italian patients with autoimmune-polyendocrinopathy-candidiasis-ectodermal-dystrophy (APECED) and in their relatives. *Clin. Endocrinol.* **70:** 421–428.
20. Anderson, M.S. *et al.* 2002. Projection of an immunological self shadow within the thymus by the aire protein. *Science* **298:** 1395–1401.
21. Ramsey, C. *et al.* 2002. Aire deficient mice develop multiple features of APECED phenotype and show altered immune response. *Hum. Mol. Genet.* **11:** 397–409.
22. Derbinski, J., A. Schulte, B. Kyewski & L. Klein. 2001. Promiscuous gene expression in medullary thymic epithelial cells mirrors the peripheral self. *Nat. Immunol.* **2:** 1032–1039.
23. Kishimoto, H. & J. Sprent. 1997. Negative selection in the thymus includes semimature T cells. *J. Exp. Med.* **185:** 263–271.
24. Derbinski, J. & B. Kyewski. 2005. Linking signalling pathways, thymic stroma integrity and autoimmunity. *Trends Immunol.* **26:** 503–506.
25. DeVoss, J. *et al.* 2006. Spontaneous autoimmunity prevented by thymic expression of a single self-antigen. *J. Exp. Med.* **203:** 2727–2735.
26. Gavanescu, I., B. Kessler H. Ploegh, *et al.* 2007. Loss of Aire-dependent thymic expression of a peripheral tissue antigen renders it a target of autoimmunity. *Proc. Natl. Acad. Sci. USA* **104:** 4583–4587.
27. Shum, A.K. *et al.* 2009. Identification of an autoantigen demonstrates a link between interstitial lung disease and a defect in central tolerance. *Sci. Transl. Med.* **1:** 9ra20–29ra20.

28. Liston, A., S. Lesage J. Wilson, *et al.* 2003. Aire regulates negative selection of organ-specific T cells. *Nat. Immunol.* **4:** 350–354.
29. Kogawa, K. 2002. Expression of AIRE gene in peripheral monocyte/dendritic cell lineage. *Immunol. Lett.* **80:** 195–198.
30. Gardner, J.M. *et al.* 2008. Deletional tolerance mediated by extrathymic Aire-expressing cells. *Science* **321:** 843–847.
31. Lee, J.W. *et al.* 2007. Peripheral antigen display by lymph node stroma promotes T cell tolerance to intestinal self. *Nat. Immunol.* **8:** 181–190.
32. Poliani, P.L. *et al.* 2010. Human peripheral lymphoid tissues contain autoimmune regulator-expressing dendritic cells. *Am. J. Pathol.* **176:** 1104–1112.
33. Bjorses, P. *et al.* 1999. Localization of the APECED protein in distinct nuclear structures. *Hum. Mol. Genet.* **8:** 259–266.
34. Koh, A.S. *et al.* 2008. Aire employs a histone-binding module to mediate immunological tolerance, linking chromatin regulation with organ-specific autoimmunity. *Proc. Natl. Acad. Sci. USA* **105:** 15878–15883.
35. Kumar, P.G. *et al.* 2001. The autoimmune regulator (AIRE) is a DNA-binding protein. *J. Biol. Chem.* **276:** 41357–41364.
36. Org, T. *et al.* 2008. The autoimmune regulator PHD finger binds to non-methylated histone H3K4 to activate gene expression. *EMBO Rep.* **9:** 370–376.
37. Bottomley, M.J. *et al.* 2001. The SAND domain structure defines a novel DNA-binding fold in transcriptional regulation. *Nat. Struct. Biol.* **8:** 626–633.
38. Derbinski, J. *et al.* 2005. Promiscuous gene expression in thymic epithelial cells is regulated at multiple levels. *J. Exp. Med.* **202:** 33–45.
39. Johnnidis, J.B. *et al.* 2005. Chromosomal clustering of genes controlled by the aire transcription factor. *Proc. Natl. Acad. Sci. USA* **102:** 7233–7238.
40. Derbinski, J., S. Pinto S. Rosch, *et al.* 2008. Promiscuous gene expression patterns in single medullary thymic epithelial cells argue for a stochastic mechanism. *Proc. Natl. Acad. Sci. USA* **105:** 657–662.
41. Villasenor, J., W. Besse C. Benoist, *et al.* 2008. Ectopic expression of peripheral-tissue antigens in the thymic epithelium: probabilistic, monoallelic, misinitiated. *Proc. Natl. Acad. Sci. USA* **105:** 15854–15859.
42. Abramson, J., M. Giraud C. Benoist, *et al.* 2010. Aire's partners in the molecular control of immunological tolerance. *Cell.* **140:** 123–135.
43. Rando, O. & H. Chang. 2009. Genome-wide views of chromatin structure. *Annu. Rev. Biochem.* **78:** 245–271.
44. Org, T. *et al.* 2009. AIRE activated tissue specific genes have histone modifications associated with inactive chromatin. *Hum. Mol. Genet.* **18:** 4699–4710.
45. Liiv, I. *et al.* 2008. DNA-PK contributes to the phosphorylation of AIRE: importance in transcriptional activity. *Biochim. Biophys. Acta.* **1783:** 74–83.
46. Oven, I. *et al.* 2007. AIRE recruits P-TEFb for transcriptional elongation of target genes in medullary thymic epithelial cells. *Mol. Cell Biol.* **27:** 8815–8823.
47. Pitkanen, J. *et al.* 2005. Cooperative activation of transcription by autoimmune regulator AIRE and CBP. *Biochem. Biophys. Res. Commun.* **333:** 944–953.
48. Nitiss, J. 2009. DNA topoisomerase II and its growing repertoire of biological functions. *Nat. Rev. Cancer.* **9:** 327–337.
49. Meloni, A. *et al.* 2008. Autoantibodies against type I interferons as an additional diagnostic criterion for autoimmune polyendocrine syndrome type I. *J. Clin. Endocrinol. Metab.* **93:** 4389–4397.
50. Zhang, L. *et al.* 2007. A robust immunoassay for anti-interferon autoantibodies that is highly specific for patients with autoimmune polyglandular syndrome type 1. *Clin. Immunol.* **125:** 131–137.
51. Kisand, K. *et al.* 2010. Chronic mucocutaneous candidiasis in APECED or thymoma patients correlates with autoimmunity to Th17-associated cytokines. *J. Exp. Med.* **207:** 299–308.
52. Puel, A. *et al.* 2010. Autoantibodies against IL-17A, IL-17F, and IL-22 in patients with chronic mucocutaneous candidiasis and autoimmune polyendocrine syndrome type I. *J. Exp. Med.* **207:** 291–297.
53. Alimohammadi, M. *et al.* 2008. Autoimmune polyendocrine syndrome type 1 and NALP5, a parathyroid autoantigen. *N. Engl. J. Med.* **358:** 1018–1028.
54. Husebye, E. *et al.* 1997. Autoantibodies against aromatic L-amino acid decarboxylase in autoimmune polyendocrine syndrome type I. *J. Clin. Endocrinol. Metab.* **82:** 147–150.
55. Soderbergh, A. *et al.* 2004. Prevalence and clinical associations of 10 defined autoantibodies in autoimmune polyendocrine syndrome type I. *J. Clin. Endocrinol. Metab.* **89:** 544–547.
56. Pugliese, A. *et al.* 1997. The insulin gene is transcribed in the human thymus and transcription levels correlated with allelic variation at the INS VNTR-IDDM2 susceptibility locus for type 1 diabetes. *Nat. Genet.* **15:** 293–297.
57. Vafiadis, P. *et al.* 1997. Insulin expression in human thymus is modulated by INS VNTR alleles at the IDDM2 locus. *Nat. Genet.* **15:** 289–292.
58. Giraud, M. *et al.* 2007. An IRF8-binding promoter variant and AIRE control CHRNA1 promiscuous expression in thymus. *Nature* **448:** 934–937.
59. Cheng, M. *et al.* 2010. Acquired autoimmune polyglandular syndrome, thymoma, and an AIRE defect. *N. Engl. J. Med.* **362:** 764–766.
60. Ochs, H., E. Gambineri & T. Torgerson. 2007. IPEX, FOXP3 and regulatory T-cells: a model for autoimmunity. *Immunol. Res.* **38:** 112–121.
61. Torgerson, T. & H. Ochs. 2007. Immune dysregulation, polyendocrinopathy, enteropathy, X-linked: forkhead box protein 3 mutations and lack of regulatory T cells. *J. Allergy Clin. Immunol.* **120:** 744–750.
62. Powell, B., N. Buist & P. Stenzel. 1982. An X-linked syndrome of diarrhea, polyendocrinopathy, and fatal infection in infancy. *J. Pediatr.* **100:** 731–737.
63. Battaglia, M. *et al.* 2006. Rapamycin promotes expansion of functional CD4+ CD25+ FOXP3+ regulatory T cells of both healthy subjects and type 1 diabetic patients. *J. Immunol.* **177:** 8338–8847.
64. Rao, A. *et al.* 2007. Successful bone marrow transplantation for IPEX syndrome after reduced-intensity conditioning. *Blood* **109:** 383–385.

65. Brunkow, M. *et al.* 2001. Disruption of a new forkhead/winged-helix protein, scurfin, results in the fatal lymphoproliferative disorder of the scurfy mouse. *Nat. Genet.* **27:** 68–73.
66. Bennett, C. *et al.* 2001. The immune dysregulation, polyendocrinopathy, enteropathy, X-linked syndrome (IPEX) is caused by mutations of FOXP3. *Nat. Genet.* **27:** 20–21.
67. Wildin, R. *et al.* 2001. X-linked neonatal diabetes mellitus, enteropathy and endocrinopathy syndrome is the human equivalent of mouse scurfy. *Nat. Genet.* **27:** 18–20.
68. Lu, L. & A. Rudensky. 2009. Molecular orchestration of differentiation and function of regulatory T cells. *Genes Dev.* **23:** 1270–1282.
69. Wing, K. & S. Sakaguchi. 2009. Regulatory T cells exert checks and balances on self tolerance and autoimmunity. *Nat. Immunol.* **11:** 7–13.
70. Sakaguchi, S., N. Sakaguchi & M. Asano. 1995. Immunologic self-tolerance maintained by activated T cells expressing IL-2 receptor alpha-chains (CD25). Breakdown of a single mechanism of self-tolerance causes various autoimmune diseases. *J. Immunol.* **155:** 1151.
71. Fontenot, J., M. Gavin & A. Rudensky. 2003. Foxp3 programs the development and function of CD4+ CD25+ regulatory T cells. *Nat. Immunol.* **4:** 330–336.
72. Hori, S., T. Nomura & S. Sakaguchi. 2003. Control of regulatory T cell development by the transcription factor Foxp3. *Science* **299:** 1057–1061.
73. Fontenot, J. *et al.* 2005. Regulatory T cell lineage specification by the forkhead transcription factor foxp3. *Immunity* **22:** 329–341.
74. Kim, J.M., J.P. Rasmussen & A.Y. Rudensky. 2007. Regulatory T cells prevent catastrophic autoimmunity throughout the lifespan of mice. *Nat. Immunol.* **8:** 191–197.
75. Williams, L. & A. Rudensky. 2007. Maintenance of the Foxp3-dependent developmental program in mature regulatory T cells requires continued expression of Foxp3. *Nat. Immunol.* **8:** 277–284.
76. Gavin, M. *et al.* 2007. Foxp3-dependent programme of regulatory T-cell differentiation. *Nature* **445:** 771–775.
77. Rubtsov, Y. *et al.* 2008. Regulatory T cell-derived interleukin-10 limits inflammation at environmental interfaces.*Immunity* **28:** 546–558.
78. Wing, K. *et al.* 2008. CTLA-4 control over Foxp3+ regulatory T cell function. *Science* **322:** 271–275.
79. Hsieh, C. *et al.* 2004. Recognition of the peripheral self by naturally arising CD25+ CD4+ T cell receptors. *Immunity* **21:** 267–277.
80. Nishikawa, H. *et al.* 2005. Definition of target antigens for naturally occurring CD4+ CD25+ regulatory T cells. *J. Exp. Med.* **201:** 681–686.
81. Kim, H. & W. Leonard. 2007. CREB/ATF-dependent T cell receptor-induced FoxP3 gene expression: a role for DNA methylation. *J. Exp. Med.* **204:** 1543–1551.
82. Mantel, P. *et al.* 2006. Molecular mechanisms underlying FOXP3 induction in human T cells. *J. Immunol.* **176:** 3593–3602.
83. Schmidt-Supprian, M. 2004. Differential dependence of CD4+ CD25+ regulatory and natural killer-like T cells on signals leading to NF-κB activation. *Proc. Natl. Acad. Sci. USA* **101:** 4566–4571.
84. Zheng, Y. *et al.* 2010. Role of conserved non-coding DNA elements in the Foxp3 gene in regulatory T-cell fate. *Nature* **463:** 808–812.
85. Fontenot, J., J. Rasmussen, M. Gavin & A. Rudensky. 2005. A function for interleukin 2 in Foxp3-expressing regulatory T cells. *Nat. Immunol.* **6:** 1142–1151.
86. Willerford, D. *et al.* 1995. Interleukin-2 receptor alpha chain regulates the size and content of the peripheral lymphoid compartment. *Immunity* **3:** 521–530.
87. Caudy, A., S. Reddy T. Chatila, *et al.* 2007. CD25 deficiency causes an immune dysregulation, polyendocrinopathy, enteropathy, X-linked-like syndrome, and defective IL-10 expression from CD4 lymphocytes. *J. Allergy Clin. Immunol.* **119:** 482–487.
88. Sharfe, N., H. Dadi, M. Shahar & C. Roifman. 1997. Human immune disorder arising from mutation of the chain of the interleukin-2 receptor. *Proc. Natl. Acad. Sci. USA* **94:** 3168–3171.
89. Liu, Y. *et al.* 2008. A critical function for TGF- signaling in the development of natural CD4+ CD25 +Foxp3+ regulatory T cells. *Nat. Immunol.* **9:** 632–640.
90. Salomon, B. *et al.* 2000. B7/CD28 costimulation is essential for the homeostasis of the CD4+ CD25 +immunoregulatory T cells that control autoimmune diabetes. *Immunity* **12:** 431–440.
91. Curotto de Lafaille, M. & J. Lafaille. 2009. Natural and adaptive foxp3+ regulatory T cells: more of the same or a division of labor. *Immunity* **30:** 626–635.
92. Apostolou, I. & H. von Boehmer. 2004. In vivo instruction of suppressor commitment in naive T cells. *J. Exp. Med.* **199:** 1401–1408.
93. Curotto de Lafaille, M. 2004. CD25-T cells generate CD25+ Foxp3+ regulatory T cells by peripheral expansion. *J. Immunol.* **173:** 7259–7268.
94. Coombes, J. *et al.* 2007. A functionally specialized population of mucosal CD103 +DCs induces Foxp3+ regulatory T cells via a TGF-beta and retinoic acid-dependent mechanism. *J. Exp. Med.* **204:** 1757–1764.
95. Sun, C. *et al.* 2007. Small intestine lamina propria dendritic cells promote de novo generation of Foxp3 T reg cells via retinoic acid. *J. Exp. Med.* **204:** 1775–1785.
96. Wu, Y. *et al.* 2006. FOXP3 controls regulatory T cell function through cooperation with NFAT. *Cell.* **126:** 375–387.
97. Chaudhry, A. *et al.* 2009. CD4+ regulatory T cells control TH17 responses in a Stat3-dependent manner. *Science* **326:** 986–991.
98. Ono, M. *et al.* 2007. Foxp3 controls regulatory T-cell function by interacting with AML1/Runx1. *Nature* **446:** 685–689.
99. Zheng, Y., A. Chaudhry & A. Kas. 2009. Regulatory T-cell suppressor program co-opts transcription factor IRF4 to control TH2 responses. *Nature* **458:** 351–356.
100. Zhou, L. *et al.* 2008. TGF-beta-induced Foxp3 inhibits TH17 cell differentiation by antagonizing RORgammat function. *Nature* **453:** 236–240.

101. Marson, A. *et al.* 2007. Foxp3 occupancy and regulation of key target genes during T-cell stimulation. *Nature* **445:** 931–935.
102. Zheng, Y. *et al.* 2007. Genome-wide analysis of Foxp3 target genes in developing and mature regulatory T cells. *Nature* **445:** 936–940.
103. Gavin, M., S. Clarke & E. Negrou. 2001. Homeostasis and anergy of CD4+ CD25+ suppressor T cells in vivo. *Nat. Immunol.* **3:** 33–41.
104. McHugh, R. *et al.* 2002. CD4+ CD25+ immunoregulatory T cells: gene expression analysis reveals a functional role for the glucocorticoid-induced TNF receptor. *Immunity* **16:** 311–323.
105. Gavin, M. *et al.* 2006. Single-cell analysis of normal and FOXP3-mutant human T cells: FOXP3 expression without regulatory T cell development. *Proc. Natl. Acad. Sci. USA* **103:** 6659–6664.
106. Walker, M. *et al.* 2003. Induction of FoxP3 and acquisition of T regulatory activity by stimulated human CD4 CD25–T cells. *J. Clin. Invest.* **112:** 1437–1443.
107. Tran, D., H. Ramsey & E. Shevach. 2007. Induction of FOXP3 expression in naive human CD4+ FOXP3 T cells by T-cell receptor stimulation is transforming growth factor-β dependent but does not confer a regulatory phenotype. *Blood* **110:** 2983–2990.
108. Walker, M., Carson B. Nepom, *et al.* 2005. De novo generation of antigen-specific CD4+ CD25 +regulatory T cells from human CD4+ CD25–cells. *Proc. Natl. Acad. Sci. USA* **102:** 4103–4108.
109. Miyara, M. *et al.* 2009. Functional delineation and differentiation dynamics of human CD4+ T cells expressing the FoxP3 transcription factor. *Immunity* **30:** 899–911.
110. Sakaguchi, S., M. Miyara, C. Costantino & D. Hafler. 2010. FOXP3+ regulatory T cells in the human immune system. *Nat. Rev. Immunol.* **10:** 490–500.
111. Allan, S., Song-Zhao G. Abraham, *et al.* 2008. Inducible reprogramming of human T cells into Treg cells by a conditionally active form of FOXP3. *Eur. J. Immunol.* **38:** 3282–3289.
112. Hoffmann, P. *et al.* 2009. Loss of FOXP3 expression in natural human CD4+ CD25 +regulatory T cells upon repetitive in vitro stimulation. *Eur. J. Immunol.* **39:** 1088–1097.
113. Kleinewietfeld, M. *et al.* 2009. CD49d provides access to" untouched" human Foxp3+ Treg free of contaminating effector cells. *Blood* **113:** 827–836.
114. Liu, W. *et al.* 2006. CD127 expression inversely correlates with FoxP3 and suppressive function of human CD4+ T reg cells. *J. Exp. Med.* **203:** 1701–1711.
115. Nadig, S. *et al.* 2010. In vivo prevention of transplant arteriosclerosis by ex vivo-expanded human regulatory T cells. *Nat. Med.* **16:** 809–813.
116. Ehrenstein, M. *et al.* 2004. Compromised function of regulatory T cells in rheumatoid arthritis and reversal by anti-TNF alpha therapy. *J. Exp. Med.* **200:** 277–285.
117. Esensten, J., D. Wofsy & J. Bluestone. 2009. Regulatory T cells as therapeutic targets in rheumatoid arthritis. *Nat. Rev. Rheumatol.* **5:** 560–565.
118. Valencia, X. *et al.* 2006. TNF downmodulates the function of human CD4+ CD25hi T-regulatory cells. *Blood* **108:** 253–261.
119. Brusko, T. *et al.* 2007. No alterations in the frequency of FOXP3+ regulatory T-cells in type 1 diabetes. *Diabetes* **56:** 604–612.
120. Kukreja, A. *et al.* 2002. Multiple immuno-regulatory defects in type-1 diabetes. *J. Clin. Invest.* **109:** 131–140.
121. Lindley, S. 2005. Defective suppressor function in CD4+ CD25+ T-cells from patients with type 1 diabetes. *Diabetes* **54:** 92–99.
122. Putnam, A., F. Vendrame, F. Dotta & P. Gottlieb. 2005. CD4+ CD25high regulatory T cells in human autoimmune diabetes. *J. Autoimmun.* **24:** 55–62.
123. Sneller, M. 1992. A novel lymphoproliferative/autoimmune syndrome resembling murine lpr/gld disease. *J. Clin. Invest.* **90:** 334–341.
124. Teachey, D., A. Seif & S. Grupp. 2009. Advances in the management and understanding of autoimmune lymphoproliferative syndrome (ALPS). *Br. J. Haematol.* **148:** 205–216.
125. Worth, A., A. Thrasher & H. Gaspar. 2006. Autoimmune lymphoproliferative syndrome: molecular basis of disease and clinical phenotype. *Br. J. Haematol.* **133:** 124–140.
126. Straus, S. *et al.* 2001. The development of lymphomas in families with autoimmune lymphoproliferative syndrome with germline Fas mutations and defective lymphocyte apoptosis. *Blood* **98:** 194–200.
127. Dimopoulou, M. *et al.* 2007. Successful treatment of autoimmune lymphoproliferative syndrome and refractory autoimmune thrombocytopenic purpura with a reduced intensity conditioning stem cell transplantation followed by donor lymphocyte infusion. *Bone Marrow Transplant.* **40:** 605–606.
128. Takahashi, T. *et al.* 1994. Generalized lymphoproliferative disease in mice, caused by a point mutation in the Fas ligand. *Cell.* **76:** 969–976.
129. Watanabe-Fukunaga, R. *et al.* 1992. Lymphoproliferation disorder in mice explained by defects in Fas antigen that mediates apoptosis.*Nature* **356:** 314–317.
130. Fisher, G. *et al.* 1995. Dominant Interfering Fas Gene Mutations Impair Apoptosis in a Human Autoimmune Lymphoproliferative Syndrome. *Cell.* **81:** 935–946.
131. Rieux-Laucat, F. *et al.* 1995. Mutations in Fas associated with human lymphoproliferative syndrome and autoimmunity. *Science* **268:** 1347–1349.
132. Wu, J. *et al.* 1996. Fas ligand mutation in a patient with systemic lupus erythematosus and lymphoproliferative disease. *J. Clin. Invest.* **98:** 1107–1113.
133. Jackson, C. *et al.* 1999. Autoimmune lymphoproliferative syndrome with defective Fas: genotype influences penetrance. *Am. J. Hum. Genet.* **64:** 1002–1014.
134. Kasahara, Y. 1998. Novel Fas (CD95/APO-1) mutations in infants with a lymphoproliferative disorder. *Int. Immunol.* **10:** 195–202.
135. Le Deist, F. *et al.* 1996. Clinical, immunological, and pathological consequences of Fas-deficient conditions. *Lancet* **348:** 719–723.

136. Bettinardi, A. *et al.* 1997. Missense mutations in the Fas gene resulting in autoimmune lymphoproliferative syndrome: a molecular and immunological analysis. *Blood* **89:** 902–909.
137. Oliveira, J. *et al.* 2007. NRAS mutation causes a human autoimmune lymphoproliferative syndrome. *Proc. Natl. Acad. Sci. USA* **104:** 8953–8958.
138. Wang, J. *et al.* 1999. Inherited human Caspase 10 mutations underlie defective lymphocyte and dendritic cell apoptosis in autoimmune lymphoproliferative syndrome type II. *Cell.* **98:** 47–58.
139. Zhu, S. *et al.* 2006. Genetic alterations in caspase-10 may be causative or protective in autoimmune lymphoproliferative syndrome. *Hum. Genet.* **119:** 284–294.
140. Chun, H. *et al.* 2002. Pleiotropic defects in lymphocyte activation caused by caspase-8 mutations lead to human immunodeficiency. *Nature* **419:** 395–399.
141. Oliveira, J. *et al.* 2010. Revised diagnostic criteria and classification for the autoimmune lymphoproliferative syndrome: report from the 2009 NIH International Workshop. *Blood* epub ahead of print June 10
142. Lenardo, M., J. Oliveira, L. Zheng & V. Rao. 2010. ALPS–ten lessons from an International Workshop on a Genetic Disease of Apoptosis. *Immunity* **32:** 291–295.
143. Holzelova, E. *et al.* 2004. Autoimmune lymphoproliferative syndrome with somatic Fas mutations. *N. Engl. J. Med.* **351:** 1409.
144. Dowdell, K. *et al.* 2010. Somatic FAS mutations are common in patients with genetically undefined autoimmune lymphoproliferative syndrome (ALPS). *Blood* **115:** 5164–5169.
145. Bleesing, J., S. Straus & T. Fleisher. 2000. Autoimmune lymphoproliferative syndrome. A human disorder of abnormal lymphocyte survival. *Pediatr. Clin. North Am.* **47:** 1291–1310.
146. Seif, A., C. Manno, S. Grupp & D. Teachey. 2008. Testing patients with Evans syndrome for the Autoimmune Lymphoproliferative Syndrome(ALPS): results of a large multinational clinical trial (ASPHO supplement). *Pediatr. Blood Cancer* **50:** S22–23.
147. Dianzani, U. 1997. Deficiency of the Fas apoptosis pathway without Fas gene mutations in pediatric patients with autoimmunity/lymphoproliferation. *Blood* **89:** 2871–2879.
148. Snow, A. *et al.* 2009. Restimulation-induced apoptosis of T cells is impaired in patients with X-linked lymphoproliferative disease caused by SAP deficiency. *J. Clin. Invest.* **119:** 2976–2989.
149. Strasser, A., P. Jost & S. Nagata. 2009. The many roles of FAS receptor signaling in the immune system. *Immunity* **30:** 180–192.
150. Goodnow, C. 2007. Multistep pathogenesis of autoimmune disease. *Cell.* **130:** 25–35.
151. Tang, Q. *et al.* 2004. In vitro-expanded antigen-specific regulatory T cells suppress autoimmune diabetes. *J. Exp. Med.* **199:** 1455–1465.
152. Brusko, T., A. Putnam & J. Bluestone. 2008. Human regulatory T cells: role in autoimmune disease and therapeutic opportunities. *Immunol. Rev.* **223:** 371–390.
153. Putnam, A. *et al.* 2009. Expansion of human regulatory T-cells from patients with type 1 diabetes. *Diabetes* **58:** 652–662.
154. Trzonkowski, P. *et al.* 2009. First-in-man clinical results of the treatment of patients with graft versus host disease with human ex vivo expanded CD4+ CD25+ CD127-T regulatory cells. *Clin. Immunol.* **133:** 22–26.

Ann. N.Y. Acad. Sci. ISSN 0077-8923

ANNALS OF THE NEW YORK ACADEMY OF SCIENCES
Issue: *The Year in Human and Medical Genetics*

Genetic regulation of human brain development: lessons from Mendelian diseases

Tracy J. Dixon-Salazar and Joseph G. Gleeson

Departments of Neurosciences and Pediatrics, Institute for Genomic Medicine, Howard Hughes Medical Institute, University of California, San Diego, La Jolla, California

Address for correspondence: Joseph G. Gleeson, M.D., University of California, San Diego, 9500 Gilman Drive 0665, La Jolla, CA 92093-0665. jogleeson@ucsd.edu

One of the fundamental goals in human genetics is to link gene function to phenotype, yet the function of the majority of the genes in the human body is still poorly understood. This is especially true for the developing human brain. The study of human phenotypes that result from inherited, mutated alleles is the most direct evidence for the requirement of a gene in human physiology. Thus, the study of Mendelian central nervous system (CNS) diseases can be an extremely powerful approach to elucidate such phenotypic/genotypic links and to increase our understanding of the key components required for development of the human brain. In this review, we highlight examples of how the study of inherited neurodevelopmental disorders contributes to our knowledge of both the "normal" and diseased human brain, as well as elaborate on the future of this type of research. Mendelian disease research has been, and will continue to be, key to understanding the molecular mechanisms that underlie human brain function, and will ultimately form a basis for the design of intelligent, mechanism-specific treatments for nervous system disorders.

Keywords: genetics; inherit; brain; development; CNS; disease; Mendelian

Introduction

The human cerebral cortex is made up of ~20 billion neurons, each of which makes an average of 7,000 synaptic contacts.[1] To further add to this complexity, each neuron contains 46 chromosomes, made up of ~3 billion nucleotides, and up to 25,000 individual genes are contained in one set of 23 chromosomes.[2] These features illustrate the potential for complex genetic regulation of the human brain and despite extensive research, much of how the brain develops and functions remains unknown. Furthermore, disorders involving the brain are among the most difficult to treat. Only with a better understanding of the nervous system will we be able to improve the quality of life of those who suffer from these disorders.

The identification of a disease-causing gene in families with inherited brain disorders is one of the most useful methods for gaining insight into human brain development, function, and pathology. Unlike other approaches, the identification of causative gene mutations provides three crucial capabilities: (1) the ability to explore the cause of disease directly in humans; (2) the ability to determine genetic disease risk, especially as it relates to penetrance (100% for homozygous null alleles), with a high level of certainty; and (3) the ability to create appropriate, translationally relevant cell-based and animal models for studying the mechanisms that underlie a given genetic disorder and the normal brain process it disrupts. Few other approaches allow for such direct insight into the human brain and identifying genes in families with inherited CNS disorders has lead to a critical understanding of the genetic control of human brain development and function.

The last 10 years has seen the advent of better technology and better diagnostic capabilities, and subsequently the rate of identification of genes in Mendelian CNS disorders has rapidly increased.[3] We know that genetic disruption of specific steps in brain development at critical time points can lead to characteristic morphological or functional

doi: 10.1111/j.1749-6632.2010.05819.x

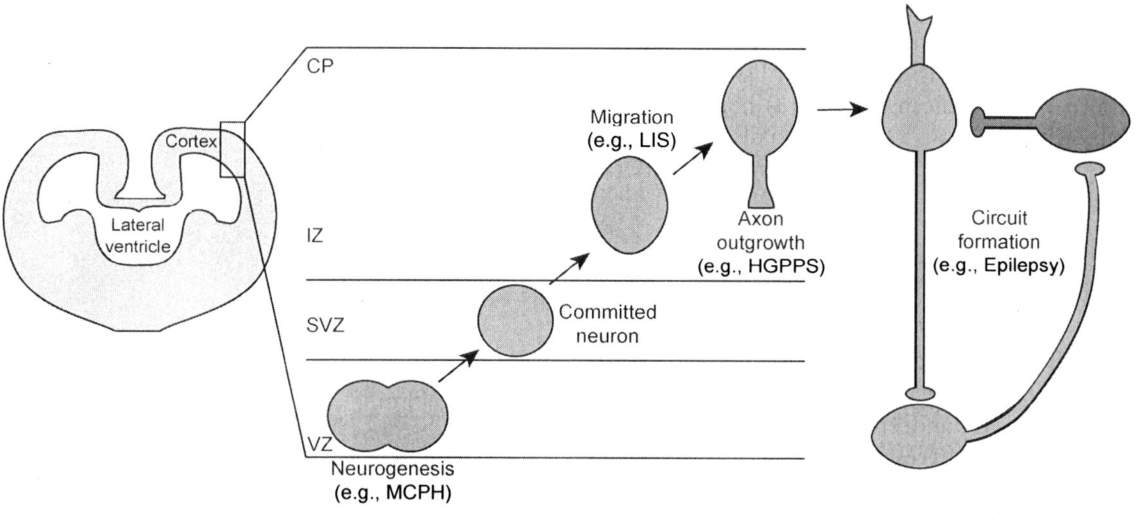

Figure 1. Overview of early cortical neuron development. Cortical neurons are born and undergo fate determination in the ventricular zone (VZ) and subventricular zone of the developing cortex. These neurons then migrate through the intermediate zone (IZ) and into the cortical plate (CP), where they come to reside in their specific cortical layer. Maturing neurons then send out axons, which grow outward until they reach their designated target. This process is repeated with each newly born neuron. Ultimately, neurons form many synaptic connections to create functional neural circuits. Mendelian disorders can occur at any of these developmental stages, and specific examples include primary microcephaly (MCPH), classical lissencephaly (LIS), horizontal gaze palsy with progressive scoliosis (HGPPS), and primary epilepsy.

anomalies, and with greater frequency brain disorders are being classified with regard to the time-dependent process they disrupt. Some developmental examples of this include disorders of cortical neurogenesis (e.g., primary microcephaly [MCPH]), disorders of neuron migration (e.g., classical lissencephaly [LIS]), disorders of axon guidance (e.g., horizontal gaze palsy with progressive scoliosis [HGPPS]), and disorders of circuit formation and function (e.g., epilepsy).[4–7] In this review, we will discuss how Mendelian genetic studies have lead to crucial insight into the pathology of these four disease examples. Further, we will explore what these findings tell us about the regulation of human brain size, cortical development, neuronal wiring, and neuronal networks, respectively. Finally, we will describe the role Mendelian genetic research is likely to play in the next 10 years, with an emphasis on whole genome sequencing and personalized medicine.

Cortical development

During development, neurons are born in distinct regions of the brain and then migrate to their specific laminar positions in a stereotyped spatiotemporal pattern. In the mammalian cerebral cortex, this process has been well characterized and the genetic mechanisms that underlie cortical development have been extensively studied.[7–9] Using a simplified model, neuronal development can be categorized into four main processes: neurogenesis, neuron migration, axon outgrowth, and circuit formation (Fig. 1). Anatomically, neural progenitors located in the ventricular zone undergo mitosis to give rise to cortical neurons. These neurons then migrate into the cortical plate where they ultimately organize into the six-layered cortex characteristic of the mammalian brain. Newborn neurons send out axons and other neurites and these axons navigate the brain milieu until locating their appropriate synaptic targets. These processes are repeated in progression by each developing neuron and, via a series of elaborate molecular and cellular mechanisms, the neurons wire themselves together to form functional neural circuits.[9,10]

Many causative genes have been identified in disorders of cortical development and the study of these genes in model systems has aided in our understanding of signaling pathways that underlie these diseases. In fact, the study of human gene mutations has formed the basis for many discoveries into development of the mammalian cortex. We next discuss four disease examples that exemplify these points: (1) primary microcephaly—a disorder of neurogenesis, (2) classical lissencephaly—a disorder of neuron migration, (3) horizontal gaze palsy with progressive scoliosis—a disorder of axon outgrowth, and (4) primary epilepsy—a disorder of circuit formation and function.

Primary microcephaly

MCPH is a neurodevelopmental disorder characterized by significantly reduced cranial circumference at birth and intellectual disability.[11–13] MCPH includes a broad spectrum of phenotypes with numerous genetic and environmental causes; however, autosomal recessive MCPH (AR-MCPH) is a distinct clinical entity. AR-MCPH is associated with normal brain architecture and nonprogressive intellectual disability in an otherwise healthy person (Fig. 2).[12,14] MRI scans of AR-MCPH patients reveal a reduced brain volume but normal brain organization, suggesting that MCPH is not a disorder of neuron migration or neuronal organization, but is rather primarily a disorder of neuronal numbers.[12,15]

Using genetic mapping techniques in consanguineous families with AR-MCPH, at least seven MCPH-causing loci[16–23] have been characterized and five genes have been identified thus far[18,24–26] (Table 1). The five AR-MCPH genes, *MCPH1*, *ASPM*, *CDK5RAP2*, *CENPJ*, and *STIL*, encode proteins that are expressed in cortical neuroprogenitor cells and, interestingly, each colocalizes with the mitotic spindle apparatus at some point during cortical neurogenesis (Fig. 3). *MCPH1* encodes the microcephalin protein, which is a component of the DNA-damage response pathway[27] and cell lines from MCPH patient fibroblasts exhibit defective cell cycle arrest, which is predicted to reduce the total number of neural progenitors.[26,28] The remaining MCPH genes, *ASPM*, *CDK5RAP2*, *CENPJ*, and *STIL*, encode centrosomal associated proteins that are predicted to regulate mitotic spindle function.[24–26] The coordinated steps of cortical development rely on numerous factors including pattern formation, cell proliferation, cell survival, and cell growth,[29,30] but studies of the MCPH genes reveal that this disorder is primarily due to a defect in one step—neuronal proliferation. Although there is currently no treatment for MCPH, the identification of a causal mutation now allows for prenatal testing for

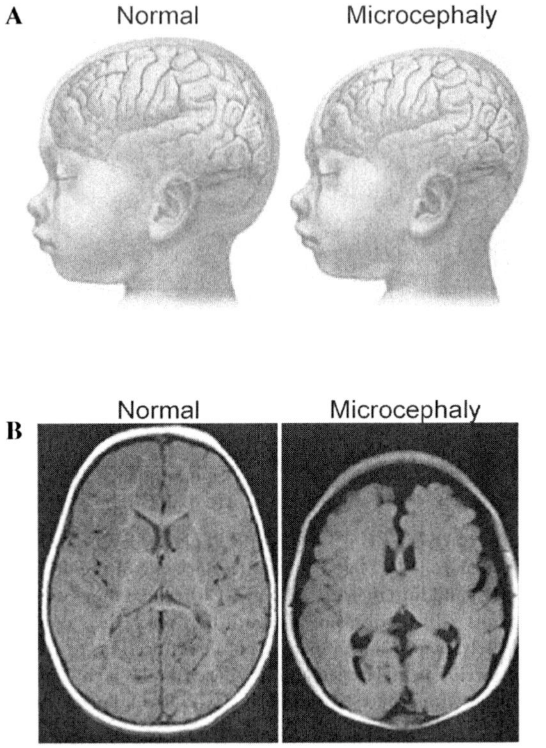

Figure 2. Anatomical features of primary microcephaly (MCPH). (A) Children with MCPH have a reduced head circumference beginning at birth and a small, but architecturally normal, brain as compared to a healthy control. Figure reprinted with permission.[101] (B) MRI scans of a healthy child and a child with microcephaly. The microcephalic brain shows reduced brain volume, with the largest volume loss seen in the cortical areas.

families. Importantly, the identification of MCPH-causing genes now provides a framework for us to expand our understanding of brain growth. These studies will lead to the development of a battery of diagnostic genetic tests, which can ultimately be used for prediction of clinical course and outcome, and studies of the molecular mechanisms that underlie microcephaly will ultimately form the basis for future therapies.

Although studies reveal that mitotic spindle regulation is important for understanding the pathogenic mechanisms of MCPH, they also may give us information about mammalian brain evolution. During human evolution, brain size, and specifically cortical surface area, has dramatically increased. Given that MCPH is a disorder of brain size, the study of the MCPH-causing genes across evolution may lead to insight into the genetic

Table 1. Genes involved in Mendelian disorders of cortical development and function

Disorder	Gene name	Localization	Cellular function
Neurogenesis			
primary microcephaly (MCPH)	*MCPH1*	Centrosome	Coordinates entry into mitosis
	ASPM	Pericentrosome	Regulates mitotic spindles
	CDK5RAP2	Centrosome	Coordinates centrosome during mitosis
	CENPJ	Centrosome	Coordinates centrosome during mitosis
	STIL	Pericentrosome	Regulates mitotic spindles
Neuron migration			
isolated lissencephaly sequence (ILS)	*DCX*	Microtubules	Bundles microtubules during migration
	LIS1	Microtubules	Component of dynein motor complex
	TUBA1A	Microtubules	Major constituent of microtubules
Axon outgrowth			
horizontal gaze palsy with progressive scoliosis (HSPPS)	*ROBO3*	Growth cone receptor	Guidance of projection axons
Congenital mirror movements (CMM)	*DCC*	Growth cone receptor	Guidance of projection axons
Circuit formation			
primary epilepsy	*SCN1A*	Axonal receptor	Action potential initiation
	SCN1B	Axonal receptor	Action potential initiation
	SCN2A	Axonal receptor	Action potential initiation
	GABRA1	Postsynaptic receptor	Synaptic inhibition
	GABRAG2	Postsynaptic receptor	Synaptic inhibition
	CHRNA2	Presynaptic receptor	Synaptic transmission
	CHRNB2	Presynaptic receptor	Synaptic transmission
	CHRNA4	Presynaptic receptor	Synaptic transmission
	CACNA1A	Presynaptic receptor	Synaptic transmission
	KCNQ2	Axonal receptor	Axon potential propagation
	KCNQ3	Axonal receptor	Axon potential propagation
	KCNA1	Presynaptic receptor	Synaptic transmission
	KCNMA1	Presynaptic receptor	Synaptic function

regulation of brain volume across species. Indeed, studies in monkeys and humans reveal that the *MCPH1* and *ASPM* genes may have experienced positive Darwinian selection over time.[12,15,31–33] Although much work is still needed to prove the extent of this selective pressure on the MCPH genes, these studies serve as an important example of how Mendelian disease genes allow us to explore other areas of brain development beyond pathogenic mechanisms.

Classical lissencephaly

LIS is a neurodevelopmental disorder characterized by the lack of cortical folds and an abnormally thick cortex (Fig. 4). Children born with LIS suffer from severe epilepsy, mental retardation, and often die in early life.[34–37] Histopathology reveals a four-layer cortical structure rather than the six layers typically seen, and most of the neurons in the abnormal cortex are found in the deeper layers.[38] Cortical lamination is a highly orchestrated process that relies

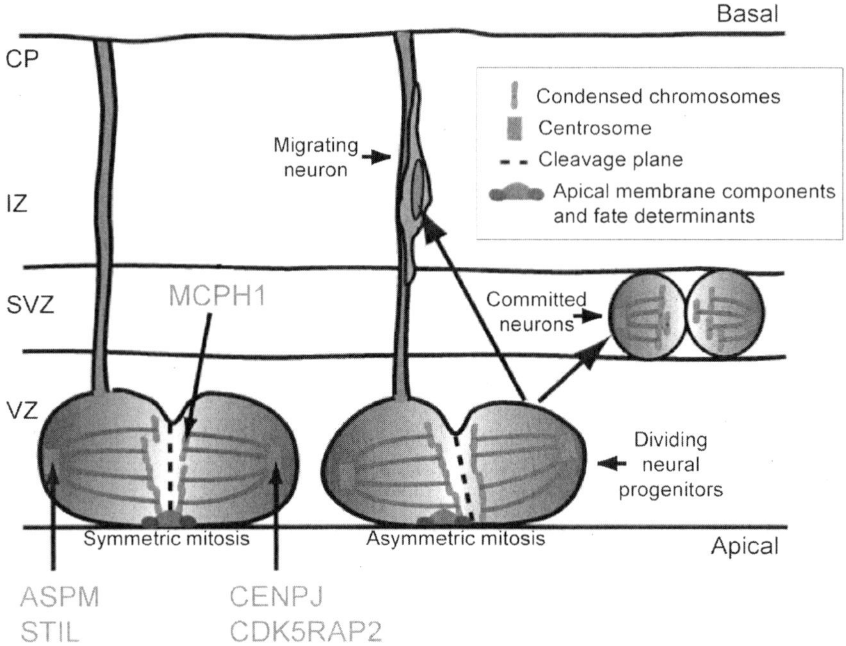

Figure 3. Genes that cause MCPH localize to the centrosome during the cell cycle and play a role in neurogenesis. During neurogenic proliferation, progenitors first undergo symmetric cell division where two daughter cells of progenitor fate are produced. This is followed by asymmetric cell division in which one daughter cell follows a progenitor cell fate and the other follows a neural fate.[29,102–104] Committed neurons then migrate to the cortical plate (CP). The cell cycle is critically regulated by the mitotic machinery,[29,105,106] and each of the five MCPH genes (*MCPH1, APSM, STIL, CENPJ*, and *CKD5RAP2)* colocalize with the mitotic apparatus during at least some part of the cell cycle. Genetic studies reveal that MCPH is primarily a disorder of cortical neurogenesis and not a disorder of migration, cell death, or cell growth and provide insight into the genes that regulate brain size in humans.

on numerous factors;[7] however, the aggregation of neurons in the deeper layers—close to where they are born—suggests that LIS is a disorder of neuronal movement.

Genetic studies reveal that lissencephaly is a multigenic disorder that can be inherited in dominant, recessive, or X-linked patterns (Table 1). At least six genes have been implicated in causing the lissencephalic phenotype,[39–46] three of which underlie LIS (lissencephaly without any other cerebral features). During development, committed neurons in the subventricular zone extend a leading process toward the cortical plate to which they ultimately will migrate. After this extension, a complex process of nucleus translocation occurs followed by retraction of the tailing edge of the migrating neuron.[7,47] The three LIS genes, *PAFAH1B1 (LIS1), DCX*, and *TUBA1A*, encode microtubule or microtubule associated proteins and each is localized to the leading process and cell body of migrating cortical neurons[7,48,49] (Fig. 5). LIS1 is a component of the cytoplasmic microtubule motor complex dynein, and studies reveal that *LIS1* may be important for translocation of the nucleus during neuron migration.[47,50–52] DCX also binds to microtubules and studies show that loss of *DCX* leads to increased microtubule dynamics.[53,54] Further, *TUBA1A* encodes a family of alpha tubulins which together with beta tubulin form the primary structure of cytoskeletal microtubules. *TUBA1A* is highly expressed during fetal brain development and disruption of this gene lead to defects in cortical lamination in animal models.[55,56] The coordinated progression of neuron migration involves many steps, including the extension of radial glia into the cortical plate to provide a guide for neurons to migrate along, the complex organization and reorganization of the cytoskeleton as the neuron migrates, and the presence of internal and external guidance cues by which neurons navigate.[7,9,47] The study of the LIS genes reveals that LIS is primarily a disorder of cytoskeletal reorganization during neuron migration. Future studies will provide a basis for the treatment, or even prevention, of this severe childhood disorder.

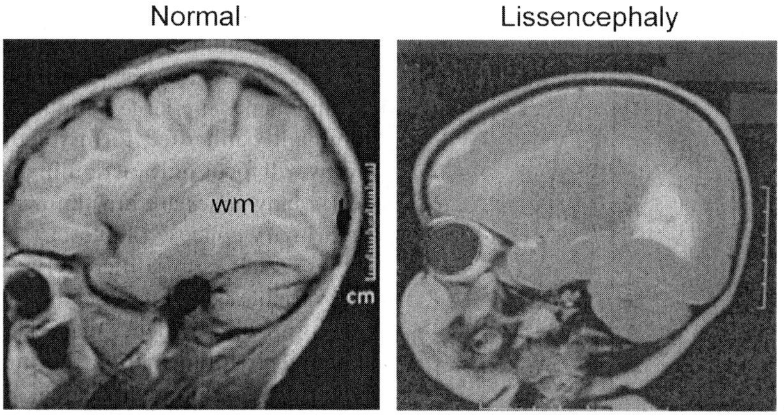

Figure 4. Anatomical features of classical lissencephaly (LIS). Children with LIS, also known as smooth brain syndrome, lack folds and ridges in the outer cortical layers and typically have a smaller head size as compared to healthy controls (wm, white matter). Figure reprinted with permission.[42]

To this end, preliminary studies in animal models reveal that it might be possible to move neurons that fail to migrate properly back into place. As proof of this concept, researchers used short hairpin RNA (shRNA) technology to knockdown *DCX* expression in the developing rat cortex.[49,57] In addition to injecting the *DCX* knockdown shRNA, a drug-inducible form of *DCX* was also injected, thereby allowing for reexpression of the *DCX* gene to occur when the drug was administered. Loss of *DCX* lead to defects in cortical neuron migration, but when *DCX* was reexpressed early postnatally, many of the delayed neurons resumed migration and ultimately reached their correct targets in the cortical plate. Although this type of treatment is far from its translation to human patients, it does provide evidence that further study of the LIS genes may lead to new treatment and prevention strategies for this disorder.

Horizontal gaze palsy with progressive scoliosis

HGPPS is an autosomal recessive disorder characterized by absence of horizontal eye movements from birth (Fig. 6A) and progressive curvature of the spine starting in the first decade of life (Fig. 6B).[58,59] Imaging studies in HGPPS patients reveal an absence of axons that cross the midline in the pons and medulla oblongata (Fig. 6C),[60,61] and functional MRI studies show that activation of the primary motor cortex occurs in an atypical ipsilateral, rather than contralateral, fashion in these patients.[60] These clinical findings reveal that HGPPS results from failure of axons to decussate during brain development, and more generally suggest that HGPPS is a disorder of axon outgrowth and guidance.

Using genetic mapping techniques in consanguineous families with HGPPS, mutations in the *ROBO3* gene were identified[61,62] (Table 1). During development, motor and sensory cortical neurons extend axons that will cross the midline in the hindbrain and travel to their ultimate destination to form synapses with their molecular targets.[5,63] *ROBO3* is a member of the *roundabout* gene family, which is well known from animal studies to regulate neurite outgrowth, growth cone guidance, and axon fasciculation.[64,65] Animal models reveal that *ROBO3* encodes a protein that is expressed on the surface of migrating growth cones and encodes a receptor for the chemorepellant molecule SLIT. Binding of the SLIT ligand to the *ROBO3* protein generates an axon repulsion signal and this, in turn, regulates axon growth and turning at or near the midline.[5,66] Axon outgrowth is a highly choreographed process that involves numerous steps including growth cone formation, axon guidance and target acquisition. The combined clinical and genetic studies of HGPPS confirm that this disorder primarily results in a failure of axon guidance during hindbrain midline crossing.

In the example of HGPPS, the identification of *ROBO3* gene mutations came after extensive characterization of the *ROBO* gene family in animal models. This example confirms two critical assumptions: (1) genetic homology between species can be effectively used to identify disease-causing genes in

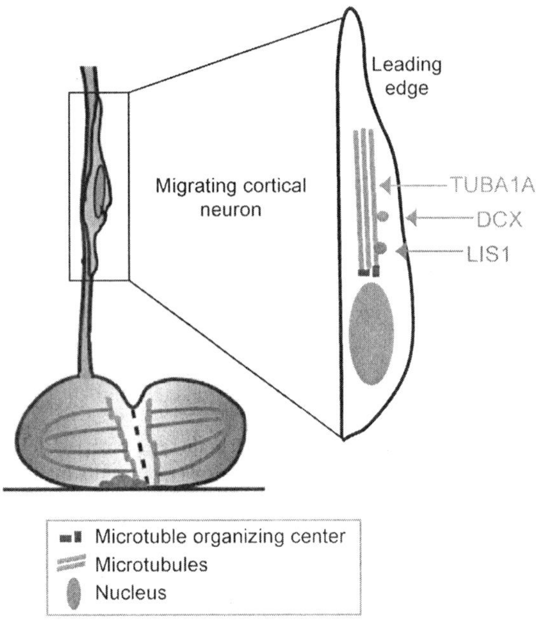

Figure 5. Genes that cause LIS localize to the leading edge of cortical neurons and play a role in neuron migration. Committed neurons migrate along radial glia to reach their specified layer in the developing cortical plate. Neuron migration involves a strictly regulated process of cytoskeletal rearrangements. Each of the LIS-causing genes (*DCX, LIS1, TUBA1A*) act as key regulators of microtubules. Genetic studies reveal that LIS is primarily a disorder of neuron migration and provide mechanistic information about the genetic regulation of cortical development in humans.

Mendelian disorders, and (2) the study of development (i.e., midline crossing) in animal models is relevant for understanding the critical determinants of brain development in humans. Although research on how *ROBO3* mutations lead to the specific HGPPS phenotype is still in its early stages, the study of this genetic disorder, and others such as the recently identified Congenital Mirror Movement Disorder associated with *DCC* mutations,[67] will undoubtedly afford us a better understanding of axon outgrowth in human brain development.

Primary epilepsy

The epilepsies are an extremely common group of brain disorders characterized by recurrent, unprovoked seizures, which manifest phenotypically as changes in motor, sensory, or cognitive function.[4] Primary epilepsy exists when seizures are not the result of a known insult (such as a brain malformation, neurodegeneration, or trauma), but rather seizures themselves are the major phenotype. Our understanding of human primary epilepsy was significantly accelerated by the identification of epilepsy-causing genes in large families with autosomal dominantly inherited forms of the disorder (this review will focus only on dominant primary epilepsies). Many genes are now known to cause heritable primary epilepsy (Table 1).[68–81] Although each of these genes causes different types of seizures that have variable ages of onset,[82] the collective data provide compelling evidence that seizures are disorders of neural circuit formation and function.[83]

During development, axonal projections form synapses with their appropriate targets and this ultimately leads to a series of interconnected neural networks in the brain and body.[63] Communication between neurons, and groups of neurons, depends primarily on the activity of transmembrane ion channels and to date mutations in over a dozen different ion channels, or their subunits, have been shown to underlie inherited primary epilepsy.[68–81] The genes identified include eight voltage-gated ion channels which are key for action potential generation and neuronal excitability and five ligand-gated ion channels which regulate synaptic transmission (Fig. 7). The coordinated steps of circuit formation and function rely on numerous factors including synapse formation of excitatory and inhibitory neurons, early experience-independent neuronal activity and later experience-dependent neuronal activity.[84] Although the repertoire of molecules involved in circuit formation has not been fully characterized, the study of the epilepsy-causing genes reveals that ion channel dysfunction is a major cause of seizures.

Unlike MCPH, LIS, or HGPPS, the pathogenic mechanisms underlying the connection between ion channel genes and epilepsy are poorly understood. In fact, the many cellular and molecular mechanisms that underlie circuit formation are only just now being unraveled. Furthermore, given that neural circuits display both intrinsic and extrinsic activity-dependent plasticity,[85] epilepsy is not a static disorder. Each abnormal burst of electrical activity has the potential to structurally and functionally change the circuit in ways that can both help or hinder the organism.[86,87] Therefore, the line between the causes and consequences of epilepsy are infinitely blurred. With the identification of genes that underlie Mendelian forms of epilepsy, our understanding of this disorder has been catapulted to new levels. Future studies of the role each of these genes plays in synaptic development will unquestionably advance our understanding of circuit formation in the brain and will have immediate applications towards epilepsy treatment.

Although there is no cure for epilepsy, the immediate benefits of identifying epilepsy-causing genes are many: (1) allowing for early and accurate diagnosis, (2) permitting research into which treatments improve or worsen seizures in a specific genetic epilepsy,(3) offering the possibility of genetic counseling to those who are carriers of the disease gene, (4) allowing for the creation of novel, gene-specific therapeutics, and (5) allowing for the generation of translationally relevant model systems to test potential therapies.[88] Furthermore, understanding the monogenic forms of epilepsy may ultimately help us understand the more complex forms of epilepsy that exist in the broader population.

In summary, the identification of genes in Mendelian brain disorders such as MCPH, LIS, HGPPS, and epilepsy has lead to a greater knowledge of the genetic machinery that underlies brain development. These studies provide direct insight into the formation and function of the human brain and are beginning to form the basis of intelligent treatment design for these devastating disorders of childhood.

Future directions

The next 10 years of Mendelian gene research look very promising. Although many genes have been identified in inherited disorders, many more remain to be discovered, as evidenced by the numerous monogenic disorders in which the causative-gene remains elusive. The identification of disease-causing genes once relied solely upon the study of large, often inbred, families containing multiple affects with the disease phenotype. Linkage analysis performed on these families determined regions of the genome that were shared by the affected individuals and this, in turn, led to the direct sequencing of candidate genes in the regions of interest.[89] With the

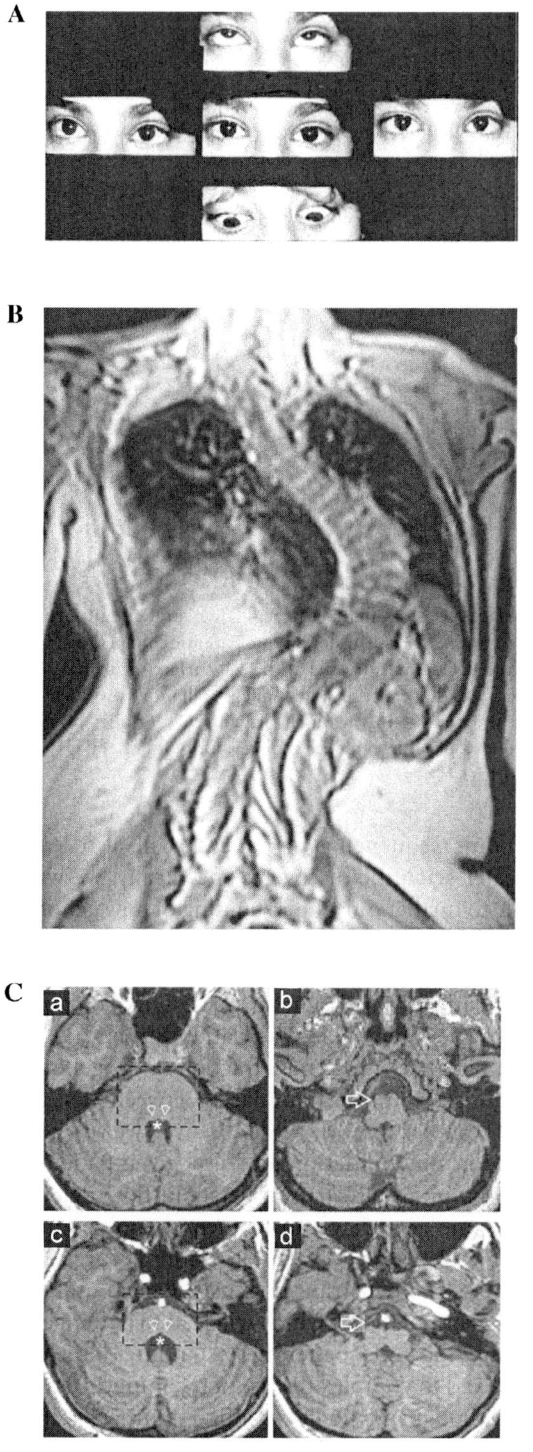

Figure 6. Anatomical features of horizontal gaze palsy with progressive scoliosis (HGPPS). (A) Photo of a patient with HGPPS showing lack of horizontal eye movements when attempting to look to either side. Upward and downward eye movements are normal. (B) MRI of the spine of a patient with HGPPS shows profound scoliosis. (C) MRI scans from a normal brain (a,b) and a patient with HGPPS (c,d). Patients with HGPPS display pons hypoplasia (boxed region in a,b); absent protrusions of the abducens nuclei (arrowheads in a,b); and flat, butterfly-like appearance of the medulla (open arrow in c,d). Figures reprinted with permission.[61]

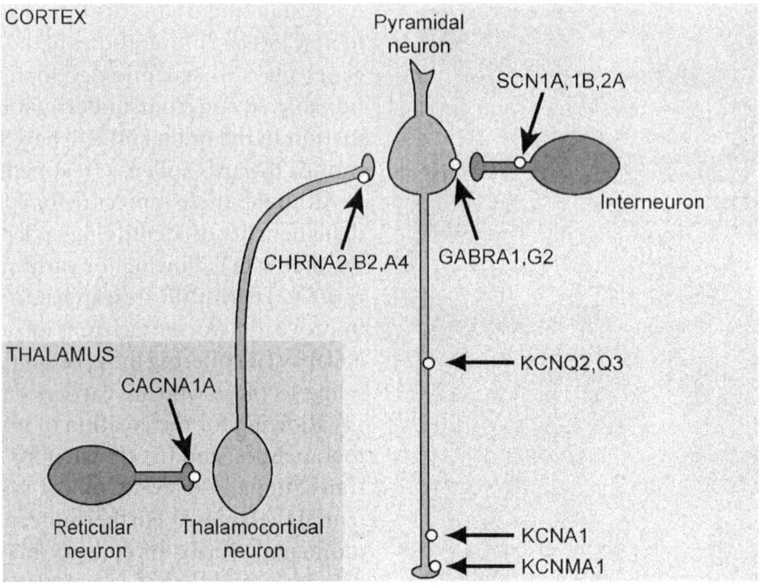

Figure 7. Many of the genes that cause primary epilepsy are cell surface receptors that act to regulate neural circuit formation and network signaling. During circuit formation, neurons from different brain areas form stereotypic synaptic connections. Circuit formation is initially neuronal activity independent, but as signaling receptors are expressed at the cell surface, neuronal activity begins to shape circuit structure and function. Thirteen of the identified genes that cause primary epilepsy fall into the voltage-gated or ligand-gated ion channel class and are expressed in the axon or at the synapse in various neuronal types. Studies reveal that epilepsy can result from a primary defect in ion channel function and also provide evidence that certain types of epilepsy result from defects in circuit formation and function in the developing neocortex.

ever dropping costs of "whole exome" and "whole genome" sequencing,[90] the identification of inherited disease-causing genes will be rapidly accelerated in the coming years. Importantly, these techniques do not require the use of large families, but instead allow for disease gene identification in much smaller families and in patients with mutations that arise *de novo* in known Mendelian disease genes. Indeed, numerous gene mutations underlying Mendelian diseases have already been identified using these "next generation" sequencing techniques.[91–98]

At some point, whole genome sequencing will ultimately translate to clinical practice, and the concept of personalized genomic medicine is on the not-so-distant horizon. Genome sequencing in patients potentially offers many benefits, including (1) the rapid diagnosis of genetic disorders, especially for those mutations that arise *de novo*; (2) the ability to determine carrier status and provide genetic counseling for Mendelian disorders of significant penetrance; and (3) the ability to determine which therapeutics are useful or harmful for a given genetic disorder, as well as allow for the design of new, gene-specific therapeutic targets. Whole genome sequencing in patients will also lead to greater scientific discovery. Specifically, we will be able to further our understanding of how genotypes regulate human phenotypes. Furthermore, we will be able to catalog human variations across many populations and begin to make sense out of the many Genome Wide Association Studies (GWAS) that targeted only a small fraction of the genome in an effort to blur the boundaries between Mendelian and complex disease.[99,100] Moreover, we will be able to determine mutation load in a given patient and disease in an effort to understand how defects in the same gene can lead to such phenotypic diversity. And, we will be able to more accurately determine the penetrance of specific gene mutations across populations, which will lead to better genetic counseling and a better understanding of the genes involved in complex disease and complex traits.

Although no single scientific approach can answer every question about the human brain and disease, Mendelian genetics has its place in future research. The study of inherited diseases provides a relatively simple, yet profoundly powerful, method for unraveling the mysteries of human genetic diseases, and study of these disorders will continue to expand our knowledge of brain development and function. It will be important to determine which genetic characteristics are unique to the various disorders and this, in turn, will elevate our understanding of how genetic mutations lead to a disease. Over the next several decades, the greatest challenges in genetic research will not be in the identification of disease-associated genes, but will rather be in defining the molecular mechanisms that explain how genetic mutations confer disease risk and phenotypic expression. By studying human Mendelian CNS disorders, we will be able to tackle these problems in their "simplest" form, and this will eventually lead to important discoveries into the genetic mechanisms that govern the formation and function of the complex human brain.

Conflicts of interest

The authors declare no conflicts of interest.

References

1. Drachman, D.A. 2005. Do we have brain to spare? *Neurology* **64:** 2004–2005.
2. International Human Genome Sequencing Consortium. 2004. Finishing the euchromatic sequence of the human genome. *Nature* **431:** 931–945.
3. Peltonen, L. *et al.* 2006. Lessons from studying monogenic disease for common disease. *Hum. Mol. Genet.* **15**(Spec. No. 1): R67–R74.
4. Engel, J. & T.A. Pedley. 1998. *Epilepsy: A Comprehensive Textbook*. Lippincott-Raven. Philadelphia.
5. Engle, E.C. 2010. Human genetic disorders of axon guidance. *Cold Spring Harb. Perspect. Biol.* **2:** a001784–a001802.
6. Guerrini, R., W.B. Dobyns & A.J. Barkovich. 2008. Abnormal development of the human cerebral cortex: genetics, functional consequences and treatment options. *Trends Neurosci.* **31:** 154–162.
7. Gupta, A., L.H. Tsai & A. Wynshaw-Boris. 2002. Life is a journey: a genetic look at neocortical development. *Nat. Rev. Genet.* **3:** 342–355.
8. Diaz, A.L. & J.G. Gleeson. 2009. The molecular and genetic mechanisms of neocortex development. *Clin. Perinatol.* **36:** 503–512.
9. Rakic, P. *et al.* 2009. Decision by division: making cortical maps. *Trends Neurosci.* **32:** 291–301.
10. Borello, U. & A. Pierani. 2010. Patterning the cerebral cortex: traveling with morphogens. *Curr. Opin. Genet. Dev.* **4:** 408–415.
11. Aicardi, J. 1998. *Diseases of the Nervous System in Childhood*. Mac Keith Press. London.
12. Cox, J. *et al.* 2006. What primary microcephaly can tell us about brain growth. *Trends Mol. Med.* **12:** 358–366.
13. Woods, C.G. 2004. Human microcephaly. *Curr. Opin. Neurobiol.* **14:** 112–117.
14. Kaindl, A.M. *et al.* 2010. Many roads lead to primary autosomal recessive microcephaly. *Prog. Neurobiol.* **90:** 363–383.
15. Thornton, G.K. & C.G. Woods. 2009. Primary microcephaly: do all roads lead to Rome? *Trends Genet.* **25:** 501–510.
16. Jackson, A.P. *et al.* 1998. Primary autosomal recessive microcephaly (MCPH1) maps to chromosome 8p22-pter. *Am. J. Hum. Genet.* **63:** 541–546.
17. Jamieson, C.R., C. Govaerts & M.J. Abramowicz. 1999. Primary autosomal recessive microcephaly: homozygosity mapping of MCPH4 to chromosome 15. *Am. J. Hum. Genet.* **65:** 1465–1469.
18. Kumar, A. *et al.* 2009. Mutations in STIL, encoding a pericentriolar and centrosomal protein, cause primary microcephaly. *Am. J. Hum. Genet.* **84:** 286–290.
19. Leal, G.F. *et al.* 2003. A novel locus for autosomal recessive primary microcephaly (MCPH6) maps to 13q12.2. *J. Med. Genet.* **40:** 540–542.
20. Moynihan, L. *et al.* 2000. A third novel locus for primary autosomal recessive microcephaly maps to chromosome 9q34. *Am. J. Hum. Genet.* **66:** 724–727.
21. Pattison, L. *et al.* 2000. A fifth locus for primary autosomal recessive microcephaly maps to chromosome 1q31. *Am. J. Hum. Genet.* **67:** 1578–1580.
22. Roberts, E. *et al.* 2002. Autosomal recessive primary microcephaly: an analysis of locus heterogeneity and phenotypic variation. *J. Med. Genet.* **39:** 718–721.
23. Roberts, E. *et al.* 1999. The second locus for autosomal recessive primary microcephaly (MCPH2) maps to chromosome 19q13.1–13.2. *Eur. J. Hum. Genet.* **7:** 815–820.
24. Bond, J. *et al.* 2002. ASPM is a major determinant of cerebral cortical size. *Nat. Genet.* **32:** 316–320.
25. Bond, J. *et al.* 2005. A centrosomal mechanism involving CDK5RAP2 and CENPJ controls brain size. *Nat. Genet.* **37:** 353–355.
26. Jackson, A.P. *et al.* 2002. Identification of microcephalin, a protein implicated in determining the size of the human brain. *Am. J. Hum. Genet.* **71:** 136–142.
27. Xu, X., J. Lee & D.F. Stern. 2004. Microcephalin is a DNA damage response protein involved in regulation of CHK1 and BRCA1. *J. Biol. Chem.* **279:** 34091–34094.
28. Alderton, G.K. *et al.* 2006. Regulation of mitotic entry by microcephalin and its overlap with ATR signalling. *Nat. Cell Biol.* **8:** 725–733.
29. Gotz, M. & W.B. Huttner. 2005. The cell biology of neurogenesis. *Nat. Rev. Mol. Cell Biol.* **6:** 777–788.
30. Fish, J.L. *et al.* 2008. Making bigger brains-the evolution of neural-progenitor-cell division. *J. Cell Sci.* **121:** 2783–2793.
31. Ali, F. & R. Meier. 2008. Positive selection in ASPM is correlated with cerebral cortex evolution across primates but not with whole-brain size. *Mol. Biol. Evol.* **25:** 2247–2250.
32. Evans, P.D., E.J. Vallender & B.T. Lahn. 2006. Molecular evolution of the brain size regulator genes CDK5RAP2 and CENPJ. *Gene* **375:** 75–79.
33. Woods, C.G., J. Bond & W. Enard. 2005. Autosomal recessive primary microcephaly (MCPH): a review of clinical, molecular, and evolutionary findings. *Am. J. Hum. Genet.* **76:** 717–728.

34. Barkovich, A.J., T.K. Koch & C.L. Carrol. 1991. The spectrum of lissencephaly: report of ten patients analyzed by magnetic resonance imaging. *Ann. Neurol.* **30:** 139–146.
35. Dobyns, W.B. *et al.* 1996. X-linked malformations of neuronal migration. *Neurology* **47:** 331–339.
36. Dobyns, W.B. *et al.* 1992. Causal heterogeneity in isolated lissencephaly. *Neurology* **42:** 1375–1388.
37. Dobyns, W.B. & C.L. Truwit. 1995. Lissencephaly and other malformations of cortical development: 1995 update. *Neuropediatrics* **26:** 132–147.
38. Gleeson, J.G. & C.A. Walsh. 2000. Neuronal migration disorders: from genetic diseases to developmental mechanisms. *Trends Neurosci.* **23:** 352–359.
39. Zaki, M. *et al.* 2007. Identification of a novel recessive RELN mutation using a homozygous balanced reciprocal translocation. *Am. J. Med. Genet. A* **143A:** 939–944.
40. Reiner, O. *et al.* 1993. Isolation of a Miller-Dieker lissencephaly gene containing G protein beta-subunit-like repeats. *Nature* **364:** 717–721.
41. Poirier, K. *et al.* 2007. Large spectrum of lissencephaly and pachygyria phenotypes resulting from de novo missense mutations in tubulin alpha 1A (TUBA1A). *Hum. Mutat.* **28:** 1055–1064.
42. Pilz, D.T. *et al.* 1998. LIS1 and XLIS (DCX) mutations cause most classical lissencephaly, but different patterns of malformation. *Hum. Mol. Genet.* **7:** 2029–2037.
43. Lo Nigro, C. *et al.* 1997. Point mutations and an intragenic deletion in LIS1, the lissencephaly causative gene in isolated lissencephaly sequence and Miller-Dieker syndrome. *Hum. Mol. Genet.* **6:** 157–164.
44. Kitamura, K. *et al.* 2002. Mutation of ARX causes abnormal development of forebrain and testes in mice and X-linked lissencephaly with abnormal genitalia in humans. *Nat. Genet.* **32:** 359–369.
45. Gleeson, J.G. *et al.* 1998. Doublecortin, a brain-specific gene mutated in human X-linked lissencephaly and double cortex syndrome, encodes a putative signaling protein. *Cell* **92:** 63–72.
46. Boycott, K.M. *et al.* 2005. Homozygous deletion of the very low density lipoprotein receptor gene causes autosomal recessive cerebellar hypoplasia with cerebral gyral simplification. *Am. J. Hum. Genet.* **77:** 477–483.
47. Bielas, S.L. & J.G. Gleeson. 2004. Cytoskeletal-associated proteins in the migration of cortical neurons. *J. Neurobiol.* **58:** 149–159.
48. Gleeson, J.G. 2001. Neuronal migration disorders. *Ment. Retard. Dev. Disabil. Res. Rev.* **7:** 167–171.
49. Kerjan, G. & J.G. Gleeson. 2009. Moving neurons back into place. *Nat. Med.* **15:** 17–18.
50. Hatten, M.E. 2005. LIS-less neurons don't even make it to the starting gate. *J. Cell Biol.* **170:** 867–871.
51. Kerjan, G. & J.G. Gleeson. 2007. Genetic mechanisms underlying abnormal neuronal migration in classical lissencephaly. *Trends Genet.* **23:** 623–630.
52. Tsai, J.W. *et al.* 2005. LIS1 RNA interference blocks neural stem cell division, morphogenesis, and motility at multiple stages. *J. Cell Biol.* **170:** 935–945.
53. Koizumi, H. *et al.* 2006. Doublecortin maintains bipolar shape and nuclear translocation during migration in the adult forebrain. *Nat. Neurosci.* **9:** 779–786.
54. Tanaka, T. *et al.* 2004. Lis1 and doublecortin function with dynein to mediate coupling of the nucleus to the centrosome in neuronal migration. *J. Cell Biol.* **165:** 709–721.
55. Jaglin, X.H. & J. Chelly. 2009. Tubulin-related cortical dysgeneses: microtubule dysfunction underlying neuronal migration defects. *Trends Genet.* **25:** 555–566.
56. Kumar, R.A. *et al.* 2010. TUBA1A mutations cause wide spectrum lissencephaly (smooth brain) and suggest that multiple neuronal migration pathways converge on alpha tubulins. *Hum. Mol. Genet.* **19:** 2817–2827.
57. Manent, J.B. *et al.* 2009. Dcx reexpression reduces subcortical band heterotopia and seizure threshold in an animal model of neuronal migration disorder. *Nat. Med.* **15:** 84–90.
58. Dretakis, E.K. & P.N. Kondoyannis. 1974. Congenital scoliosis associated with encephalopathy in five children of two families. *J. Bone Joint Surg. Am.* **56:** 1747–1750.
59. Sharpe, J.A., J.L. Silversides & R.D. Blair. 1975. Familial paralysis of horizontal gaze. Associated with pendular nystagmus, progressive scoliosis, and facial contraction with myokymia. *Neurology* **25:** 1035–1040.
60. Haller, S., S.G. Wetzel & J. Lutschg. 2008. Functional MRI, DTI and neurophysiology in horizontal gaze palsy with progressive scoliosis. *Neuroradiology* **50:** 453–459.
61. Jen, J.C. *et al.* 2004. Mutations in a human ROBO gene disrupt hindbrain axon pathway crossing and morphogenesis. *Science* **304:** 1509–1513.
62. Chan, W.M. *et al.* 2006. Horizontal gaze palsy with progressive scoliosis can result from compound heterozygous mutations in ROBO3. *J. Med. Genet.* **43:** e11–e14.
63. Sanes, J.R. & M. Yamagata. 2009. Many paths to synaptic specificity. *Annu. Rev. Cell Dev. Biol.* **25:** 161–195.
64. Marillat, V. *et al.* 2004. The slit receptor Rig-1/Robo3 controls midline crossing by hindbrain precerebellar neurons and axons. *Neuron* **43:** 69–79.
65. Sabatier, C. *et al.* 2004. The divergent Robo family protein rig-1/Robo3 is a negative regulator of slit responsiveness required for midline crossing by commissural axons. *Cell* **117:** 157–169.
66. Chen, Z. *et al.* 2008. Alternative splicing of the Robo3 axon guidance receptor governs the midline switch from attraction to repulsion. *Neuron* **58:** 325–332.
67. Srour, M. *et al.* 2010. Mutations in DCC cause congenital mirror movements. *Science* **328:** 592.
68. Aridon, P. *et al.* 2006. Increased sensitivity of the neuronal nicotinic receptor alpha 2 subunit causes familial epilepsy with nocturnal wandering and ictal fear. *Am. J. Hum. Genet.* **79:** 342–350.
69. Baulac, S. *et al.* 2001. First genetic evidence of GABA(A) receptor dysfunction in epilepsy: a mutation in the gamma2-subunit gene. *Nat. Genet.* **28:** 46–48.
70. Biervert, C. *et al.* 1998. A potassium channel mutation in neonatal human epilepsy. *Science* **279:** 403–406.
71. Charlier, C. *et al.* 1998. A pore mutation in a novel KQT-like potassium channel gene in an idiopathic epilepsy family. *Nat. Genet.* **18:** 53–55.

72. Cossette, P. *et al.* 2002. Mutation of GABRA1 in an autosomal dominant form of juvenile myoclonic epilepsy. *Nat. Genet.* **31:** 184–189.
73. De Fusco, M. *et al.* 2000. The nicotinic receptor beta 2 subunit is mutant in nocturnal frontal lobe epilepsy. *Nat. Genet.* **26:** 275–276.
74. Du, W. *et al.* 2005. Calcium-sensitive potassium channelopathy in human epilepsy and paroxysmal movement disorder. *Nat. Genet.* **37:** 733–738.
75. Escayg, A. *et al.* 2000. Mutations of SCN1A, encoding a neuronal sodium channel, in two families with GEFS+2. *Nat. Genet.* **24:** 343–345.
76. Heron, S.E. *et al.* 2002. Sodium-channel defects in benign familial neonatal-infantile seizures. *Lancet* **360:** 851–852.
77. Jouvenceau, A. *et al.* 2001. Human epilepsy associated with dysfunction of the brain P/Q-type calcium channel. *Lancet* **358:** 801–807.
78. Singh, N.A. *et al.* 1998. A novel potassium channel gene, KCNQ2, is mutated in an inherited epilepsy of newborns. *Nat. Genet.* **18:** 25–29.
79. Steinlein, O.K. *et al.* 1995. A missense mutation in the neuronal nicotinic acetylcholine receptor alpha 4 subunit is associated with autosomal dominant nocturnal frontal lobe epilepsy. *Nat. Genet.* **11:** 201–203.
80. Wallace, R.H. *et al.* 1998. Febrile seizures and generalized epilepsy associated with a mutation in the Na+-channel beta1 subunit gene SCN1B. *Nat. Genet.* **19:** 366–370.
81. Zuberi, S.M. *et al.* 1999. A novel mutation in the human voltage-gated potassium channel gene (Kv1.1) associates with episodic ataxia type 1 and sometimes with partial epilepsy. *Brain* **122**(Pt. 5): 817–825.
82. Ottman, R. *et al.* 2010. Genetic testing in the epilepsies—report of the ILAE Genetics Commission. *Epilepsia* **51:** 655–670.
83. Steinlein, O.K. 2004. Genetic mechanisms that underlie epilepsy. *Nat. Rev. Neurosci.* **5:** 400–408.
84. Kullmann, D.M. 2010. Neurological channelopathies. *Annu. Rev. Neurosci.* **33:** 151–172.
85. Malenka, R.C. & M.F. Bear. 2004. LTP and LTD: an embarrassment of riches. *Neuron* **44:** 5–21.
86. Sayin, U., T.P. Sutula & C.E. Stafstrom. 2004. Seizures in the developing brain cause adverse long-term effects on spatial learning and anxiety. *Epilepsia* **45:** 1539–1548.
87. Sutula, T.P., J. Hagen & A. Pitkanen. 2003. Do epileptic seizures damage the brain? *Curr. Opin. Neurol.* **16:** 189–195.
88. Mullen, S.A. & I.E. Scheffer. 2009. Translational research in epilepsy genetics: sodium channels in man to interneuronopathy in mouse. *Arch. Neurol.* **66:** 21–26.
89. Dawn Teare, M. & J.H. Barrett. 2005. Genetic linkage studies. *Lancet* **366:** 1036–1044.
90. Bonetta, L. 2010. Whole-genome sequencing breaks the cost barrier. *Cell* **141:** 917–919.
91. Choi, M. *et al.* 2009. Genetic diagnosis by whole exome capture and massively parallel DNA sequencing. *Proc. Natl. Acad. Sci. USA* **106:** 19096–101.
92. Hoischen, A. *et al.* 2010. De novo mutations of SETBP1 cause Schinzel-Giedion syndrome. *Nat. Genet.* **42:** 483–485.
93. Lupski, J.R. *et al.* 2010. Whole-genome sequencing in a patient with Charcot-Marie-Tooth neuropathy. *N. Engl. J. Med.* **362:** 1181–1191.
94. Ng, S.B. *et al.* 2010. Exome sequencing identifies the cause of a Mendelian disorder. *Nat. Genet.* **42:** 30–35.
95. Nikopoulos, K. *et al.* 2010. Next-generation sequencing of a 40 Mb linkage interval reveals TSPAN12 mutations in patients with familial exudative vitreoretinopathy. *Am. J. Hum. Genet.* **86:** 240–247.
96. Reversade, B. *et al.* 2009. Mutations in PYCR1 cause cutis laxa with progeroid features. *Nat. Genet.* **41:** 1016–1021.
97. Roach, J.C. *et al.* 2010. Analysis of genetic inheritance in a family quartet by whole-genome sequencing. *Science* **328:** 636–639.
98. Volpi, L. *et al.* 2010. Targeted next-generation sequencing appoints c16orf57 as clericuzio-type poikiloderma with neutropenia gene. *Am. J. Hum. Genet.* **86:** 72–76.
99. Cirulli, E.T. & D.B. Goldstein. 2010. Uncovering the roles of rare variants in common disease through whole-genome sequencing. *Nat. Rev. Genet.* **11:** 415–425.
100. Manolio, T.A. *et al.* 2009. Finding the missing heritability of complex diseases. *Nature* **461:** 747–753.
101. A.D.A.M., I. 2009. Medline Plus Microcephaly (image), NIH.
102. Caviness, V.S. Jr., T. Takahashi & R.S. Nowakowski. 1995. Numbers, time and neocortical neuronogenesis: a general developmental and evolutionary model. *Trends Neurosci.* **18:** 379–383.
103. Noctor, S.C. *et al.* 2001. Neurons derived from radial glial cells establish radial units in neocortex. *Nature* **409:** 714–720.
104. Reid, C.B., S.F. Tavazoie & C.A. Walsh. 1997. Clonal dispersion and evidence for asymmetric cell division in ferret cortex. *Development* **124:** 2441–2450.
105. Buchman, J.J. & L.H. Tsai. 2007. Spindle regulation in neural precursors of flies and mammals. *Nat. Rev. Neurosci.* **8:** 89–100.
106. Takahashi, T., R.S. Nowakowski & V.S. Caviness, Jr. 1995. The cell cycle of the pseudostratified ventricular epithelium of the embryonic murine cerebral wall. *J. Neurosci.* **15:** 6046–6057.

Ann. N.Y. Acad. Sci. ISSN 0077-8923

ANNALS OF THE NEW YORK ACADEMY OF SCIENCES
Issue: *The Year in Human and Medical Genetics*

Hearing loss: a common disorder caused by many rare alleles

Dorith Raviv, Amiel A. Dror, and Karen B. Avraham

Department of Human Molecular Genetics and Biochemistry, Sackler Faculty of Medicine, Tel Aviv University, Tel Aviv, Israel

Address for correspondence: Karen B. Avraham, Department of Human Molecular Genetics & Biochemistry, Sackler School of Medicine, Tel Aviv University, Tel Aviv 69978, Israel. karena@post.tau.ac.il

Perception of sound is a fundamental role of the auditory system. Traveling with the force of their mechanical energy, sound waves are captured by the ear and activate the sensory pathway of this complex organ. The hair cells, specialized sensory cells within the inner ear, transmit the mechanical energy into electrical nerve stimuli that reach the brain. A large number of proteins are responsible for the overarching tasks required to maintain the complex mechanism of sound sensation. Many hearing disorders are due to single gene defects inherited in a Mendelian fashion, thus enabling clinical diagnostics. However, at the same time, hearing impairment is genetically heterogeneous, with both common and rare forms occurring due to mutations in over 100 genes. The crosstalk between human and mouse genetics has enabled comprehensive studies on gene identification and protein function, taking advantage of the tools animal models have to offer. The aim of the following review is to provide background and examples of human deafness genes and the discovery of their function in the auditory system.

Keywords: deafness; Usher syndrome; myosin VIIA; cadherin 23; stereocilin; microRNA-96

Hearing loss: a Mendelian disorder

Hearing loss is the most common sensory loss, affecting 1–2 out of 1,000 newborns and over half of the population by the age of 80 (U.S. National Institute on Deafness and Other Communication Disorders, http://www.nidcd.nih.gov/health/statistics/hearing.html). Severe hearing loss that occurs during infancy can have dramatic effects on speech acquisition and literacy. Deafness with a later onset might have a deleterious effect on the quality of life of an affected individual. Hearing loss may be sensorineural, caused by damage to the inner ear or auditory nerve, or conductive, due to reduction of sound waves passing through the outer and middle ear (reviewed in Ref. 1). Sensorineural hearing loss can be present at birth or acquired at later stages in life. Among the factors causing acquired hearing loss are effects from different infections and diseases and ototoxic drugs, as well as exposure to excessive noise, either social or industrial. Approximately 60–70% of hearing loss is due to genetic factors and is genetically heterogeneous, with single gene defects in many different genes (Fig. 1).[4] Genetic hearing impairment can occur either as syndromic hearing loss (SHL), associated with other abnormalities, or nonsyndromic hearing loss (NSHL) as an isolated phenotype, which accounts for about 70% of the cases. Vestibular dysfunction may be associated with NSHL as well. Hearing loss can be inherited either in a dominant or recessive mode, whether it is syndromic or nonsyndromic.

A comprehensive listing of all genes and the proteins they encode is no longer practical to complete in a review, as it was just a few years ago. The breadth and variability of phenotypes, mutations, mechanisms, and pathways is remarkable. Because the inner ear is an organ with diverse cell types, many cellular targets are susceptible to damage caused by genetic mutations. In this review, we concentrate on one of the most sensitive targets to genetic mutations, the hair cell, a sensory cell that functions to detect and amplify the mechanical signal evoked by sound, translating it into electrical signals that are transmitted to the brain. We define several characteristics and qualities of hereditary hearing loss. Four genes are described as examples of Mendelian phenotypes that helped decipher

doi: 10.1111/j.1749-6632.2010.05868.x

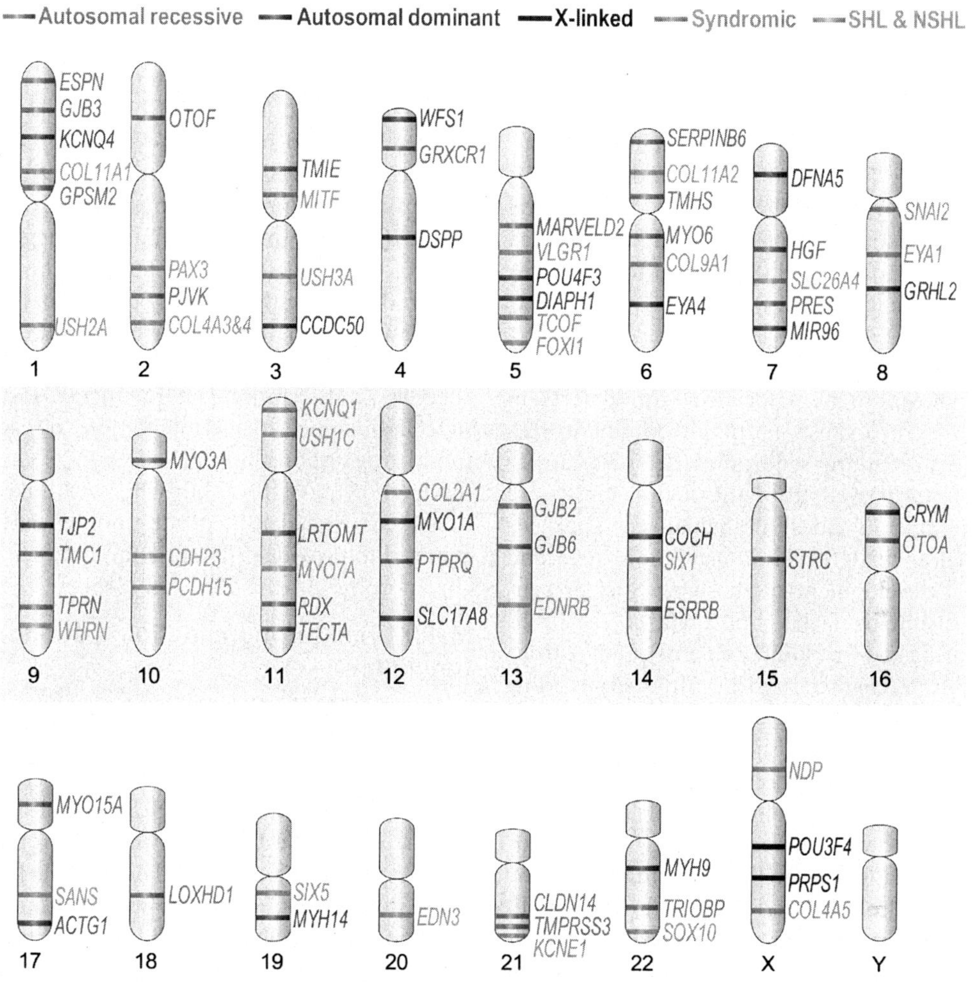

Figure 1. A scheme representing the chromosomal location of genes with mutations that cause hearing loss. The genes are classified as nonsyndromic autosomal recessive (red), nonsyndromic autosomal dominant (blue), X-linked genes (black), syndromic (green), and genes associated with both syndromic and nonsyndromic hearing loss (light blue). Previously published in Refs. 2 and 3, with permission by authors.

major mechanisms or functions of the inner ear and have shed light on auditory pathways. In each case, the mouse was instrumental in defining the mechanism. Each gene represents a different prototype of the variability and heterogeneity of hereditary hearing loss due to a single gene defect.

Hearing loss: from past to present

Descriptions of hereditary hearing loss go back as far as the 16th century.[5] In the early days of the 17th century, Paulus Zacchias (1584–1659), an Italian physician, stated that marriage of deaf people should be avoided, as they would give birth to deaf children like themselves. This statement suggests that Zacchias was aware of the hereditary factor in congenital deafness. In those days, however, it was still believed that hearing was the "gate of the mind," and therefore congenital deafness lead to defective intelligence.[6] By 1880, evidence for both autosomal dominant and autosomal recessive inheritances of hearing loss was presented, although the terms "autosomal dominant" and "autosomal recessive" were not yet determined. X-linked inherited deafness is presumed to have been first described in 1863, although it was not fully characterized until 1930 by Dow and Poynter.[7]

Prevention of marriage between deaf individuals was formally proposed in legislation in 1883, by Alexander Graham Bell. The approach of legally banning hearing-impaired individuals from society was taken even further under Nazi Germany and the embrace of eugenics. During Nazi control, 1,600 deaf individuals were murdered.[5] Today, there are many specialized schools and communities around the world for the deaf providing cultural centers and institutions of higher education, such as Gallaudet University in Washington DC, USA (http://www.gallaudet.edu/).

Syndromic hearing loss

SHL is known to be associated with defects in many different organs.[8] Forms of dominant SHL include Wardenburg syndrome (deafness, structural defects derived from the neural crest, and pigmentation anomalies); Branchial-oto-renal syndrome (Branchial fistulas, renal anomalies, and abnormal development of the ear); Stickler syndrome (distinctive facial abnormalities, eye problems, hearing loss, and joint problems), and neurofibromatosis 2 (characterized by tumors, including acoustic neuromas). Recessive SHL includes the deafness–blindness of Usher syndrome, Pendred syndrome (inner ear and enlargement of the thyroid gland, known as goiter), Jervell and Lange-Nielsen syndrome (characterized by electrocardiographic changes as well as hearing loss), and Cockayne syndrome (described with dwarfism, retinal atrophy, and deafness). X-linked syndromic deafness includes Alport syndrome (also known as hereditary nephritis) and Norrie disease (congenital ocular symptoms with progressive hearing loss). It is difficult to state how many forms of SHL there are. Hearing impairment is often overlooked because of other disorders present in patients. Toriello once estimated that "new" syndromes are being described weekly.[9] At least 32 genes have been identified for SHL (Hereditary Hearing Loss Homepage, http://hereditaryhearingloss.org).[2]

Nonsyndromic hearing loss

Since the early 1990s, 149 NSHL loci have been identified and are marked as DFNA, for autosomal dominant, DFNB, for autosomal recessive, and DFNX, for X-linked loci (Hereditary Hearing loss Homepage).[2] The first dominant locus, DFNA1, was identified in 1992[10] in a large kindred from Costa Rica. Five years later, the gene *DIAPH1*, which encodes the protein diaphanous that regulates actin polymerization, was found to be responsible for the hearing loss in this family.[11] DFNB1, the first recessive deafness locus identified, was revealed in 1994 in two consanguineous families from Tunisia that were affected by this form of deafness.[12] The responsible gene, *GJB2*, identified soon after in 1997, encodes the gap-junction protein connexin 26 (Cx26).[13] Today, *GJB2* is the most common deafness-causing gene, accounting for up to 50% of the cases of nonsyndromic sensorineural hearing loss in some populations.[14] In 1995 the first sex-linked hearing loss-causing gene was identified, *POU3F4*,[15] which encodes a transcription factor with a POU domain; *POU3F4* mapped to the interval of Xq21 and was found to be responsible for DFNX2. DFNX2 was formerly known as DFN3, but when a Y-linked deafness locus was identified[16] the nomenclature was changed to provide differentiation between the sex chromosomes.[17]

Major breakthroughs defining auditory function

Several major discoveries uncovered mechanisms of auditory function, combining gene and protein networks. Myosin VIIA mutations, associated with both Usher syndrome and NSHL, encode a protein that is part of the Usher network of proteins that interact with one another and help define functions of the stereocilia. Cadherin 23 and protocadherin 15 mutations are associated with both syndromic and nonsyndromic forms of hearing loss. Most compelling, these proteins together form the tip link between stereocilia, a major site of mechanotransduction. Mutations in stereocilin are associated with NSHL, and the loss of this protein leads to structural changes in the outer hair cells, implicating this protein in hair cell function. Finally, an entirely new paradigm in hearing loss was demonstrated with the discovery of mutations in a microRNA (miRNA). Together, these examples provide a snapshot of several different scenarios of this hereditary disease, which together form the complex function of the auditory system.

DFNB2/USH1B

An understanding of Usher syndrome on a molecular level has been achieved by defining part of the protein network in the inner ear that is responsible for development and mechanisms of the hair bundle. Usher syndrome is the most common cause of combined deafness–blindness known in humans, affecting 1:25,000 children.[18,19] It is defined by congenital, bilateral deafness followed by a later onset of vision loss, caused by retinitis pigmentosa (RP). The retinal degeneration is a consequence of photoreceptor cell death beginning from the periphery of the retina. Patients suffer from night blindness, followed by narrowing of the visual field ("tunnel vision"), leading to absolute blindness.[20] Usher syndrome is heterogeneous, both clinically and genetically, with at least 12 chromosomal loci involved in the three types of USH syndrome, Usher type 1 (USH1), USH2, and USH3 (further subdivided, when relevant, to A–H). The different types are distinguishable from each other mainly by the severity and onset of deafness and the variability of vestibular defects, while all have RP. USH1 patients are profoundly deaf from birth and have vestibular dysfunction. The deafness in USH2 is less severe and the vestibular function is normal. USH3 is characterized by mild, progressive hearing loss with occasional vestibular dysfunction.[21]

The first *USH1* gene identified, for USH1B, was the gene encoding myosin VIIa (*MYO7A*).[22] Myosin VIIa is an unconventional myosin that moves toward the plus end of actin, from the stereocilia base to its tip; it has been described as a dimeric motor and is known to require high concentrations of ATP for its motile activity along actin filaments.[23] As part of the myosin superfamily, myosin VIIa has the common structure of myosins, a motor head domain that includes the actin-binding and the ATP-binding sites, a neck domain with five IQ motifs that are expected to bind myosin-light chains, and the variable tail domain. The tail region contains the functional domains that determine myosin VIIa specificity,[24] including two large tandem repeats that are separated by SH3 domains. These repeats consist of a MyTH4 (myosin tail homology 4) domain and a FERM domain (4.1, ezrin, readixin, moesin), which is thought to be responsible for the attachment of the protein to the plasma membrane.[19] The tail portion of myosin VIIa is known to bind the scaffolding proteins harmonin[25] and SANS,[19] as well as protocadherin15,[26] which are all part of the USH protein network, in addition to binding to the F-actin filaments (Fig. 2).[23,28] These interactions form the protein network of Usher syndrome. Myosin VIIa is also known to interact with vezatin,[29] which bridges myosin VIIa to the cadherin–catenin complex, thus increasing cell–cell adhesion. In the inner ear these two proteins have a membrane-to-membrane interaction in the links that interconnect the bases of adjacent stereocilia. Recently, a direct interaction between cadherin 23 and myosin VIIa was reported.[30] This finding enhances the relationship within the Usher protein network, as all of its components have been shown to interact with one another.

In addition to its involvement in Usher syndrome, mutations in myosin VIIA causes nonsyndromic deafness, either dominant (DFNA11[31]) or recessive (DFNB2[32,33]). Myosin VIIa was also found to be mutated in the mutant Shaker1 (*sh1*) mouse, causing progressive degeneration of the cochlea (Fig. 2) together with vestibular dysfunction.[34] Myosin VIIa was found to be expressed in the cochlea as early as embryonic day 16.5.[35] The protein is expressed in the stereocilia and cuticular plate of the hair cells (Fig. 2).[36] The Shaker-1 (*sh1*) gene has seven different alleles, and in some of the mutated alleles ($Myo7a^{6J}$ and $Myo7a^{816SB}$ in particular) the mutation causes abnormal stereocilia development in both inner and outer hair cells.[35] There is a high correlation between the location of the mutation within the protein and its severity. Mutations causing ultrastructural changes in the stereocilia were either deletions or missense mutations in the myosin motor-head core. Mutated forms of myosin VIIa fail to localize correctly to the stereocilia.[37] These mutated forms are responsible for the more severe phenotypes.

DFNB12/USH1D and DFNB23/USH1F

Mutations in *CDH23* (cadherin-related 23) are responsible for both DFNB12 and USH1D.[38] Mutations were identified in a number of consanguineous families of Pakistanian and Indian origin[38] as well as a Cuban family.[39] The protein encoded by *CDH23* comprises 3,354 amino acids and is predicted to have 27 cadherin repeats that are involved in Ca^{2+}-dependent cell adhesion, an N′ terminal signal peptide, and one transmembrane domain.[40] Null mutations in this gene are associated with USH1D, whereas missense mutations are associated with DFNB12.[38,41] The *waltzer* mouse phenotype is due to mutations in *Cdh23* and serves as a mouse model for USH1D and DFNB12.[42,43] These mice have hearing loss and vestibular dysfunction, caused by disorganization of their stereocilia (Fig. 3).[43] Another mouse that suffers from progressive

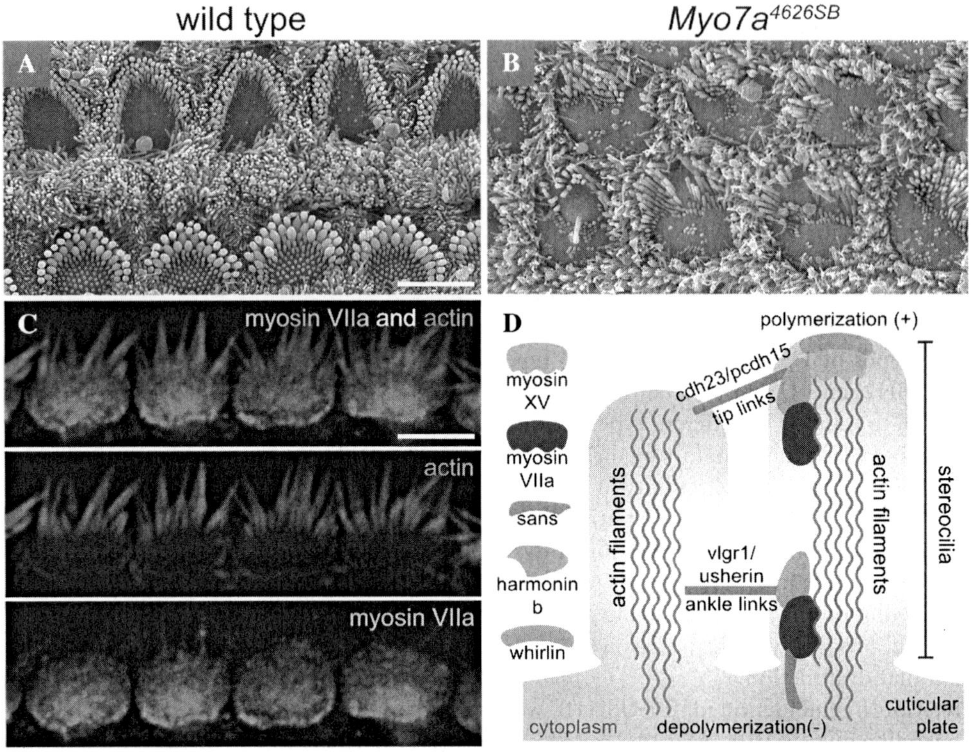

Figure 2. Phenotypic effect of mutations in *MYO7A*. (A and B) Scanning electron micrographs of wild type (A) and (B) null allele of myosin VIIa (*Myo7a*^{4626SB}) that expresses less than 1% of myosin VIIA protein, showing a near complete loss of stereocilia structure and disorganization of the remaining hair bundle. (C) Immunolocalization of myosin VIIa is shown in the cuticular plate. (D) Schematic representation of the network of proteins that are involved in Usher syndrome within the stereocilia. Myosin VIIA is connected to the actin filaments as well as to harmonin b and to the tip links that are composed of cadherin 23 (cdh23) and protocadherin (pcdh15). Scale bars: 5 μm in A applies to B; 5 μm in C. Figures A and B were received from Tim Self and Karen Steel, and D was previously published in Friedman *et al.*,[27] with permission by the authors.

hearing loss and carries a mutation in *Cdh23* is the *N*-ethyl-*N*-nitrosourea(ENU)-induced mouse *salsa*.[41] As opposed to the *waltzer* mouse, the hair bundle in the *salsa* mouse is intact and only the stereocilia tip-links are lost, presumably causing loss of tension within the stereocilia. Based on these two mouse models, it was suggested that USH1D, caused by null mutations in *CDH23,* is due to hair bundle developmental defects, while DFNB12, with the missense mutations, is caused by tip link defects. Cadherin 23 was shown to have a role in maintaining the hair bundle during development[43] but in adults has the role of comprising the tip links, transmitting the force to the ion channels. The tip links are extracellular filaments connecting between adjacent stereocilia and control the opening of mechanosensory transduction channels (Fig. 3).[46] The mechanosensory channels are located near the stereocilia tips. Sound vibrations causes deflection of the hair bundle toward the longest stereocilia. This deflection increases the tension in the tip link and the hair cells will respond by opening or closing the ion transduction channels,[47] leading to an influx of cations into the stereocilia and resulting in depolarization of the hair cell.

Another component of the tip links is encoded by *PCDH15* (protocadherin-15),[48] which has 11 calcium-binding cadherin repeats and is responsible for USH1F and DFNB23, as was identified in two Pakistani families.[49] In the syndromic form, affected members are profoundly deaf and suffer from progressive RP. The mouse model for USH1F is Ames waltzer, which is profoundly deaf with disorganized abnormal stereocilia (Fig. 3).[50,44] Protocadherin 15 forms the lower part of the tip link and interacts with cadherin 23, which forms the upper

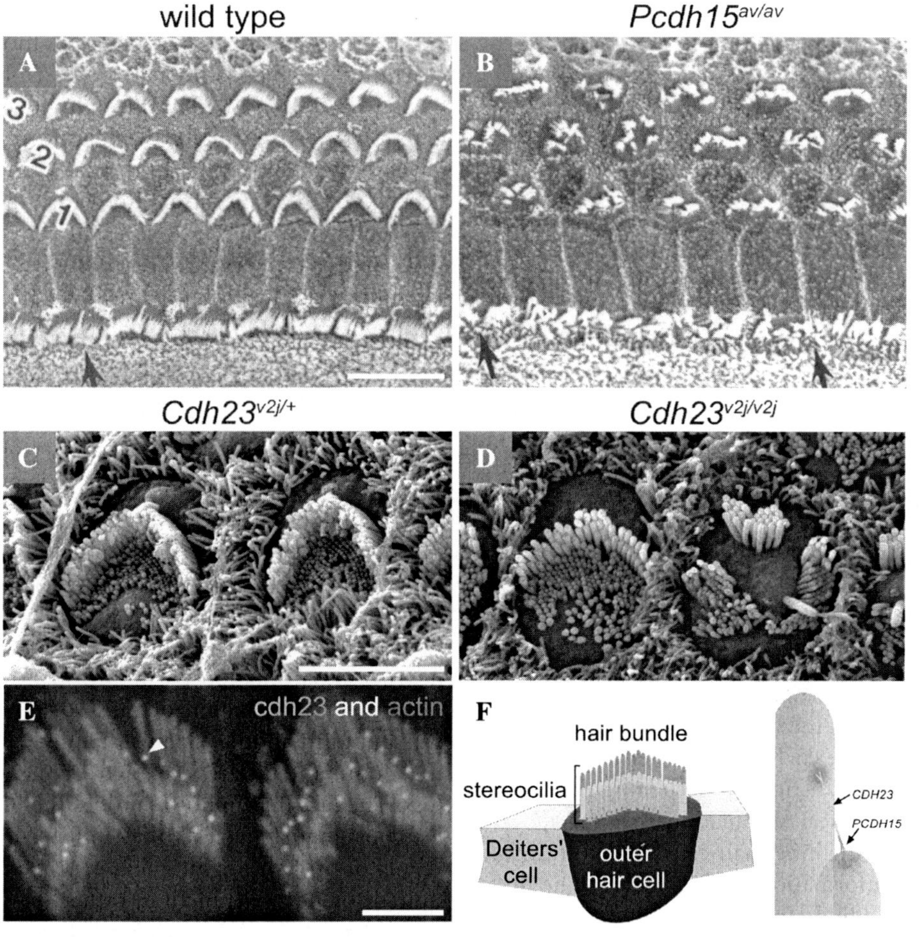

Figure 3. The tip link of adjacent stereocilia is assembled by cadherin 23 and protocadherin 15. (A–D) Scanning electron micrographs of cochlear sensory epithelium of Ames waltzer (*av*) and waltzer (*v*) mice carrying mutations in *Pcdh15* and *Cdh23*, respectively. (A) Wild-type sensory epithelium is organized with one row of inner hair cell and three rows of outer hair cells with a typical V shape arrangement of their stereocilia. (B) A spontaneous mutation of *Pcdh15* leads to severely impaired organization of the hair bundles of both inner (indicated by arrows) and outer hair cells that lost the V shape structure within their apical surface. (C) High-power images of outer hair cells highlights the polarity nature of these cells characterized by hair bundles that point toward the same direction to the kinocilium. (D) The outer hair cells of $Cdh23^{v2j/v2j}$ mutant mice lost their planar cell polarity, showing groups of stereocilia of the surface of the cell facing different directions. Moreover, the essential staircase arrangement of the stereocilia is no longer maintained within the bundle. (E) Immunolocalization of cadherin 23 (green) is defined to the tip of the stereocilia (red) in the adult guinea pig. (F) Schematic representation of an outer hair cell surrounded by two supporting cells. The hair bundle is composed of rows of stereocilia in gradual heights, which create the typical staircase structure of the bundle (yellow). The tip links that connect adjacent stereocilia of lower and higher rows are composed of cadherin 23 and protocadherin 15. Ages: (A and B) P10; (C and D) P4. Scale bars: 10 μm in A, B, 5 μm C, D, 2 μm E. Figures (A and B) are reprinted with permission from Alagramam *et al.*[44]; figures (C and D) from Di Palma *et al.*[43]; and figures (E and F) from Kazmierczak *et al.*[45]

part, together creating the tip-link filament,[45] which is about 180nm long. The interaction between these two proteins is made through their extracellular cadherin domains, at their N′ terminus, and is Ca^{2+} dependent. The *salsa* mutation also affects the interaction between cadherin 23 and protocadherin 15, even though the mutation affects an amino acid that is outside the interaction site.[41] The studies of these two proteins revealed the structure of the tip links, one of the most important components of the auditory hair bundle, which is important for both maintaining the highly organized structure of the stereocilia and playing a role in the mechanoelectrical (MET) process in the inner ear.

DFNB16

Mutation of the gene encoding stereocilin, *Strc*, is responsible for DFNB16, recessively inherited NSHL, found in two families of Pakistani and French origin.[51] The affected individuals have postlingual hearing loss from early childhood, which is more severe at higher frequencies. The stereocilin gene encodes an ~1800 amino acid long protein that is localized at the tips of the stereocilia of the outer hair cells (Fig. 4). Stereocilin was first hypothesized to be responsible for the attachment of the tectorial membrane to the sensory cells in the inner ear.[53] A mouse model lacking stereocilin[52] provided a clearer picture of the role of stereocilin in the inner ear. The tectorial membrane in these mice is intact but hair bundle imprinting, which represents the anchoring of outer hair cells stereocilia to the tectorial membrane, was not detectable. Outer hair cells derived from $Strc^{-/-}$ mice lack the horizontal top connectors. The distal ends of the stereocilia are less connected in the mutants than in wild-type mice, suggesting that the top connectors have a role in the cohesion of the outer hair cell hair bundle of adult mice. In the cochlea, the outer hair cells acts as an amplifier, assisting in sharpening the sound-induced mechanical stimulation that reaches the cochlea, while the inner hair cells serve to transmit the sensory information into the brain. The outer hair cells, in addition to sound amplification, generate mechanical and electrical waveform distortions, essential for hearing in noisy environments.[54] This distortion creates a masking effect that contributes to speech understanding and simplicity. The studies of $Strc^{-/-}$ mutant mice suggest that the top connectors may either generate distortion by themselves, due to nonlinear stiffness, or they may alter hair-bundle displacements, leading to distortion of sound waveforms. This work suggests a role for stereocilin in a major mechanism of outer hair cell function, revealed through genetic mutations.

DFNA50

MiRNAs are small noncoding RNAs, 21–23 nucleotides long, which regulate gene expression through the RNA interference (RNAi) mechanism, known to affect proliferation, differentiation, and developmental processes in eukaryotes.[55] Mir genes are included within introns of protein-coding genes or in intragenic regions. Intronic miRNAs are transcribed along with the host protein-coding gene. A miR is first transcribed to a long RNA molecule, the pri-miRNA, which is further processed by the nucleases Drosha and Dicer1 through the pre-miRNA intermediate to generate the mature miRNA.[56,57] The seed sequence (nucleotides 2–8) of the mature miRNA is a key determinant of which mRNAs will be regulated by each miRNA. MiR genes are often organized in clusters that are cotranscribed as polycistronic genes to a single pri-miRNA from which a couple of mature miRNAs will originate.[58]

MiR-96 is part of the miR-183/96/182 cluster, which is expressed in the mouse inner ear, retina, and additional tissues.[59–61] In the inner ear, it is mainly expressed in developing hair cells. The locus *DFNA50* segregated in a Spanish family with postlingual, progressive, NSHL at all frequencies.[62] Further analysis revealed a mutation in miR-96 (+13 G > A), causing replacement of the fourth nucleotide within the conserved seed region of the mature miR-96 sequence, which segregated with the hearing impairment in the family.[63] Another mutation in the seed region of miR-96 (+14 C > A) was found in another Spanish family with autosomal dominant, progressive, high-frequency hearing loss. The fact that in both families the hearing loss was postlingual suggests that miR-96 may have a regulatory role in maintaining gene expression required for normal function, rather than inner ear development. Notably, this is the first study showing a nucleotide substitution in the seed region of a miRNA that is responsible for a human Mendelian disease.[63]

Diminuendo (*Dmdo*) is an ENU-induced mouse model, with an A > T substitution in the miR-96 seed[64] (Fig. 5). These mice show progressive hearing impairment in heterozygotes and profound deafness in homozygotes, associated with hair cell defects. In homozygotes, most of the hair cells are degenerated by 4 weeks of age. Luciferase assays in transfected non-inner ear cells demonstrated that five mRNAs, *Aqp5, Celsr2, Myrip, Odf2,* and *Ryk*, are regulated by miR-96.[63,64] These targets are suggested to be overexpressed in the inner ears of humans or mice with mutations in the miR-96 seed. However, while the level of *Aqp5* and *Celsr2* mRNAs was notably upregulated in the cochlear organ of Corti of the mutant *Dmdo* mice compared to wildtype, the differences in protein expression levels were small.[64] MiR-96 may function to fine-tune the expression level of these

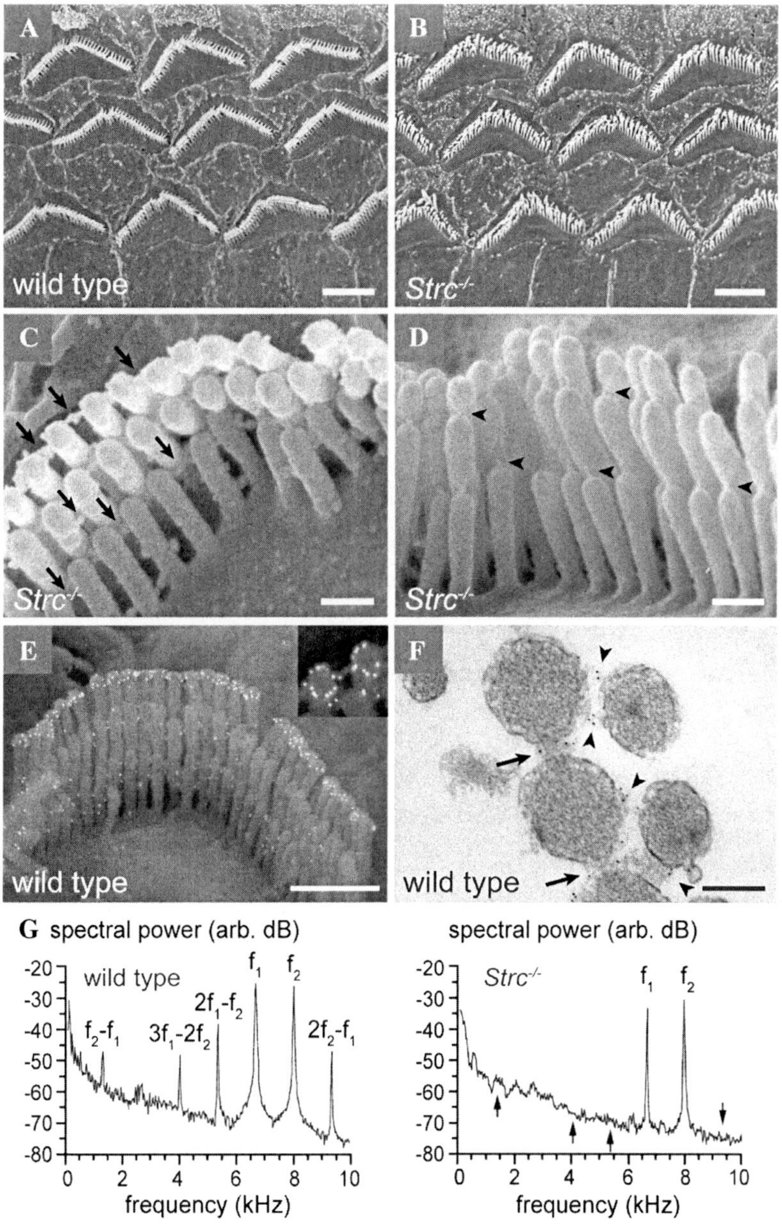

Figure 4. Stereocilin assembles the horizontal top connectors and is suggested to be essential for the nonlinearity underlying cochlear waveform distortions. (A–D) Scanning electron micrographs (SEM) of cochlear sensory outer hair cells of stereocilin-deficient mice, $Strc^{-/-}$. (A) Low power image of the three rows of outer hair cells in a wild-type mouse. (B) The stereocilia of $Strc^{-/-}$ outer hair cells retain the typical V shape but are dispersed from one another. (C) Higher magnification image of a single outer hair cell of $Strc^{-/-}$ mice shows that the top connectors between stereocilia (arrows) are absent in the mutant hair bundles. (D) Despite the missing top connectors, the tip links of $Strc^{-/-}$ hair bundles are maintained (arrow heads). (E and F) The localization of stereocilin is detected by immunogold labeling on scanning electron micrographs of a wild-type outer hair cell (E) in a ring-shaped labeling around the tips of the tallest stereocilia row. Labeling can also be detected between the stereocilia within the top connectors in a transmission electron micrograph of a stereocilin-labeled wild-type outer hair cell hair bundle (F) marked with arrows. (G) Cochlear microphonic (CM) shown by frequency spectra. The black arrows within the $Strc^{-/-}$ mice spectra indicate the frequency points where prominent intermodulations are apparent. Ages: (A–E) P14; (F) P22. Scale bars: 2.5 μm in A and B, 0.25 μm C and D, 1 μm (E), and 0.2 μm in F. Figures (A–G) are reprinted with permission from Verpy *et al.*[52]

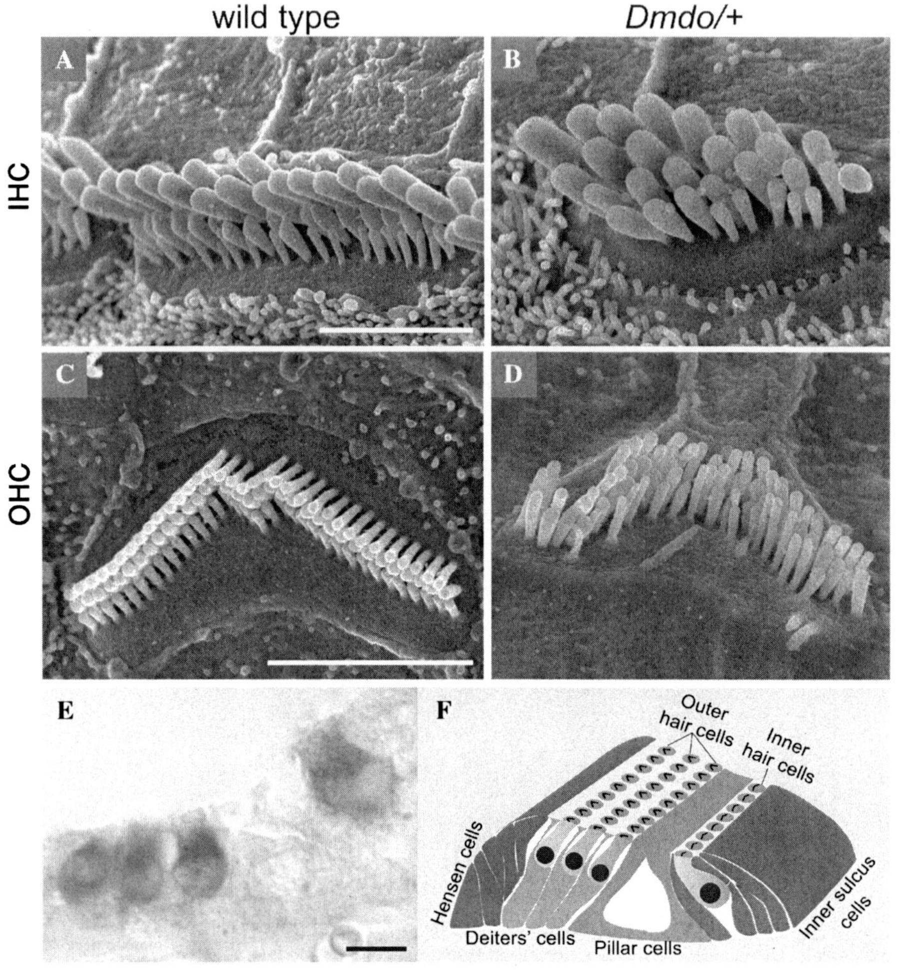

Figure 5. Mir-96 is required for normal morphology and function of the auditory system. (A–D) Scanning electron microscopy (SEM) images of cochlear sensory hair cells of diminuendo mice depleted of miR-96. Wild-type inner (A) and outer (C) hair cells exhibit normal morphology of the hair bundles with actin-rich stereocilia protrusions arranged in a typical staircase manner. (B and D) Created by ENU mutagenesis, a point mutation within the seed of miR-96 in *Dmdo* mice leads to altered morphology of the auditory hair cells. (B) The inner hair cells of the *Dmdo/+* mutant mice show additional rows of stereocilia with a bulged and rounded cell surface. (D) *Dmdo/+* outer hair cells lost their staircase arrangement, while stereocilia of the first row of the bundle are taller than the middle rows. (E) *In situ* hybridization (ISH) revealed the spatial expression of miR-96 within the auditory apparatus defined to the cochlear inner and outer hair cells (dark blue). (F) Schematic illustration combining the top and cross view of the cochlear sensory epithelium, also known as the organ of Corti. The hair cells (green) are shown in their typical arrangement of one inner and three outer rows. The hair bundle of each hair cell is apparent on its surface (black). A wide array of supporting cells are essential, surrounding the hair cells, including Dieters' (blue) Hensens and inner sulcus cells (purple). The pillar cells (brown) create the tunnel of Corti that separates between the inner and outer rows of hair cells. Ages: (A–D); P28 (E) P0. Scale bars: 3 μm in A–D, 5 μm in E. Figures (A–E) are reprinted with permission from Lewis *et al.*,[64] and Figure F was previously published in Dror and Avraham.[65]

putative targets. It is also possible that other post translational regulation might play a role in regulating AQP5 and CELSR2 protein expression levels. These changes may indirectly affect the expression of many other genes, which are not direct targets of mir-96. A genome-wide approach revealed nearly 100 affected transcripts that were either down- or upregulated in the organ of Corti of mutant mice compared to wild-type mice. In particular, five genes that are well known in the inner ear were down-regulated (*prestin, oncomodulin, Pitpnm1, Gfi1,* and *Ptprq*), although none of them has the seed region for mir-96. This suggested a downstream effect of epigenetic down-regulation caused by the miRNA that may help explain the mechanism behind the hearing impairment of these mice.

Conclusion

The advancements in the field of hereditary hearing loss have been dramatic since the mapping of the first chromosomal loci for X-linked deafness in 1988, SHL in 1990, and NSHL in 1992. Since then the discovery of a new gene or allele for hearing loss has occurred frequently, seemingly a new discovery every month. Along with the fact that connexin 26 accounts for a majority of congenital NSHL, these advances have provided new opportunities for diagnostics. Early identification of specific mutations using molecular diagnostics alleviates the need for other clinical tests.[66] It also facilitates predictions regarding the potential progressivity of the hearing loss (if not early onset profound) and other abnormalities, in the case of SHL. For example, children born with the *PCDH15* R245X mutation are profoundly deaf, but the RP does not arise until about the age of 10.[67] Early diagnosis allows the family and physicians to be aware of impending blindness, which may dictate future treatment, for example, cochlear implantation.

Efforts are currently underway worldwide to develop therapeutics based on gene therapy, stem cells, and/or regenerative medicine. Caution must be taken, however, in raising immediate hopes in these areas for patients, since while there has been progress in basic research toward inner ear therapeutics,[68,69] major technological and other hurdles remain.[70]

Most important, there are many single-gene defects for inherited hearing loss yet to be discovered. While over 50 genes have been found for NSHL since the discovery of connexin 26, most mutations are rare and occur in one or a few families, making diagnostics today limited to only a handful of genes. Massively parallel sequencing (or deep sequencing) has been applied in a few cases toward gene discovery for hearing loss.[71,72] As targeted genome capture, whole exome sequencing, and deep sequencing of entire genomes become more affordable, these techniques will be applied more to hereditary hearing loss as well, enabling further gene and mutation discovery. Most compelling, the complex genetic and protein interactions of the auditory system will be elucidated through studies of classical Mendelian phenotypes.

Acknowledgments

Research in the Avraham laboratory is funded by the Israel Science Foundation Grant 1486/07, the National Institutes of Health (NIDCD) R01DC005641, the Israel Ministry of Health, and the European Commission FP6 Integrated Project Eumodic 037188. We would like to thank Kumar Algaramam, Morag Lewis, Konrad Noben-Trauth, Ulrich Mueller,Elizabeth Verpy, Christine Petit, Tim Self, and Karen Steel for sharing figures with us.

Conflicts of interest

The authors declare no conflicts of interest.

References

1. Petit, C. & G. P. Richardson. 2009. Linking genes underlying deafness to hair-bundle development and function. *Nat. Neurosci.* **12:** 703–710.
2. Van Camp, G. & R. J. H. Smith. 2010. Hereditary Hearing Loss Homepage. Available at http://hereditary hearingloss.org. Update: 1.10.2010
3. Dror, A. A. & K. B. Avraham. 2010. Hearing impairment: a panoply of genes and functions. *Neuron* **68:** 293–308.
4. Petit, C., J. Levilliers & J. P. Hardelin. 2001. Molecular genetics of hearing loss. *Annu. Rev .Genet.* **35:** 589–646.
5. Gorlin, R. J. 1995. Genetic Hearing Loss-A Brief History. In *Hereditary Hearing Loss and its Syndromes*.R. J. Gorlin, H. V. Toriello & M. M. Cohen, Eds.: 3–4. Oxford University Press. New York.
6. Cranefield, P. F. & W. Federn. 1970. Paulus Zacchias on mental deficiency and on deafness. *Bull. N. Y. Acad. Med.* **46:** 3–21.
7. Dow, G. S. & C. I. Poynter. 1930. The Dar family. *Eugen. News.* **15:** 128–130.
8. R. J.,Gorlin, H. V. Toriello & M. M. Cohen, Eds. *Hereditary Hearing Loss and its Syndromes.* Oxford University Press. New York.
9. Toriello, H. V. 1988. New syndromes from old: evaluation of heterogeneity and variability in syndrome definition and delineation. *Am. J. Med. Genet. Suppl.* **4:** 55–70.
10. Leon, P. E., H. Raventos, E. Lynch, *et al.* 1992. The gene for an inherited form of deafness maps to chromosome 5q31. *Proc. Natl. Acad. Sci. USA* **89:** 5181–5184.
11. Lynch, E. D., M. K. Lee, J. E. Morrow, *et al.* 1997. Nonsyndromic deafness DFNA1 associated with mutation of a human homolog of the *Drosophila* gene *diaphanous. Science* **278:** 1315–1318.
12. Guilford, P., S. Ben Arab, S. Blanchard, *et al.* 1994. A nonsyndrome form of neurosensory, recessive deafness maps to the pericentromeric region of chromosome 13q. *Nat. Genet.* **6:** 24–28.

13. Kelsell, D. P., J. Dunlop, H. P. Stevens, *et al.* 1997. Connexin 26 mutations in hereditary non-syndromic sensorineural deafness. *Nature* **387:** 80–83.
14. Kenneson, A., K. Van Naarden Braun & C. Boyle. 2002. *GJB2* (connexin 26) variants and nonsyndromic sensorineural hearing loss: a HuGE review. *Genet. Med.* **4:** 258–274.
15. de Kok, Y.J., S. M. van der Maarel, M. Bitner-Glindzicz, *et al.* 1995. Association between X-linked mixed deafness and mutations in the POU domain gene *POU3F4*. *Science* **267:** 685–688.
16. Wang, Q. J., C. Y. Lu, N. Li, *et al.* 2004. Y-linked inheritance of non-syndromic hearing impairment in a large Chinese family. *J. Med. Genet.* **41:** e80.
17. Petersen, M. B., Q. Wang & P. J. Willems. 2008. Sex-linked deafness. *Clin. Genet.* **73:** 14–23.
18. El-Amraoui, A. & C. Petit. 2005. Usher I syndrome: unravelling the mechanisms that underlie the cohesion of the growing hair bundle in inner ear sensory cells. *J. Cell. Sci.* **118:** 4593–4603.
19. Reiners, J., K. Nagel-Wolfrum, K. Jurgens, *et al.* 2006. Molecular basis of human Usher syndrome: deciphering the meshes of the Usher protein network provides insights into the pathomechanisms of the Usher disease. *Exp. Eye. Res.* **83:** 97–119.
20. van Soest, S., A. Westerveld, P. T. de Jong, *et al.* 1999. Retinitis pigmentosa: defined from a molecular point of view. *Surv. Ophthalmol.* **43:** 321–334.
21. Williams, D. S. 2008. Usher syndrome: animal models, retinal function of Usher proteins, and prospects for gene therapy. *Vision. Res.* **48:** 433–441.
22. Weil, D., S. Blanchard, J. Kaplan, *et al.* 1995. Defective myosin VIIA gene responsible for Usher syndrome type 1B. *Nature* **374:** 60–61.
23. Inoue, A. & M. Ikebe. 2003. Characterization of the motor activity of mammalian myosin VIIA. *J. Biol. Chem.* **278:** 5478–5487.
24. Sellers, J. R. 2000. Myosins: a diverse superfamily. *Biochim. Biophys. Acta.* **1496:** 3–22.
25. Adato, A., V. Michel, Y. Kikkawa, *et al.* 2005. Interactions in the network of Usher syndrome type 1 proteins. *Hum. Mol. Genet.* **14:** 347–356.
26. Senften, M., M. Schwander, P. Kazmierczak, *et al.* 2006. Physical and functional interaction between protocadherin 15 and myosin VIIa in mechanosensory hair cells. *J. Neurosci.* **26:** 2060–2071.
27. Friedman, L. M., A. A. Dror & K. B. Avraham. 2007. Mouse models to study inner ear development and hereditary hearing loss. *Int. J. Dev. Biol.* **51:** 609–631.
28. Petit, C. 2001. Usher syndrome: from genetics to pathogenesis. *Annu. Rev. Genomics. Hum. Genet.* **2:** 271–297.
29. Kussel-Andermann, P., A. El-Amraoui, S. Safieddine, *et al.* 2000. Vezatin, a novel transmembrane protein, bridges myosin VIIA to the cadherin-catenins complex. *EMBO J.* **19:** 6020–6029.
30. Bahloul, A., V. Michel, J. P. Hardelin, *et al.* 2010. Cadherin-23, myosin VIIa and harmonin, encoded by Usher syndrome type I genes, form a ternary complex and interact with membrane phospholipids. *Hum. Mol. Genet.* **19:** 3557–3565.
31. Liu, X. Z., J. Walsh, Y. Tamagawa, *et al.* 1997. Autosomal dominant non-syndromic deafness caused by a mutation in the myosin VIIA gene. *Nat. Genet.* **17:** 268–269.
32. Liu, X. Z., J. Walsh, P. Mburu, *et al.* 1997. Mutations in the myosin VIIA gene cause non-syndromic recessive deafness. *Nat. Genet.* **16:** 188–190.
33. Weil, D., P. Kussel, S. Blanchard, *et al.* 1997. The autosomal recessive isolated deafness, DFNB2, and the Usher 1B syndrome are allelic defects of the myosin-VIIA gene. *Nat. Genet.* **16:** 191–193.
34. Gibson, F., J. Walsh, P. Mburu, *et al.* 1995. A type VII myosin encoded by the mouse deafness gene shaker-1. *Nature* **374:** 62–64.
35. Self, T., M. Mahony, J. Fleming, *et al.* 1998. Shaker-1 mutations reveal roles for myosin VIIA in both development and function of cochlear hair cells. *Development* **125:** 557–566.
36. Hasson, T., J. Walsh, J. Cable, *et al.* 1997. Effects of shaker-1 mutations on myosin-VIIa protein and mRNA expression. *Cell. Motil. Cytoskeleton.* **37:** 127–138.
37. Riazuddin, S., S. Nazli, Z. M. Ahmed, *et al.* 2008. Mutation spectrum of *MYO7A* and evaluation of a novel nonsyndromic deafness *DFNB2* allele with residual function. *Hum. Mutat.* **29:** 502–511.
38. Bork, J. M., L. M. Peters, S. Riazuddin, *et al.* 2001. Usher syndrome 1D and nonsyndromic autosomal recessive deafness DFNB12 are caused by allelic mutations of the novel cadherin-like gene *CDH23*. *Am. J. Hum. Genet.* **68:** 26–37.
39. Bolz, H., B. von Brederlow, A. Ramirez, *et al.* 2001. Mutation of *CDH23*, encoding a new member of the cadherin gene family, causes Usher syndrome type 1D. *Nat. Genet.* **27:** 108–112.
40. Ahmed, Z. M., S. Riazuddin, S. Riazuddin & E. R. Wilcox. 2003. The molecular genetics of Usher syndrome. *Clin. Genet.* **63:** 431–444.
41. Schwander, M., W. Xiong, J. Tokita, *et al.* 2009. A mouse model for nonsyndromic deafness (DFNB12) links hearing loss to defects in tip links of mechanosensory hair cells. *Proc. Natl. Acad. Sci. USA* **106:** 5252–5257.
42. Wilson, S. M., D. B. Householder, V. Coppola, *et al.* 2001. Mutations in *Cdh23* cause nonsyndromic hearing loss in *waltzer* mice. *Genomics* **74:** 228–233.
43. Di Palma, F., R. H. Holme, E. C. Bryda, *et al.* 2001. Mutations in *Cdh23*, encoding a new type of cadherin, cause stereocilia disorganization in waltzer, the mouse model for Usher syndrome type 1D. *Nat. Genet.* **27:** 103–107.
44. Alagramam, K. N., C. L. Murcia, H. Y. Kwon, *et al.* 2001. The mouse Ames waltzer hearing-loss mutant is caused by mutation of *Pcdh15*, a novel protocadherin gene. *Nat. Genet.* **27:** 99–102.
45. Kazmierczak, P., H. Sakaguchi, J. Tokita, *et al.* 2007. Cadherin 23 and protocadherin 15 interact to form tip-link filaments in sensory hair cells. *Nature* **449:** 87–91.
46. Siemens, J., C. Lillo, R. A. Dumont, *et al.* 2004. Cadherin 23 is a component of the tip link in hair-cell stereocilia. *Nature* **428:** 950–955.
47. Gillespie, P. G. & U. Muller. 2009. Mechanotransduction by hair cells: models, molecules, and mechanisms. *Cell* **139:** 33–44.

48. Ahmed, Z. M., R. Goodyear, S. Riazuddin, *et al.* 2006. The tip-link antigen, a protein associated with the transduction complex of sensory hair cells, is protocadherin-15. *J. Neurosci.* **26:** 7022–7034.
49. Ahmed, Z. M., S. Riazuddin, S. L. Bernstein, *et al.* 2001. Mutations of the protocadherin gene *PCDH15* cause Usher syndrome type 1F. *Am. J. Hum. Genet.* **69:** 25–34.
50. Raphael, Y., K. N. Kobayashi, G. A. Dootz, *et al.* 2001. Severe vestibular and auditory impairment in three alleles of Ames waltzer (*av*) mice. *Hear. Res.* **151:** 237–249.
51. Verpy, E., S. Masmoudi, I. Zwaenepoel, *et al.* 2001. Mutations in a new gene encoding a protein of the hair bundle cause non-syndromic deafness at the *DFNB16* locus. *Nat. Genet.* **29:** 345–349.
52. Verpy, E., D. Weil, M. Leibovici, *et al.* 2008. Stereocilin-deficient mice reveal the origin of cochlear waveform distortions. *Nature* **456:** 255–258.
53. Jovine, L., J. Park & P. M. Wassarman. 2002. Sequence similarity between stereocilin and otoancorin points to a unified mechanism for mechanotransduction in the mammalian inner ear. *BMC Cell. Biol.* **3:** 28.
54. Goldstein, J. L. 1967. Auditory nonlinearity. *J. Acoust. Soc. Am.* **41:** 676–689.
55. Bartel, D. P. 2004. MicroRNAs: genomics, biogenesis, mechanism, and function. *Cell* **116:** 281–297.
56. Friedman, L. M. & K. B. Avraham. 2009. MicroRNAs and epigenetic regulation in the mammalian inner ear: implications for deafness. *Mamm. Genome.* **20:** 581–603.
57. Carthew, R. W. & E. J. Sontheimer. 2009. Origins and Mechanisms of miRNAs and siRNAs. *Cell* **136:** 642–655.
58. Saini, H. K., S. Griffiths-Jones & A. J. Enright. 2007. Genomic analysis of human microRNA transcripts. *Proc. Natl. Acad. Sci. USA* **104:** 17719–17724.
59. Weston, M. D., M. L. Pierce, S. Rocha-Sanchez, *et al.* 2006. MicroRNA gene expression in the mouse inner ear. *Brain. Res.* **1111:** 95–104.
60. Ryan, D. G., M. Oliveira-Fernandes & R. M. Lavker. 2006. MicroRNAs of the mammalian eye display distinct and overlapping tissue specificity. *Mol. Vis.* **12:** 1175–1184.
61. Sacheli, R., L. Nguyen, L. Borgs, *et al.* 2009. Expression patterns of miR-96, miR-182 and miR-183 in the development inner ear. *Gene. Expr. Patterns.* **9:** 364–370.
62. Modamio-Hoybjor, S., M. A. Moreno-Pelayo, *et al.* 2004. A novel locus for autosomal dominant nonsyndromic hearing loss, DFNA50, maps to chromosome 7q32 between the DFNB17 and DFNB13 deafness loci. *J. Med. Genet.* **41:** e14.
63. Mencia, A., S. Modamio-Hoybjor, N. Redshaw, *et al.* 2009. Mutations in the seed region of human miR-96 are responsible for nonsyndromic progressive hearing loss. *Nat. Genet.* **41:** 609–613.
64. Lewis, M. A., E. Quint, A. M. Glazier, *et al.* 2009. An ENU-induced mutation of miR-96 associated with progressive hearing loss in mice. *Nat. Genet.* **41:** 614–618.
65. Dror, A. A. & K. B. Avraham. 2009. Hearing loss: mechanisms revealed by genetics and cell biology. *Annu. Rev. Genet.* **43:** 411–437.
66. Brownstein, Z. & K. B. Avraham. 2009. Deafness genes in Israel: implications for diagnostics in the clinic. *Pediatr. Res.* **66:** 128–134.
67. Ben-Yosef, T., S. L. Ness, A. C. Madeo, *et al.* 2003. A mutation of PCDH15 among Ashkenazi Jews with the type 1 Usher syndrome. *N. Engl. J. Med.* **348:** 1664–1670.
68. Gubbels, S. P., D. W. Woessner, J. C. Mitchell, *et al.* 2008. Functional auditory hair cells produced in the mammalian cochlea by in utero gene transfer. *Nature* **455:** 537–541.
69. Oshima, K., K. Shin, M. Diensthuber, *et al.* 2010. Mechanosensitive hair cell-like cells from embryonic and induced pluripotent stem cells. *Cell* **141:** 704–716.
70. Shibata, S. B. & Y. Raphael. 2010. Future approaches for inner ear protection and repair. *J. Commun. Disord.* **43:** 295–310.
71. Rehman, A. U., R. J. Morell, I. A. Belyantseva, *et al.* 2010. Targeted capture and next-generation sequencing identifies *C9orf75*, encoding taperin, as the mutated gene in nonsyndromic deafness *DFNB79*. *Am. J. Hum. Genet.* **86:** 378–388.
72. Walsh, T., H. Shahin, T. Elkan-Miller, *et al.* 2010. Whole exome sequencing and homozygosity mapping identify mutation in the cell polarity protein GPSM2 as the cause of nonsyndromic hearing loss DFNB82. *Am. J. Hum. Genet.* **87:** 90–94.

Ann. N.Y. Acad. Sci. ISSN 0077-8923

ANNALS OF THE NEW YORK ACADEMY OF SCIENCES
Issue: *The Year in Human and Medical Genetics*

Investigation of Mendelian forms of obesity holds out the prospect of personalized medicine

Alexandra I. F. Blakemore and Philippe Froguel

Department of Genomics of Common Disease, School of Public Health, Imperial College London, London, United Kingdom

Address for correspondence: Alexandra I. F. Blakemore, Department of Genomics of Common Disease, School of Public Health, Imperial College London, London, UK. a.blakemore@imperial.ac.uk

Mendelian forms of obesity are already known to account for approximately 5% of the severely obese but are currently underinvestigated. In contrast, there has been much recent concentration on genome-wide single nucleotide polymorphism (SNP) associations in obesity, with particular emphasis given to the role of the fat mass and obesity associated *(FTO)* gene. Unfortunately, despite the enormous resources devoted to this work, none of the SNP markers in the ~30 genes discovered to have associations with common obesity have meaningful predictive power. This is very different from the situation for Mendelian obesity, where mutations have very clear effects on phenotype. Study of Mendelian obesity has also added significantly to our understanding of mechanisms of appetite regulation, with all known causative genes being active in the brain and most forming part of the leptin–melanocortin signaling pathway. Investigation of genomic structural variation has also recently revealed deletions causing obesity, sometimes with concomitant neurocognitive dysfunction. Advances in next-generation sequencing are expected to uncover additional highly penetrant causes of obesity. Screening for Mendelian forms of obesity is rarely carried out but holds considerable promise for improved clinical care of these high-risk patients.

Keywords: obesity; Mendelian genetics; appetite regulation

Introduction

The genetic analysis of obesity has two major goals. First, scientists aim to understand the basic science of regulation of adiposity and the pathophysiology of obesity to design new approaches to prevention, therapy, and disease management. Second, for the first time, such knowledge holds out the prospect of personalized medicine for both molecular diagnosis and treatment. For obese individuals, whose condition results from carriage of a rare variant of strong effect, and their affected relatives, the predictive power of genetic analysis is very high, and the potential for improving the quality of the medicine is very real.

While individually uncommon, a number of single-gene or monolocus forms of obesity exist, and there is increasing evidence that together they account for a nontrivial proportion (around 1 in 20 so far) of the dozens of millions of severely obese individuals.[1] Characterization of Mendelian forms of obesity has already contributed enormously to our understanding of pathways for appetite and satiety, and through this the mechanisms of body weight dysregulation and obesity development.[2,3] In the rare cases of leptin deficiency, this increased understanding has also resulted in dramatic improvements in treatment.[4]

By 2015, 700 million adults worldwide are expected to be obese (BMI >30 kg/m^2); and the prevalence of childhood obesity is also rising rapidly.[5] This has profound implications for public health and health care economics, as obesity is a major risk factor for type 2 diabetes (T2D) and is also associated with increased risk of cardiovascular and respiratory disease, osteoarthritis, hypertension, stroke, and certain cancers. Although obesity prevalence is rising in Europe and North America, obesity is

doi: 10.1111/j.1749-6632.2010.05880.x

also a serious cause for concern in the developing world, with a predicted sharp rise in obesity and consequent T2D in regions such as the Middle East, sub-Saharan Africa, Southeast Asia, and India.[6,7] Developing nations now make up a significant proportion of the top 10 countries with the highest proportions of obese individuals,[8] placing a significant burden on the health-care systems and economies of affected countries.[9]

Identification of the drivers of this "obesity epidemic" is critically important to effectively target economic resources to combat them; indeed, all past and current strategies to prevent or to reverse obesity at the level of entire populations have proven dramatically unsuccessful. Although it is clear that the increasingly sedentary, urbanized lifestyles and the widespread availability of highly palatable calorie-dense foodstuffs have contributed to the global rise in obesity, in addition to these "obesogenic" environmental influences there is also a strong genetic component to obesity that affects differences in individual risk within populations. Some people carry genetic variants that make them more sensitive to Westernized obesogenic environments so that they become obese, whereas others in the population maintain a healthy weight range despite experiencing the same unhealthy environmental pressures.

The evidence that common obesity has a significant genetic component is strong. Comparison of obesity concordance rates between monozygotic and dizygotic twin pairs, and family segregation studies, have demonstrated that obesity, BMI, and other measures of adiposity are highly heritable traits. Heritability estimates are high with over 70% of the variance in BMI attributed to genetic factors.[10–14] The relative contribution of Mendelian forms of obesity to this overall heritability remains to be determined, but it is not unreasonable to postulate that a large proportion of individuals with early-onset severe obesity carry highly functionally deleterious variants of genes having a key role in appetite.

The first indications of how obesity might be inherited came from murine models with mutations in the genes encoding the adipokine leptin and its receptor. Leptin is produced primarily by adipocytes and acts as an indicator of stored fat levels. The long, signaling isoform of the leptin receptor is expressed in the hypothalamus where it initiates a signaling pathway regulating feeding behavior and energy expenditure.[15]

Study of consanguineous families revealed the first recessive single-gene forms of obesity

The first two cases of recessive single-gene obesity in humans—mutations in the human genes encoding leptin and the leptin receptor (*LEP* and *LEPR*, respectively)—were originally identified by sequencing these genes in obese children of consanguineous families with multiple affected family members.[16,17] A consanguineous mating is one between related people, most frequently cousins, but may also be defined as "unions of individuals with at least one common ancestor such as occurring within population isolates, small towns, and tribes; intracommunity or endogamous marriages."[18] In populations where consanguineous unions are frequent they are often perceived as having social and economic benefits but they also increase the likelihood of having children with recessive disorders (often with an unusual pattern of prevalence specific to the particular population) because both parents are more likely to carry the same deleterious gene mutation. As outlined above, leptin acts within the hypothalamic leptin–melanocortin pathway, which regulates appetite and satiety, and homozygosity for inactivating *LEP* mutations results in hyperphagia, with extreme childhood-onset obesity and hypogonadism.[17] Subsequent identification of mutations in the leptin receptor gene that cause monogenic obesity and hypogonadism also confirmed the central role of this pathway in appetite regulation.[16]

In the examples above, murine models of obesity had suggested candidate genes for sequence analysis. Where no candidates suggest themselves, however, consanguineous families may still provide useful information through hypothesis-free homozygosity mapping. High-density single-nucleotide polymorphism (SNP) analysis has been used to identify cosegregation of regions of homozygosity with phenotype of interest in highly consanguineous families affected by rare recessive conditions.[19] These regions of homozygosity can pinpoint critical genomic regions, leading to rapid identification of disease loci.[20–22] Highly polymorphic microsatellite markers can also be used to exclude or fine-map the regions of interest. Similar homozygosity mapping may be carried out at a population level in highly consanguineous communities. Following definition of the minimal critical genomic region, candidate genes within it can be sequenced to identify the disease-causing mutation. An additional potential benefit to this study design is that genetic risk variants in non-European populations may be different from those already identified in populations of European descent, so that such multiethnic analyses may reveal entirely new loci and/or underlying biological pathways.

Mutations in other genes involved in the hypothalamic appetite regulation systems also result in monogenic obesity

Identification of the *LEP* and *LEPR* genes highlighted the role of the hypothalamus in regulating feeding behavior. Further investigation of neurosignaling pathways defined the leptin–melanocortin axis, which provided further candidate genes to be investigated in familial obesity. Because the melanocortin 4 receptor gene (*MC4R*) is small and intronless it is amenable to sequencing. Mutations in the *MC4R* gene cause autosomal dominant forms of monogenic obesity. This is the most prevalent single-gene cause of obesity known, causing around 6% of childhood and 2% of adult obesity cases.[2,23,24] Haploinsufficiency of *MC4R* causes hyperphagia and severe obesity with onset in childhood, increased lean mass, increased height, and severe hyperinsulinemia. In most cases each family had its own "private" mutation, so that the implications of each sequence change had to be established directly by biological assay. In each case, where homozygous individuals were seen, they were usually more severely affected than heterozygotes. Patients carrying mutations that do not completely abolish MC4R signaling capacity had less severe phenotypes. Genes for which mutations cause monogenic obesity may also contain common variants associated with less severe phenotypes: SNPs in the *MC4R* region (including the rs17782313 SNP 188 kb downstream of the *MC4R* gene) are associated with common forms of obesity.[25–28] In contrast, infrequent gain-of-function *MC4R* mutations are associated with a significant reduction in risk of obesity.[29]

Another partner in the leptin–melanocortin pathway, the preproopiomelanocortin (*POMC*) gene, encodes a preprohormone that is processed to produce several hormones, one of which is alpha melanocyte-stimulating hormone, which interacts with the melanocortin-4 receptor.[30] Mutations in *POMC* cause monogenic obesity in association with adrenal insufficiency and red hair.[31]

The cleavage of the *POMC* gene product into separate hormones (and also the processing of proinsulin in pancreatic beta cells) is carried out by an enzyme called prohormone convertase, which is encoded by the *PCSK1* gene. Mutations in *PCSK1* cause extreme childhood obesity, abnormal glucose homeostasis, hypogonadotropic hypogonadism, hypocortisolism, and elevated plasma proinsulin and POMC concentrations, but very low insulin levels, suggestive of defective prohormone processing,[32,33] sometimes with persistent diarrhea.[34] MC4R signaling takes place in the paraventricular (PVN) nuclei of the hypothalamus. The gene *SIM1* (single-minded homologue 1) is essential for proper development of PVN. Patients with chromosomal aberrations resulting in disruption or deletion of the *SIM1* gene have early-onset obesity.[35,36] *SIM1* heterozygous knockout mice develop early-onset obesity, with increased linear growth, hyperinsulinemia, and hyperleptinemia; the mice are also hyperphagic, but unlike leptin- or MC4R-deficient mice, their energy expenditure is not decreased.[37]

Most monogenic obesity cases have private mutations[38] so (with the exception of *MC4R*, which is easily sequenced) they have largely been investigated only in individual families: their true contribution to "common" obesity in the general population is, therefore, largely unknown and probably underestimated. It is likely that the current development of whole-exome and whole-genome sequencing of patients with putative Mendelian forms of obesity will find additional mutations in known obesity genes as well as novel gene mutations. So far, however, the investigation of such monogenic forms of obesity has been critical to our understanding of the regulation of feeding behavior.

Investigation of patients with complex "obesity-plus" phenotypes reveals novel obesity genes

In addition to monogenic forms of obesity, a number of syndromes have been identified, including obesity and developmental delay, as presenting clinical features. The archetypal example of this is Prader–Willi syndrome (PWS), which includes, in addition to severe obesity resulting from hyperphagia, a range of dysmorphic features, developmental delay, and hypogonadism. PWS results from loss of activity of the paternal copy of a region of chromosome 15.[39] Both paternal deletions and imprinting defects in this region result in PWS.[40,41] For several years, loss of the *SNRPN* gene was thought to be critical to PWS pathology, but the recent identification of PWS patients with unusually small deletions has allowed tighter definition of the critical region to the *HBII-85* snoRNA family, although the full details of how these genes affect feeding behavior remains to be elucidated.[42,43]

Similar phenotype mapping approaches in other genomic disorders have resulted in the identification of further novel genes and genomic regions involved in the pathogenesis of obesity. One example is Wilms tumor-aniridia-genitourinary anomalies-mental retardation (WAGR) syndrome.[44] A subset of individuals suffering from WAGR syndrome who, in addition to the classic clinical features of WAGR syndrome, also develop severe childhood obesity were demonstrated to have longer deletions of the 11p12p14 region including the brain-derived neurotrophic factor gene (*BDNF*).[45] Further support for the role of the BDNF pathway in syndromes encompassing both obesity and developmental delay is provided by the fact that mutations in the BDNF receptor (encoded by the *NTRK2* gene) result in the development of a phenotype characterized by severe obesity and cognitive impairment.[46] Mice that express decreased amounts of the BDNF receptor, encoded by *NTRK2*, have hyperphagia and maturity-onset obesity.[47]

Other complex conditions that have implicated genomic regions that may be involved in the development of obesity include Bardet–Biedl syndrome (BBS): a highly heterogeneous set of ciliopathy-based disorders sharing the clinical features of postaxial polydactly, obesity, developmental delay, retinal dystrophy, anosmia, and genital and renal abnormalities.[48] There have been a number of descriptions of BBS as an example of "triallelic inheritance"[49] because patients carry three mutations (homozygosity or compound heterozygosity for deleterious mutations in one of the six BBS genes, plus another mutation in one of the other BBS genes), but it is also sometimes described as a recessive disease with a modifier of penetrance.[50]

Obesity is also a feature of pseudohypoparathyroidism, also known as Albright's hereditary osteodystrophy, a disorder characterized by short stature, obesity, and skeletal defects, including brachydactyly.[51] The disease is caused by heterozygous mutations of *GNASI*, but obesity develops only when the mutation is inherited maternally. *GNASI* is imprinted in the PVN nucleus of the hypothalamus, and, in mouse models, loss of the maternal allele resulted in impaired melanocortin stimulation of energy expenditure, without affecting melanocortin's regulation of food intake.

Recently, important advances have been made by investigation of large cohorts of obese patients with no known genetic diagnosis but demonstrate complex "obesity-plus" phenotypes. Walters *et al.* identified a rare deletion of over 593 kilobases at chromosome 16p11.2 in 3.7% of children with "obesity-plus" phenotypes.[1] This study is important in that it is the first to demonstrate that genomic structural variants (GSVs) identified in such children are also found in obese patients from the general population, but not associated with developmental delay or cognitive dysfunction.[1] Deletions and duplications of the same 16p11.2 region had previously been observed in neuropsychiatric disorders (autism, schizophrenia, and bipolar disorder[52–54]), but Walters *et al.* demonstrated that although the presence of neuropsychiatric phenotype was variable, the deletion was highly penetrant for obesity. The deletion was also present in 0.7% of all "obesity-alone" individuals studied (these were ascertained only for obesity, not for any neuropsychiatric disorder) but not in any normal weight controls from an obesity case control study in children and adults.[26] The deletion accounts, therefore, for a nontrivial number of subjects with presumed common obesity: in fact, the deletion is associated with a 43-fold increase in risk of morbid obesity.

It is likely that several more rare GSVs of similar strong effect remain to be identified, both for obesity and other conditions. In a parallel study, Bochukova *et al.*[55] had reported identification of a number of candidate GSVs for obesity, including another 16p11 deletion (including the *SH2B1* gene among several other genes), which is close to but distinct from that identified by Walters *et al.* Limited replication was carried out in that study so the implications of the GSVs reported by Bochukova *et al.* for obesity in the general population have yet to be clearly established.

The full role of GSVs in obesity is only now beginning to be elucidated. As with other types of mutation, GSVs that influence adiposity may include both rare, highly penetrant variants of strong influence (as described above) and common variants with more subtle effects. There are already reported associations of common GSVs with conditions such as psoriasis[56,57] and autoimmune glomerulonephritis[58,59] and rare GSVs resulting in autism[60] and schizophrenia.[53,61] In addition to the direct benefit in diagnosis of subgroups of GSV-carrier individuals, the identification of rare, highly penetrant structural variants is also useful in that it can highlight genomic regions harboring novel obesity loci for targeted sequence analyses.

SNPs identified from genome-wide association studies of common obesity have poor predictive value

Although they have provided vital biological insight and even led to improved management of affected individuals in a small number of cases, currently known monogenic, monolocus, and syndromic forms of obesity are generally assumed to account for only a modest proportion of obesity cases. In an attempt to identify novel loci contributing to common obesity, a range of genome-wide association studies (GWAS) have been carried out, bringing the number of obesity-related genes to > 30.[62] The strongest and most widely replicated association is with a region encompassing the *FTO* gene.[26–28,63–67] As previously discussed, SNPs in the *MC4R* region are also associated with obesity in GWAS studies.[25–28] In addition to *FTO* and *MC4R* regions, many additional susceptibility loci have been identified through individual GWAS and subsequent meta-analyses.[26–28,68,69] Many of these loci are relevant to known hypothalamic regulatory pathways and one maps close to *GIPR*, which encodes an incretin receptor.[69]

Despite the success of GWAS in identifying novel obesity loci, the SNPs identified to date confer only very modest effects on risk, and the majority of the observed heritability of obesity remains unexplained.[69–71] The issue of "missing heritability" is by no means unique to obesity: it is observed in almost all of the common, complex diseases for which GWAS have so far been carried out, and there is much speculation about possible explanations for the phenomenon.[72] At the most basic level, it may be that current heritability estimates are inflated, and so even if all variants contributing to the genetic risk are identified much of the estimated heritability will remain unexplained. Alternatively, it could be that gene–gene and gene–environment interactions could explain more of the heritability than a simple additive model of individual risk variants; that is, the overall effect of all common SNP variants identified so far may be much greater than their summed individual effects: for example, 45% of the variance for human height can be reportedly explained by considering all genotyped GWAS SNPs simultaneously.[73] Investigation of the potential additional effects of environmental interactions in humans will require analysis of very large-scale prospective population cohorts with careful phenotyping and standardized data collection.

Another strong possibility, particularly pertinent to this review, is that the genetic architecture of many complex conditions, including obesity, may be partly refractory to dissection using the GWAS approach because of the contribution of rare highly penetrant variants. GWAS are predicated on the "common disease, common variant" model of complex disease, which assumes that SNPs present in >5% of the population will account for most of the genetic risk for a particular condition. However, it is now clear that individually infrequent variants of much stronger effect may collectively explain a nontrivial proportion of the genetic risk for a disease such as obesity. One important consequence of this is that for carriers of such variants, the predictive power of genetic analysis is high. In contrast, analysis of GWAS-identified SNPs is not useful for assessing disease risk in particular individuals: a combination of 20 obesity susceptibility variants failed to provide sufficient discriminative power to be of clinical use.[74] The identification of Mendelian forms of obesity is, therefore, raising, for the first time, the real prospect of personalized medicine in obesity. Thus, there is an urgent need to identify and characterize rare obesity-causing genetic variants for clinical use.

The potential impact of next generation sequencing

Infrequent (or recurrent) sequence or structural variants are unlikely to be effectively tagged by

SNPs. Recent rapid progress in the technology available for massively parallel DNA sequencing or next generation sequencing (NGS) could allow identification of new highly penetrant variants, causing Mendelian obesity. Costs of NGS are falling rapidly, and the bioinformatics capacity to store and interpret the resulting data is becoming more easily available, meaning that exome or full genome sequencing is likely to become increasingly accessible as a routine research tool for study of human disease.[75,76] Initial results have already demonstrated that exome sequencing can be an effective strategy for revealing novel mutations in recessive disorders, for example, identification of *de novo* mutations in the *SETBP1* gene as the cause of Schinzel–Giedion syndrome, and examples of this approach being successfully applied, are now accumulating in the literature.[77–79]

Sequencing of carefully phenotyped obese subjects has not yet been systematically undertaken. Initial large-scale sequencing initiatives, such as the 1000 Genomes Project,[80] have aimed to catalogue sequence variants without reference to phenotype. More recently, there have been initiatives to begin sequencing subjects with disease, including T2D. These large-scale NGS projects will undoubtedly add to our knowledge of human diversity, but there are significant problems yet to be addressed: some population groups are under-represented, information on phenotype may be absent or variable in quality, and there may be a large number of artifacts due to genetic changes occurring during cell culture (unfortunately, the majority of sequencing projects so far have used DNA from cell lines).

In the case of obesity, several NGS strategies might be adopted, including sequencing members of families with apparently Mendelian obesity (especially, where there is consanguinity), subjects at the extremes of the population variance for BMI, or subjects with complex "obesity-plus" phenotypes to identify new candidate genes and mutations for obesity. The major advantage of sequencing consanguineous families for recessive forms of obesity is that one expects homozygosity for exactly the same mutation (as opposed to nonconsanguineous families which might exhibit compound heterozygosity), but the disadvantage is that regions of general homozygosity might be sufficiently frequent and extensive as to obscure the true causative mutation. It is advantageous to sequence a number of family members (typically quartets of three affected and one nonaffected) and, if possible, also to compare results between different families with the same phenotype.

Sequencing of subjects who are at the extremes of population variation or "obesity-plus" patients will present its own challenges, notably in careful selection of patients with very similar phenotypes to maximize the chances of picking up mutations of the same genes or regions, to increase the likelihood of being able to distinguish benign from pathogenic variants (see below). A particular issue with the "obesity-plus" patient group is that these subjects may be more likely to have a GSV as the causative mutation, and methods of detecting GSVs from NGS data are still in active development.[81–86]

The main challenge of NGS projects will be in interpretation of a very large body of data: the intensive filtering of the thousands of apparently novel mutations found in every exome is mandatory, as it is not trivial to determine which particular genetic variant is responsible for the obesity in each subject. Whole-genome sequencing will magnify this problem significantly. Definitive identification of causative variants will require a combination of approaches including comparison with genomes of nonobese subjects; family studies to determine which variants cosegregate with obesity and bioinformatic prediction of the functional implications of the variant. In the end, there may be no substitute for basic biological study of altered gene and protein functions *in vitro* or in animal models.

Prospects for personalized medicine

As the cost of high-throughput sequencing continues to fall, it will become increasingly feasible to carry out whole-exome or whole-genome NGS studies of individuals with monogenic obesity or "obesity-plus" phenotypes. Such samples are likely to be enriched for rare, highly penetrant variants, with the potential to reveal novel genes and pathways involved in these conditions.

This type of research will help to paint a clearer picture of the genetic landscape of obesity. Identification of further Mendelian forms of obesity will allow affected individuals and their families to be offered genetic counseling and targeted support, including early intervention with the intention of preventing the worst deleterious effects of extreme obesity on metabolism and on the cardiovascular system, which may even occur in teenagers with obesity.

In addition to lifestyle intervention, it may be possible to target specific therapies to those most likely to benefit. As genetics continues to elucidate important pathogenic mechanisms and pathways, it opens new potential therapeutic targets. One area of very active research is in development of MC4R agonists. Several groups have already assessed the potential of these agents in treatment of obese patients with *MC4R* defects, with generally encouraging results.[87–91] Unfortunately, a previous centrally acting therapeutic agent for obesity (the cannabinoid receptor antagonist, rimonabant) caused serious neuropsychiatric adverse events, resulting in its withdrawal from the market. All new centrally active antiobesity drugs will, therefore, need to be carefully assessed for similar adverse events.[92] A range of other therapeutic agents for appetite control are currently in development, and it may be possible in the future to use genetics to stratify patients for optimal prescribing.[93,94]

Another obvious example is bariatric surgery (currently the only long-term effective treatment for morbid obesity). Understanding of the basis of a particular patient's obesity might be a useful prognostic indicator or be used to guide choice of surgery type (e.g., a patient with monogenic disease resulting in hyperphagia might be a poor candidate for a purely restrictive procedure, such as gastric banding, but might be offered a surgical procedure resulting in both restriction and malabsorption). Such concepts are already being explored in the literature, with early reports of the most common form of monogenic obesity, *MC4R* mutations, affecting the outcome of obesity surgery.[95–97]

Additionally, knowledge of a person's genetic risk of obesity could guide prescription of a range of therapeutic agents known to result in weight gain. For example, for patients with neuropsychiatric conditions, identification of patients strongly predisposed to weight gain might inform choice of therapy; antipsychotic-mediated weight gain further increases the burden of ill health for psychiatric patients and is a frequently cited reason for noncompliance with drug regimens. This might be particularly relevant because there is a clear overlap between neuropsychiatric disease and obesity in some patients carrying GSVs.[1] It may be pertinent to consider genotyping for GSVs having profound effects on obesity risk in patients with autism, for example. It may be better to avoid obesogenic antipsychotics in these cases. Because the obesity phenotype of the autism-associated chromosome 16p11.2 deletion does not always manifest before the age of ten,[1] intensive lifestyle intervention strategies might also be employed to try to prevent or delay obesity in deletion carriers.

Conclusion

Although obesity is considered by the general public and, sadly, by many clinicians to be simply the result of poor lifestyle choices, in a nontrivial number of subjects it is in fact a Mendelian genetic disorder. Sufferers, even those with very disabling extreme obesity, are not usually offered genetic advice or investigation. These patients, and their families, deserve the same degree of care offered to sufferers of other monogenic disorders. In the near future, we expect identification of an increasing number of rare, highly penetrant genetic causes of obesity, and because these have high predictive power, they may have immediate clinical utility. There is a very real prospect that obesity will be one of the earliest areas where true personalized medicine might be applied.

Conflicts of interest

The authors declare no conflicts of interest.

References

1. Walters, R.G. *et al*. 2010. A new highly penetrant form of obesity due to deletions on chromosome 16p11.2. *Nature* **463:** 671–675. doi:nature08727 [pii] 10.1038/nature08727.
2. Blakemore, A.I. & P. Froguel. 2008. Is obesity our genetic legacy? *J. Clin. Endocrinol. Metab.* **93:** S51–56. doi:93/11_Supplement_1/s51 [pii] 10.1210/jc.2008-1676.
3. O'Rahilly, S. 2009. Human genetics illuminates the paths to metabolic disease. *Nature* **462:** 307–314. doi:nature08532 [pii] 10.1038/nature08532.
4. Farooqi, I.S. *et al*. 2002. Beneficial effects of leptin on obesity, T cell hyporesponsiveness, and neuroendocrine/metabolic dysfunction of human congenital leptin deficiency. *J. Clin. Invest.* **110:** 1093–1103. doi:10.1172/JCI15693.
5. World Health Organization. 2006. *Obesity and overweight factsheet*. Retrieved August 10, 2010 from http://www.who.int/mediacentre/factsheets/fs311/en/index.html.
6. Hossain, P., B. Kawar & El Nahas M. 2007. Obesity and diabetes in the developing world–a growing challenge. *N. Engl. J. Med.* **356:** 213–215. doi:356/3/213 [pii] 10.1056/NEJMp068177.
7. Shaw, J.E., R.A. Sicree & P.Z. Zimmet. 2010. Global estimates of the prevalence of diabetes for 2010 and 2030. *Diabetes Res. Clin. Pract.* **87:** 4–14. doi:S0168-8227(09)00432-X [pii] 10.1016/j.diabres.2009.10.007.

8. Ono, T.G.R., R. Guthold & K. Strong. 2005. *WHO Global Comparable Estimates*. Geneva, Switzerland: World Health Organization.
9. Kelly, T., W. Yang, C.S. Chen, *et al.* 2008. Global burden of obesity in 2005 and projections to 2030. *Int. J. Obes. (Lond.)* **32:** 1431–1437.
10. Stunkard, A.J., T.T. Foch & Z. Hrubec. 1986. A twin study of human obesity. *AMA* **256:** 51–54.
11. Stunkard, A.J., J.R. Harris, N.L. Pedersen & G.E. McClearn. 1990. The body-mass index of twins who have been reared apart. *N. Engl. J. Med.* **322:** 1483–1487.
12. Stunkard, A.J. *et al.* 1986. An adoption study of human obesity. *N. Engl. J. Med.* **314:** 193–198.
13. Maes, H.H., M.C. Neale & L.J. Eaves. 1997. Genetic and environmental factors in relative body weight and human adiposity. *Behav. Genet.* **27:** 325–351.
14. Wardle, J., S. Carnell, C.M. Haworth & R. Plomin. 2008. Evidence for a strong genetic influence on childhood adiposity despite the force of the obesogenic environment. *Am. J. Clin. Nutr.* **87:** 398–404.
15. Farooqi, I.S. & S. O'Rahilly. 2009. Leptin: a pivotal regulator of human energy homeostasis. *Am. J. Clin. Nutr.* **89:** 980S–984S. doi:ajcn.2008.26788C [pii] 10.3945/ajcn.2008.26788C.
16. Clement, K. *et al.* 1998. A mutation in the human leptin receptor gene causes obesity and pituitary dysfunction. *Nature* **392:** 398–401.
17. Montague, C.T. *et al.* 1997. Congenital leptin deficiency is associated with severe early-onset obesity in humans. *Nature* **387:** 903–908.
18. Tadmouri, G.O. *et al.* 2009. Consanguinity and reproductive health among Arabs. *Reprod. Health* **6:** 17. doi:1742-4755-6-17 [pii] 10.1186/1742-4755-6-17.
19. Campbell, H., I. Rudan, A.H. Bittles & A.F. Wright. 2009. Human population structure, genome autozygosity and human health. *Genome Med.* **1:** 91. doi:gm91 [pii] 10.1186/gm91.
20. White, D.R. *et al.* 2007. Autozygosity mapping of Bardet–Biedl syndrome to 12q21.2 and confirmation of FLJ23560 as BBS10. *Eur. J. Hum. Genet.* **15:** 173–178. doi:5201736 [pii] 10.1038/sj.ejhg.5201736.
21. Khaliq, S. *et al.* 2007. Mapping of a novel type III variant of Knobloch syndrome (KNO3) to chromosome 17q11.2. *Am. J. Med. Genet A.* **143A:** 2768–2774. doi:10.1002/ajmg.a.31739.
22. Kurian, M.A. *et al.* 2009. Homozygous loss-of-function mutations in the gene encoding the dopamine transporter are associated with infantile parkinsonism-dystonia. *J. Clin. Invest.* **119:** 1595–1603. doi:39060 [pii] 10.1172/JCI39060.
23. Farooqi, I.S. *et al.* 2003. Clinical spectrum of obesity and mutations in the melanocortin 4 receptor gene. *N. Engl. J. Med.* **348:** 1085–1095.
24. Stutzmann, F. *et al.* 2008. Prevalence of melanocortin-4 receptor deficiency in Europeans and their age-dependent penetrance in multigenerational pedigrees. *Diabetes* **57:** 2511–2518. doi:db08-0153 [pii] 10.2337/db08-0153.
25. Loos, R.J. *et al.* 2008. Common variants near MC4R are associated with fat mass, weight and risk of obesity. *Nat. Genet.* **40:** 768–775.
26. Meyre, D. *et al.* 2009. Genome-wide association study for early-onset and morbid adult obesity identifies three new risk loci in European populations. *Nat. Genet.* **41:** 157–159. doi:ng.301 [pii] 10.1038/ng.301.
27. Thorleifsson, G. *et al.* 2009. Genome-wide association yields new sequence variants at seven loci that associate with measures of obesity. *Nat. Genet.* **41:** 18–24. doi:ng.274 [pii] 10.1038/ng.274.
28. Willer, C.J. *et al.* 2009. Six new loci associated with body mass index highlight a neuronal influence on body weight regulation. *Nat. Genet.* **41:** 25–34. doi:ng.287 [pii] 10.1038/ng.287.
29. Stutzmann, F. *et al.* 2007. Non-synonymous polymorphisms in melanocortin-4 receptor protect against obesity: the two facets of a Janus obesity gene. *Hum. Mol. Genet.* **16:** 1837–1844.
30. Lu, D. *et al.* 1994. Agouti protein is an antagonist of the melanocyte-stimulating-hormone receptor. *Nature* **371:** 799–802. doi:10.1038/371799a0.
31. Krude, H. *et al.* 1998. Severe early-onset obesity, adrenal insufficiency and red hair pigmentation caused by POMC mutations in humans. *Nat. Genet.* **19:** 155–157.
32. O'Rahilly, S. *et al.* 1995. Brief report: impaired processing of prohormones associated with abnormalities of glucose homeostasis and adrenal function. *N. Engl. J. Med.* **333:** 1386–1390. doi:10.1056/NEJM199511233332104.
33. Jackson, R.S. *et al.* 1997. Obesity and impaired prohormone processing associated with mutations in the human prohormone convertase 1 gene. *Nat. Genet.* **16:** 303–306.
34. Farooqi, I.S. *et al.* 2007. Hyperphagia and early-onset obesity due to a novel homozygous missense mutation in prohormone convertase 1/3. *J. Clin. Endocrinol. Metab.* **92:** 3369–3373. doi:jc.2007-0687 [pii] 10.1210/jc.2007-0687.
35. Holder, J.L., N.F. Butte Jr. & A.R. Zinn. 2000. Profound obesity associated with a balanced translocation that disrupts the SIM1 gene. *Hum. Mol. Genet.* **9:** 101–108.
36. Faivre, L. *et al.* 2002. Deletion of the SIM1 gene (6q16.2) in a patient with a Prader–Willi-like phenotype. *J. Med. Genet.* **39:** 594–596.
37. Michaud, J.L. *et al.* 2001. Sim1 haploinsufficiency causes hyperphagia, obesity and reduction of the paraventricular nucleus of the hypothalamus. *Hum. Mol. Genet.* **10:** 1465–1473.
38. Farooqi, I.S. & S. O'Rahilly. 2008. Mutations in ligands and receptors of the leptin–melanocortin pathway that lead to obesity. *Nat. Clin. Pract. Endocrinol. Metab.* **4:** 569–577. doi:ncpendmet0966 [pii] 10.1038/ncpendmet0966.
39. Goldstone, A.P. 2004. Prader–Willi syndrome: advances in genetics, pathophysiology and treatment. *Trends Endocrinol. Metab.* **15:** 12–20. doi:S1043276003002443 [pii].
40. Ledbetter, D.H. *et al.* 1981. Deletions of chromosome 15 as a cause of the Prader–Willi syndrome. *N. Engl. J. Med.* **304:** 325–329.
41. Mascari, M.J. *et al.* 1992. The frequency of uniparental disomy in Prader–Willi syndrome. Implications for molecular diagnosis. *N. Engl. J. Med.* **326:** 1599–1607.
42. Sahoo, T. *et al.* 2008. Prader–Willi phenotype caused by paternal deficiency for the HBII-85 C/D box small nucleolar RNA cluster. *Nat. Genet.* **40:** 719–721. doi:ng.158 [pii] 10.1038/ng.158.

43. de Smith, A.J. *et al.* 2009. A deletion of the HBII-85 class of small nucleolar RNAs (snoRNAs) is associated with hyperphagia, obesity and hypogonadism. *Hum. Mol. Genet.* **18:** 3257–3265. doi:ddp263 [pii] 10.1093/hmg/ddp263.
44. Fischbach, B.V., K.L. Trout, J. Lewis, *et al.* 2005. WAGR syndrome: a clinical review of 54 cases. *Pediatrics* **116:** 984–988. doi:116/4/984 [pii] 10.1542/peds.2004-0467.
45. Han, J.C. *et al.* 2008. Brain-derived neurotrophic factor and obesity in the WAGR syndrome. *N. Engl. J. Med.* **359:** 918–927.
46. Yeo, G.S. *et al.* 2004. A de novo mutation affecting human TrkB associated with severe obesity and developmental delay. *Nat. Neurosci.* **7:** 1187–1189.
47. Xu, B. *et al.* 2003. Brain-derived neurotrophic factor regulates energy balance downstream of melanocortin-4 receptor. *Nat. Neurosci.* **6:** 736–742. doi:10.1038/nn1073 nn1073 [pii].
48. Badano, J.L., N. Mitsuma, P.L. Beales & N. Katsanis. 2006. The ciliopathies: an emerging class of human genetic disorders. *Annu. Rev. Genomics Hum. Genet.* **7:** 125–148. doi:10.1146/annurev.genom.7.080505.115610.
49. Katsanis, N. *et al.* 2001. Triallelic inheritance in Bardet–Biedl syndrome, a Mendelian recessive disorder. *Science* **293:** 2256–2259.
50. Burghes, A.H., H.E. Vaessin & A. de La Chapelle. 2001. Genetics. The land between Mendelian and multifactorial inheritance. *Science* **293:** 2213–2214. doi:10.1126/ science.1065930 293/5538/2213 [pii].
51. Weinstein, L.S., J. Liu, A. Sakamoto, *et al.* 2004. Minireview: GNAS: normal and abnormal functions. *Endocrinology* **145:** 5459–5464. doi:10.1210/en.2004-0865 en.2004-0865 [pii].
52. Weiss, L.A. *et al.* 2009. Association between microdeletion and microduplication at 16p11.2 and autism. *Nat. Genet.* **41:** 1223–1227.
53. McCarthy, S.E. *et al.* 2009. Microduplications of 16p11.2 are associated with schizophrenia. *Nat. Genet.* **41:** 1223–1227. doi:ng.474 [pii] 10.1038/ng.474.
54. Kumar, R.A. *et al.* 2008. Recurrent 16p11.2 microdeletions in autism. *Hum. Mol. Genet.* **17:** 628–638. doi:ddm376 [pii] 10.1093/hmg/ddm376.
55. Bochukova, E.G. *et al.* 2010. Large, rare chromosomal deletions associated with severe early-onset obesity. *Nature* **463:** 666–670. doi:nature08689 [pii] 10.1038/nature08689.
56. de Cid, R. *et al.* 2009. Deletion of the late cornified envelope LCE3B and LCE3C genes as a susceptibility factor for psoriasis. *Nat. Genet.* **41:** 211–215. doi:ng.313 [pii] 10.1038/ng.313.
57. Hollox, E.J. *et al.* 2008. Psoriasis is associated with increased beta-defensin genomic copy number. *Nat. Genet.* **40:** 23–25. doi:ng.2007.48 [pii] 10.1038/ng.2007.48.
58. Aitman, T.J. *et al.* 2006. Copy number polymorphism in Fcgr3 predisposes to glomerulonephritis in rats and humans. *Nature* **439:** 851–855. doi:nature04489 [pii] 10.1038/nature04489.
59. Fanciulli, M. *et al.* 2007. FCGR3B copy number variation is associated with susceptibility to systemic, but not organ-specific, autoimmunity. *Nat. Genet.* **39:** 721–723. doi:ng2046 [pii] 10.1038/ng2046.
60. Sebat, J. *et al.* 2007. Strong association of de novo copy number mutations with autism. *Science* **316:** 445–449. doi:1138659 [pii] 10.1126/science.1138659.
61. Walsh, T. *et al.* 2008. Rare structural variants disrupt multiple genes in neurodevelopmental pathways in schizophrenia. *Science* **320:** 539–543. doi:1155174 [pii] 10.1126/science.1155174.
62. Walley, A.J., J.E. Asher & P. Froguel. 2009. The genetic contribution to non-syndromic human obesity. *Nat. Rev. Genet.* **10:** 431–442. doi:nrg2594 [pii] 10.1038/nrg2594.
63. Cotsapas, C. *et al.* 2009. Common body mass index-associated variants confer risk of extreme obesity. *Hum. Mol. Genet.* **18:** 3502–3507. doi:ddp292 [pii] 10.1093/hmg/ddp292.
64. Dina, C. *et al.* 2007. Variation in FTO contributes to childhood obesity and severe adult obesity. *Nat. Genet.* **39:** 724–726. doi:ng2048 [pii] 10.1038/ng2048.
65. Hinney, A. *et al.* 2007. Genome wide association (GWA) study for early onset extreme obesity supports the role of fat mass and obesity associated gene (FTO) variants. *PLoS ONE* **2:** e1361. doi:10.1371/journal.pone.0001361.
66. Loos, R.J. & C. Bouchard. 2008. FTO: the first gene contributing to common forms of human obesity. *Obes. Rev.* **9:** 246–250. doi:OBR481 [pii] 10.1111/j.1467-789X.2008.00481.x.
67. Scuteri, A. *et al.* 2007. Genome-wide association scan shows genetic variants in the FTO gene are associated with obesity-related traits. *PLoS Genet.* **3:** e115.
68. Benzinou, M. *et al.* 2008. Common nonsynonymous variants in PCSK1 confer risk of obesity. *Nat. Genet.* **40:** 943–945.
69. Speliotes, E.K. *et al.* Association analyses of 249,796 individuals reveal 18 new loci associated with body mass index. *Nat. Genet.* **42:** 937–948. doi:ng.686 [pii] 10.1038/ng.686.
70. Bogardus, C. 2009. Missing heritability and GWAS utility. *Obesity* **17:** 209–210.
71. Stolerman, E.S. *et al.* 2009. TCF7L2 variants are associated with increased proinsulin/insulin ratios but not obesity traits in the Framingham Heart Study. *Diabetologia* **52:** 614–620. doi:10.1007/s00125-009-1266-2.
72. Manolio, T.A. *et al.* 2009. Finding the missing heritability of complex diseases. *Nature* **461:** 747–753. doi:nature08494 [pii] 10.1038/nature08494.
73. Yang, J. *et al.* 2010. Common SNPs explain a large proportion of the heritability for human height. *Nat. Genet.* **42:** 565–569. doi:ng.608 [pii] 10.1038/ng.608.
74. Sandholt, C.H. *et al.* 2010. Combined analyses of 20 common obesity susceptibility variants. *Diabetes* **59:** 1667–1673. doi:db09–1042 [pii] 10.2337/db09–1042.
75. Mardis, E.R. 2008. Next-generation DNA sequencing methods. *Annu. Rev. Genomics Hum. Genet.* **9:** 387–402. doi:10.1146/annurev.genom.9.081307.164359.
76. Mardis, E.R. 2008. The impact of next-generation sequencing technology on genetics. *Trends Genet.* **24:** 133–141. doi:S0168-9525(08)00023-1 [pii] 10.1016/j.tig.2007.12.007.
77. Hoischen, A. *et al.* 2010. De novo mutations of SETBP1 cause Schinzel–Giedion syndrome. *Nat. Genet.* **42:** 483–485. doi:ng.581 [pii] 10.1038/ng.581.
78. Walsh, T. *et al.* 2010. Whole exome sequencing and homozygosity mapping identify mutation in the cell polarity protein GPSM2 as the cause of nonsyndromic hearing loss DFNB82.

Am. J. Hum. Genet. **87:** 90–94. doi:S0002-9297(10)00257-0 [pii] 10.1016/j.ajhg.2010.05.010.
79. Pierce, S.B. *et al.* 2010. Mutations in the DBP-deficiency protein HSD17B4 cause ovarian dysgenesis, hearing loss, and ataxia of Perrault Syndrome. *Am. J. Hum. Genet.* **87:** 282–288. doi:S0002-9297(10)00363-0 [pii] 10.1016/j.ajhg.2010.07.007.
80. *1,000 genomes. A deep catalogue of human genetic variation.* Available at www.1000genomes.org
81. Korbel, J.O. *et al.* 2009. PEMer: a computational framework with simulation-based error models for inferring genomic structural variants from massive paired-end sequencing data. *Genome Biol.* **10:** R23. doi:gb-2009-10-2-r23 [pii] 10.1186/gb-2009-10-2-r23.
82. Albers, C.A. *et al.* 2010. Dindel: accurate indel calls from short-read data. *Genome Res.* doi:gr.112326.110 [pii] 10.1101/gr.112326.110.
83. Abel, H.J. *et al.* 2010. SLOPE: a quick and accurate method for locating non-SNP structural variation from targeted next-generation sequence data. *Bioinformatics* **26:** 2684–2688. doi:btq528 [pii] 10.1093/bioinformatics/btq528.
84. Zeitouni, B. *et al.* 2010. SVDetect: a tool to identify genomic structural variations from paired-end and mate-pair sequencing data. 2010. *Bioinformatics* **26:** 1895–1896. doi:btq293 [pii] 10.1093/bioinformatics/btq293.
85. Hajirasouliha, I. *et al.* 2010. Detection and characterization of novel sequence insertions using paired-end next-generation sequencing. *Bioinformatics* **26:** 1277–1283. doi:btq152 [pii] 10.1093/bioinformatics/btq152.
86. Chen, K. *et al.* 2009. BreakDancer: an algorithm for high-resolution mapping of genomic structural variation. *Nat. Methods* **6:** 677–681. doi:nmeth.1363 [pii] 10.1038/nmeth.1363.
87. Roubert, P. *et al.* 2010. Novel pharmacological MC4R agonists can efficiently activate mutated MC4R from obese patient with impaired endogenous agonist response. *J. Endocrinol.* **207:** 177–183. doi:JOE-09-0336 [pii] 10.1677/JOE-09-0336.
88. Krishna, R. *et al.* 2009. Potent and selective agonism of the melanocortin receptor 4 with MK-0493 does not induce weight loss in obese human subjects: energy intake predicts lack of weight loss efficacy. *Clin. Pharmacol. Ther.* **86:** 659–666. doi:clpt2009167 [pii] 10.1038/clpt.2009.167.
89. Kumar, K.G. *et al.* 2009. Analysis of the therapeutic functions of novel melanocortin receptor agonists in MC3R- and MC4R-deficient C57BL/6J mice. *Peptides* **30:** 1892–1900. doi:S0196-9781(09)00303-9 [pii] 10.1016/j.peptides.2009.07.012.
90. Fan, Z.C. & Y.X. Tao. 2009. Functional characterization and pharmacological rescue of melanocortin-4 receptor mutations identified from obese patients. *J. Cell. Mol. Med.* **13:** 3268–3282. doi:JCMM726 [pii] 10.1111/j.1582-4934.2009.00726.x.
91. Xiang, Z., B. Proneth, M.L. Dirain, *et al.* 2010. Pharmacological characterization of 30 human melanocortin-4 receptor polymorphisms with the endogenous proopiomelanocortin-derived agonists, synthetic agonists, and the endogenous agouti-related protein antagonist. *Biochemistry* **49:** 4583–4600. doi:10.1021/bi100068u.
92. Nathan, P.J., B.V. O'Neill, A. Napolitano & E.T. Bullmore. 2010. Neuropsychiatric adverse effects of centrally acting antiobesity drugs. *CNS Neurosci. Ther.* doi:CNS172 [pii] 10.1111/j.1755-5949.2010.00172.x.
93. Halford, J.C., E.J. Boyland, J.E. Blundell, *et al.* 2010. Pharmacological management of appetite expression in obesity. *Nat. Rev. Endocrinol.* **6:** 255–269. doi:nrendo.2010.19 [pii] 10.1038/nrendo.2010.19.
94. Vetter, M.L., L.F. Faulconbridge, V.L. Webb & T.A. Wadden. 2010. Behavioral and pharmacologic therapies for obesity. *Nat. Rev. Endocrinol.* **6:** 578–588. doi:nrendo.2010.121 [pii] 10.1038/nrendo.2010.121.
95. Potoczna, N. *et al.* 2004. Gene variants and binge eating as predictors of comorbidity and outcome of treatment in severe obesity. *J. Gastrointest. Surg.* **8:** 97–981, discussion 981–972. doi:S1091-255X(04)00380-4 [pii] 10.1016/j.gassur.2004.09.032.
96. Aslan, I.R. *et al.* 2010. Bariatric surgery in a patient with complete MC4R deficiency. *Int. J. Obes. (Lond).* [Epub ahead of print] PubMed PMID: 20733581.
97. Aslan, I.R. *et al.* 2010. Weight loss afer Roux-en-Y gastric bypass in obese patients heterozygous for MC4R mutations. *Obes. Surg.* [Epub ahead of print] PubMed PMID: 20957447.

Ann. N.Y. Acad. Sci. ISSN 0077-8923

ANNALS OF THE NEW YORK ACADEMY OF SCIENCES
Issue: *The Year in Human and Medical Genetics*

Congenital disorders of glycosylation

Jaak Jaeken
University Hospital Gasthuisberg, Leuven, Belgium

Address for correspondence: Jaak Jaeken, University Hospital Gasthuisberg, Herestraat 49, BE-3000 Leuven, Belgium. jaak.jaeken@uzleuven.be

Congenital (genetic) disorders of glycosylation (CDG) are a rapidly growing disease family, with some 45 members reported since its first clinical description in 1980. Most of these are protein hypoglycosylation diseases, but recently three defects in lipid glycosylation have been identified. Most protein hypoglycosylation diseases are due to defects in the *N*-glycosylation pathway (16 diseases). The remaining ones affect the *O*-glycosylation pathway (8 diseases), both the *N*- and the *O*-glycosylation pathways, or other pathways (17 diseases). CDG can affect nearly all organs and systems, but there is often an important neurological component. The first-line screening for the *N*-glycosylation diseases is serum transferrin isoelectrofocusing (IEF), and for the *O*-glycosylation disorders apo CIII IEF. It has to be stressed that a normal test result does by no means exclude a CDG. In case of an abnormal result and as long as the basic defect has not been elucidated, the disease is labeled CDG-x (CDG-Ix when the transferrin IEF shows a type 1 pattern, and CDG-IIx when it shows a type 2 pattern).

Keywords: apo–CIII isoelectrofocusing; lipid glycosylation; *N*-glycosylation; *O*-glycosylation; transferrin isoelectrofocusing

Introduction

Congenital (genetic) disorders of glycosylation (CDG) are a rapidly growing family of genetic diseases first reported in 1980.[1] They are due to defects in the synthesis of the glycan moiety of glycoproteins or glycolipids, and in the attachment of these glycans to proteins and lipids.[2,3] There are two main types of protein glycosylation: *N*-glycosylation and *O*-glycosylation. *N*-glycosylation (*N*-glycans attached to an aminogroup of asparagines of proteins) comprises an assembly part and a processing part and extends over three cellular compartments: the cytosol, the endoplasmic reticulum (ER), and the Golgi. The assembly part of the *N*-glycosylation starts on the cytosolic part of the ER with the transfer of *N*-acetylglucosamine (GlcNAc) phosphate from UDP-GlcNAc to membrane-bound dolichyl monophosphate (Dol-P), forming GlcNAc-pyrophosphatedolichol (GlcNAc-PP-Dol). One GlcNAc and five mannose (Man) residues are subsequently attached to this dolichol-linked monosaccharide in a stepwise manner. The donor of these mannoses is a nucleotide-activated sugar, guanosine diphosphate (GDP)-Man, which is synthesized from fructose 6-phosphate, an intermediate of the glycolytic pathway. The lipid-linked heptasaccharide $Man_5GlcNAc_2$ is translocated by a flippase across the ER membrane and is elongated at the lumenal side by the attachment of four further mannose residues and, subsequently, three glucose residues. The four mannosyltransferases and three glucosyltransferases involved require dolichylphosphate-bound monosaccharides (Dol-P-Man and Dol-P-Glc). The completed $Glc_3Man_9GlcNAc_2$ oligosaccharide is then transferred to selected asparagine residues of nascent proteins by the oligosaccharyltransferase complex. The processing part of the *N*-glycosylation starts in the ER by trimming the glucoses (catalyzed by glucosidases I and II) and one mannose (catalyzed by alpha-mannosidase I). The residual glycoprotein intermediate is directed to the cis-Golgi where the processing pathway branches. A minor branch targets glycoproteins to the lysosomes (after the action of a

doi: 10.1111/j.1749-6632.2010.05840.x

GlcNAc-phosphotransferase and removal of the GlcNAc residues, leaving high-mannose glycoproteins capped with Man 6-P). The main branch leads to further trimming of mannoses (leaving a trimannosyl core) and addition of, respectively, GlcNAc, galactose, and eventually, sialic acid in the medial- and trans-Golgi. Another modification of many *N*-glycoproteins in the Golgi is the attachment of fucose.

O-glycosylation (*O*-glycans attached to the hydroxyl group of threonine or serine of proteins) has no processing and thus only consists of assembly. Contrary to *N*-glycosylation, this assembly mainly occurs in the Golgi. *O*-glycans show a greater diversity than *N*-glycans. Examples of important *O*-glycans are *O*-*N*-acetylglucosaminylglycans (mucin-type glycans), *O*-xylosylglycans (glycosaminylglycans), *O*-mannosylglycans, and *O*-fucosylglycans. There is a growing group of CDG with combined *N*- and *O*-glycosylation defects; a number of these are defects in proteins with a broader range of functions besides glycosylation. Recently, defects have been identified in glycosphingolipid glycosylation (ST3GAL5-CDG or Amish infantile epilepsy) and in glycosylphosphatidylinositol (GPI)-anchor glycosylation (PIGM-CDG and PIGV-CDG). This text uses the novel nomenclature, that is, the gene symbol followed by "-CDG."[4,5] The old designation is given between brackets.

Defects of protein *N*-glycosylation

Sixteen diseases are known in protein N-glycosylation: 14 assembly defects (CDG-I group) and 2 processing defects (CDG-II group). Only those CDG will be described here with more than two reported patients.

PMM2-CDG (CDG-Ia)

Phosphomannomutase (PMM)2 deficiency is a (cytosolic) defect in the second step of the mannose pathway (transforming mannose 6-phosphate into mannose 1-phosphate), which normally leads to the synthesis of GDP-mannose. This nucleotide sugar is the donor of the mannoses used in the ER to assemble the dolichol-pyrophosphate oligosaccharide precursor. Deficiency of GDP-mannose causes hypoglycosylation of numerous glycoproteins, including serum proteins, lysosomal enzymes, and membranous glycoproteins.

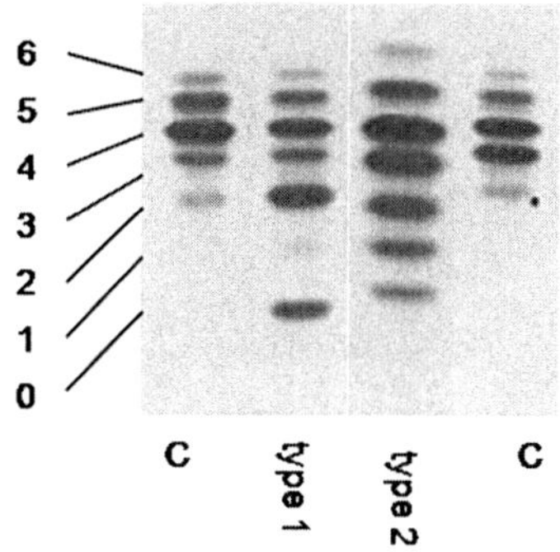

Figure 1. Serum transferrin isoelectrofocusing. Numbers 0 to 6 on the left indicate different sialotransferrins. C: control; type 1: type 1 pattern; type 2: type 2 pattern.

The diagnosis of PMM2-CDG (and of congenital disorders of *N*-glycosylation in general) is usually made by isoelectrofocusing (IEF) and immunofixation of serum transferrin[6] (Fig. 1) or by capillary zone electrophoresis of total serum.[7] Normal serum transferrin is mainly composed of tetrasialotransferrin and small amounts of mono-, di-, tri-, penta-, and hexasialotransferrins. The partial deficiency of sialic acid (a negatively charged and end-standing sugar) in CDG causes a cathodal shift. Two main types of cathodal shift can be recognized: type 1 is characterized by an increase of both disialo- and asialotransferrin, and a decrease of tetrasialotransferrin; in type 2 there is also an increase of the tri- and/or monosialotransferrin bands. In PMM2-CDG a type 1 pattern is found. In addition to the above-mentioned serum glycoprotein abnormalities, laboratory findings include increase of serum transaminases, hypoalbuminemia, hypocholesterolemia, and tubular proteinuria. The diagnosis is confirmed by finding a decreased activity of PMM2 in leukocytes or fibroblasts. Prenatal diagnosis is possible by enzymatic analysis of amniocytes or chorionic villus cells. This should be combined with mutation analysis of the PMM2 gene.

PMM2-CDG is by far the most prevalent protein *N*-glycosylation disorder (more than 700 patients known). The clinical spectrum is very large. The nervous system is affected in all patients, and most other organs are involved in a variable way. The neurological picture comprises alternating internal strabism and other abnormal eye movements, axial hypotonia, psychomotor retardation, ataxia, and hyporeflexia. After infancy, symptoms include retinitis pigmentosa, often stroke-like episodes, and sometimes

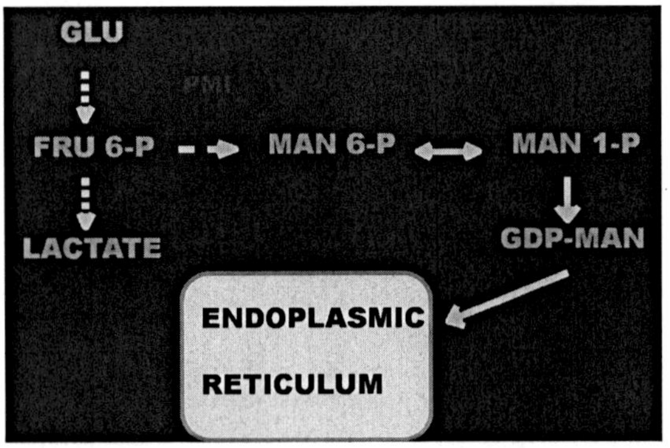

Figure 2. Scheme of the (cytosolic) mannose pathway with indication of the defect (vertical bar) in MPI-CDG. GLU, glucose; FRU, fructose; MAN, mannose; P, phosphate; GDP, guanosine diphosphate.

epilepsy. During the first year(s) of life, there are variable feeding problems such as anorexia, vomiting, and diarrhea. These can lead to severe failure to thrive. Other features are a variable dysmorphy (large hypoplastic/dysplastic ears, abnormal subcutaneous adipose tissue distribution), hepatomegaly, skeletal abnormalities, and hypogonadism. Some infants develop pericardial effusion and/or cardiomyopathy. At the other end of the spectrum are patients with a very mild phenotype (no dysmorphy, mild psychomotor retardation). Patients often have an extroverted and happy appearance. There is an increased mortality in the first years of life due to vital organ involvement or severe infection. Some 90 mutations have been identified in the PMM2 gene. The mutation leading to the R141H substitution is present in about 75% of the alleles of Central European patients.[8]

MPI-CDG (CDG-Ib)

The defect in phosphomannose-isomerase (MPI) deficiency is in the first step of the biosynthesis of the nucleotide sugar GDP-mannose (Fig. 2). The biochemical abnormalities are indistinguishable from those found in PMM2-CDG. Some 20 patients have been reported with this mainly hepatic-intestinal disorder. It is the only known *N*-glycosylation disorder without or with only minor neurological involvement. Symptoms start between 1 and 11 months of age and consist of various combinations of recurrent vomiting, abdominal pain, protein-losing enteropathy, recurrent thromboses, gastrointestinal bleeding, liver disease, and symptoms of hypoglycemia. Several patients have died from this disorder. This is the only known CDG that is efficiently treatable (see Treatment).[9]

ALG6-CDG (CDG-Ic)

ALG6 encodes glucosyltransferase I. Its deficiency causes a defect in the attachment of the first glucose (of three) to the dolichol-linked $Man_9GlcNAc_2$ ER intermediate. This is the second most common protein *N*-glycosylation disease with at least 30 patients identified. As in PMM2-CDG, patients show hypotonia, strabism, and seizures but psychomotor development is less retarded, there is less dysmorphy and usually no retinitis pigmentosa or cerebellar hypoplasia. Remarkably, some glycoproteins have unusually low blood levels (particularly of factor XI, and coagulation inhibitors antithrombin and protein C). Analysis of the dolichol-linked oligosaccharides in fibroblasts shows an accumulation of the glycan intermediate $Man_9GlcNAc_2$.[10]

ALG3-CDG (CDG-Id)

Six patients have been reported with this mannosyltransferase VI deficiency. Symptomatology included severe psychomotor retardation, hypsarrhytmia, postnatal microcephaly, optic atrophy, iris coloboma, hyperinsulinemic hypoglycemia with islet cell hyperplasia, and brain and corpus callosum atrophy. The hallmark biochemical feature of this CDG is an accumulation in fibroblasts of dolichyl pyrophosphate-$Man_5GlcNAc_2$.[11]

ALG12-CDG (CDG-Ig)

In the six reported patients with this mannosyltransferase VIII deficiency, the phenotypes showed various combinations of facial dysmorphy, psychomotor retardation, hypotonia, inverted nipples, subcutaneous fat pads, skeletal dysplasia, and decreased serum IgG levels. The typical biochemical feature of this disorder is an accumulation of dolichyl pyrophosphate-$Man_7GlcNAc_2$.[12]

ALG8-CDG (CDG-Ih)

This CDG is due to glucosyltransferase II deficiency. Five patients have been reported from four families. Three showed a severe disease with dysmorphy and multiorgan failure resulting in early death. The fourth patient had a milder phenotype with hepatomegaly and protein-losing enteropathy. In the fifth patient there was prominent central nervous system involvement besides enterohepatic and renal disease. Dolichyl pyrophosphate-$Glc_1Man_9GlcNAc_2$ accumulated in fibroblasts.[13]

ALG1-CDG (CDG-Ik)

The four reported patients with mannosyltransferase I deficiency showed epilepsy, severe psychomotor retardation, and variable features such as dysmorphy, liver dysfunction, cardiomyopathy, nephrotic syndrome, hypogonadism, and depletion of beta-cells. Lipid-linked oligosaccharide analysis showed an accumulation of $GlcNAc_2$-PP-dolichol in fibroblasts.[14]

RFT1-CDG (CDG-In)

This is a defect in the flippase that transfers $Man_5GlcNAc_2$-PP-Dol from the cytoplasmic to the lumenal side of the ER. The six reported patients showed mainly severe psychomotor retardation, hypotonia, drug-resistant epilepsy, and sensorineural deafness. This is the first CDG with sensorineural deafness as a consistent feature. Lipid-linked oligosaccharide analysis showed an accumulation of the above-mentioned intermediate in fibroblasts.[15]

TUSC3-CDG

This is a defect in one of the subunits of the oligosaccharyltransferase complex. It has been reported in two families with autosomal recessive, nonsyndromic mental retardation and suggests a crucial role of this complex in cognitive functioning. Remarkably, IEF of serum transferrin showed a normal pattern.[16]

MAGT1-CDG

Like TUSC3-CDG, this is a disorder of one of the subunits of the oligosaccharyltransferase complex, and has been reported in a family with X-linked, nonsyndromic, and nonprogressive mental retardation. This is the only known CDG with X-linked inheritance.[17]

MGAT2-CDG (CDG-IIa)

Four patients have been reported with *N*-acetylglucosaminyltransferase II deficiency, the first identified *N*-glycan processing defect. Besides neurological involvement (psychomotor retardation, epilepsy, behavioral disturbances), these patients present with craniofacial dysmorphy, skeletal abnormalities, gastrointestinal disturbances, and growth retardation. As in other CDG-II group patients, serum glutamic oxaloacetic transaminase is increased but glutamic pyruvic transminase is normal. IEF of serum transferrin shows a type 2 pattern and structural analysis of the transferrin oligosaccharides indicates an abnormal glycan structure (a monoantennary *N*-acetyllactosamine type glycan) as seen in some other CDG-II diseases.[18]

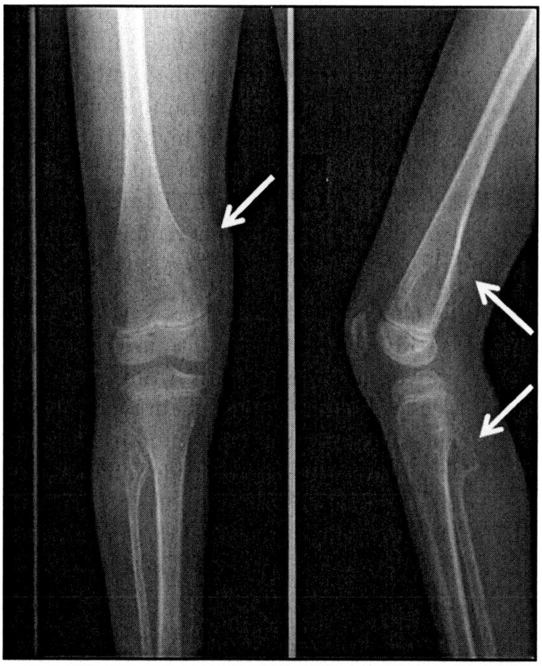

Figure 3. Radiological picture of lower limb exostoses in EXT1/EXT2-CDG. The exostoses are indicated by white arrows.

Defects of protein *O*-glycosylation

Defects have been identified in the synthesis of *O*-xylosylglycans (EXT1/EXT2-CDG and B4GALT7-CDG), of *O*-*N*-acetylgalactosaminylglycans (GALNT3-CDG), of O-xylosyl/N-acetylgalactosaminylglycans (SLC35D1-CDG), of *O*-mannosylglycans (POMT1/POMT2-CDG and POMGNT1-CDG), and of *O*-fucosylglycans (SCDO3-CDG and B3GALTL-CDG).

EXT1/EXT2-CDG

This is the defect of multiple cartilaginous exostoses and the only known CDG with autosomal dominant inheritance. It is characterized by osteochondromas of the ends of long bones (Fig. 3) and is thus a monosystemic CDG. These tumors are often present at birth, and their growth slows during adolescence and stops in adulthood. A small percentage of these lesions shows malignant degeneration. Complications may arise from compression of peripheral nerves and blood vessels. The EXT1/EXT2 complex is localized in the Golgi and has both glucuronyltransferase and *N*-acetyl-D-hexosaminyltransferase activities involved in the polymerization of heparan sulfate. It has been hypothesized that mutations in these glycosyltransferases impair the synthesis of a glycosaminoglycan that exerts a tumor-suppressor function. This would explain the higher risk of patients to develop chondrosarcomas and osteosarcomas. Mutations in *EXT1* and in *EXT2* are responsible for over 70% of the cases of hereditary multiple exostoses.[19]

B4GALT7-CDG

This defect in beta-1,4-galactosyltransferase 7 has been reported in three patients from two families with a premature aging phenotype, hyperelastic skin, microcephaly, and joint hyperlaxity. The defect disrupts the trisaccharide linker region of glycosaminoglycans (*O*-linked xylose-galactose-galactose), specifically in the attachment of the first galactose to xylose.[20]

GALNT3-CDG

Deficiency of isoform 3 of *N*-acetylgalactosaminyltransferase causes recurrent, painful calcified subcutaneous masses known as familial, hyperphosphatemic tumoral calcinosis. This can be complicated by secondary infections and incapacitating mutilations. The hyperphosphatemia is due to increased renal phosphate retention.[21]

SLC35D1-CDG

The solute carrier 35D1 encodes an ER UDP-glucuronic acid/UDP-*N*-acetylgalactosamine dual transporter needed for chondroitin sulfate biosynthesis. Loss-of-function mutations cause Schneckenbecken dysplasia, a rare, severe skeletal dysplasia comprising mainly platyspondyly, extremely short long bones, and small ilia with snail-like appearance.[22]

POMT1/POMT2-CDG

POMT1 mutations can cause Walker-Warburg syndrome, a rare neuronal migration disorder characterized by brain and eye involvement associated with congenital muscular dystrophy. The brain lesions consist of "cobblestone" lissencephaly, agenesis of the corpus callosum, cerebellar hypoplasia, hydrocephaly, and sometimes encephalocoele. This disease usually runs a fatal course before the age of 1 year. Psychomotor development is absent. In this disorder there is an aberrant glycosylation of alpha-dystroglycan, an external membrane protein expressed in muscle, brain, and other tissues. The enzyme catalyses the first step in the synthesis of the *O*-mannose-linked core Galbèta1-4GlcNAcbèta1-2Man-*O*-Ser/Thr. Some patients with Walker-Warburg syndrome have mutations in the protein *O*-mannosyltransferase 2 gene (*POMT2*), in the fukutin gene, or in the fukutin-related protein gene.[23]

POMGNT1-CDG

This disease is also known as muscle-eye-brain disease. It is a neuronal migration/congenital muscular dystrophy syndrome similar to but less severe than Walker-Warburg syndrome. These patients have mutations in the gene encoding protein *O*-mannosyl-bèta1,2-*N*-acetylglucosaminyltransferase 1, catalyzing the second step in the synthesis of the *O*-mannosylglycan core.[23]

SCDO3-CDG

Spondylocostal dysostosis type 3 is a notch pathway defect in lunatic fringe, an *O*-fucose-specific bèta1,3-*N*-acetylglucosaminyltransferase. Patients show a severe vertebral phenotype with malsegmentation due to disruption of somitogenesis. Somites are precursors of the axial skeleton and associated musculature. It should be noted that spondylocostal dysostosis type 1 and type 2 are also due to defects in the notch pathway.[24]

B3GALTL-CDG

This so-called Peters'-plus syndrome is characterized by peculiar eye malformations including corneal opacities and iridocorneal adhesions besides growth retardation and variable abnormalities in other organs. Mutations are in a beta1,3-glucosyltransferase that adds glucose to *O*-linked fucose. This disaccharide modification is specific to thrombospondin type 1 repeats, found in extracellular proteins that function in cell–cell and cell–matrix interactions.[25]

Defects of glycosphingolipid and GPI-anchor glycosylation

Only three disorders have been reported in this CDG group: SIAT9-CDG or Amish infantile epilepsy, and GPI deficiency due to PIGM deficiency and to PIGV deficiency.

SIAT9-CDG

This defect causes Amish infantile epilepsy, an infantile-onset epilepsy syndrome associated with developmental retardation and blindness. The

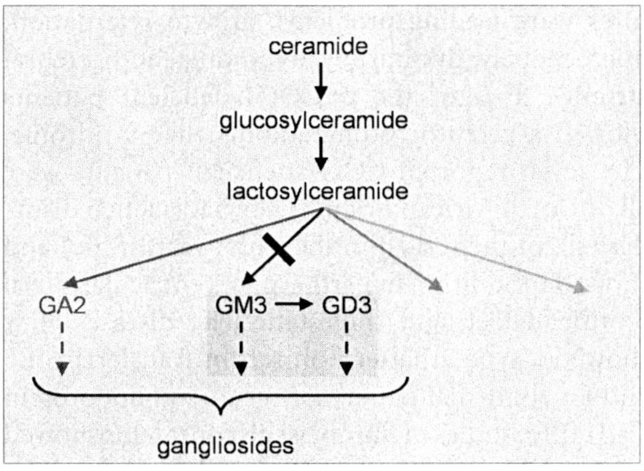

Figure 4. Scheme of the ganglioside synthesis pathway with indication of the defect (horizontal bar) in SIAT9-CDG.

SIAT9 gene encodes GM3 synthase, a sialyltransferase (lactosylceramide alpha-2,3 sialyltransferase). This enzyme catalyzes the initial step in the biosynthesis of most complex gangliosides from lactosylceramide (Fig. 4). The defect causes an accumulation of lactosylceramide associated with a decrease of the gangliosides of the GM3 and GD3 series.[26]

PIGM-CDG

This is a new form of inherited (autosomal recessive) GPI deficiency presenting in infancy. It is characterized by splanchnic vein thrombosis and epilepsy. As compared with paroxysmal nocturnal hemoglobinuria, it does not result in clinically significant hemolysis and bone marrow failure. The defect results in histone hypoacetylation at the promoter of the PIGM gene. This gene codes for the first mannosyltransferase in the GPI-anchor biosynthesis pathway. The GPI anchor is synthesized in the ER through at least nine sequential reaction steps mediated by at least 18 proteins. Thus a number of other disorders have to be expected in this pathway. A partial treatment is available (see Treatment).[27]

PIGV-CDG

PIGV codes for the second mannosyltransferase in the GPI anchor biosynthesis. Mutations in this gene have recently been identified in the hyperphosphatasia mental retardation syndrome, also known as Mabry syndrome. More particularly, mutations have been found in a subgroup of this syndrome with facial dysmorphism (hypertelorism, long palpebral fissures, a broad nasal bridge and tip, thin upper lip), and variable neurological features. The hyperphosphatasia can be explained by the fact that alkaline phosphatase (as more than 100 other proteins) is a GPI-anchored protein.[28]

Defects of multiple glycosylation and other pathways

This group comprises combined defects in *N*- and *O*-glycosylation (DPM1-CDG, MPDU1-CDG, B4GALT1-CDG, GNE-CDG, SLC35A1-CDG), a fucosylation defect (SLC35C1-CDG), two dolichol synthesis defects (DK1-CDG, SRD5A3-CDG), defects in subunits of the COG complex (COG7-CDG, COG1-CDG, COG8-CDG-COG4-CDG-COG5-CDG), a defect in a V-ATPase (ATP6V0A2-CDG), and a defect in SEC23B, a COPII component.

DPM1-CDG (CDG-Ie)

Six children been reported with this defect in the catalytic subunit of Dol-P-Man synthase. They showed a severe neurological involvement as in NOT56L-CDG and also an accumulation of $Man_5GlcNAc_2$-PP-dolichol in fibroblasts.[29]

MPDU1-CDG (CDG-If)

The deficient protein is considered to be a chaperone involved in the use of both Dol-P-Man and Dol-P-Glc. The four reported patients showed a severe encephalopathy, and three of them had a scaly, erythematous skin disorder. One patient had dwarfism with growth hormone deficiency. Fibroblasts accumulated $Man_5GlcNAc_2$-PP-dolichol as well as $Man_9GlcNAc_2$-PP-dolichol.[30]

GNE-CDG

Mutations in UDP-GlcNAc 2-epimerase/*N*-acetylmannosamine kinase cause sialuria, recessive hereditary inclusion body myopathy and Nonaka myopathy. The gene encodes a bifunctional enzyme that catalyses the first two (and rate-limiting) steps in sialic acid biosynthesis. This myopathy has an adult onset with progressive distal and proximal muscle weakness. A peculiar feature is that it spares the quadriceps muscles. Muscle histology shows rimmed vacuoles on Gomori's trichrome stain, small fibers in groups, and tubulofilaments without evidence of inflammation. IEF of serum transferrin is normal in this disorder.[31]

SLC35C1-CDG (CDG-IIc)

A few patients have been reported with this so-called leukocyte adhesion deficiency type II syndrome (LAD II). This syndrome comprises craniofacial dysmorphy, severe growth and psychomotor retardation, as well as recurrent bacterial infections with unusually high leukocytosis. Due to hypofucosylation, neutrophils of these patients lack sialyl-Lewis X, a fucose-containing carbohydrate ligand of the selectin family of cell adhesion molecules. This ligand is required for the recruitment of neutrophils to infection sites. A partial treatment is available for some patients, depending on the mutation (see Treatment).[32]

DK1-CDG (CDG-Im)

This is a defect in dolichol kinase catalyzing the final step of the *de novo* biosynthesis of dolichol phosphate. This compound is involved in several glycosylation reactions: *N*-glycosylation, GPI-anchor biosynthesis, and *C*- and *O*-mannosylation. Two families with two affected sibs each have been reported. All four had ichthyosis of the skin and died before the age of nine months. Inconstant features included dilated cardiomyopathy, hair problems, failure to thrive, postnatal microcephaly, epilepsy, hypotonia, bilateral nystagmus, and hypoketotic hypoglycaemia.[33]

SRD5A3-CDG

This is the second reported defect in the synthesis of dolichol phosphate, more specifically in the conversion of polyprenol to dolichol. Clinical data are from 11 patients belonging to seven families. Besides psychomotor retardation the most prominent symptoms are cerebellar (atrophy, vermis malformations), ophtalmological (hypoplasia or coloboma of the iris, retina, choroid, optic disc; nystagmus, optic atrophy, microphtalmia, glaucoma, cataract), and cutaneous (ichthyosis, erythroderma, dry skin, atopic dermatisis).[34]

COG-CDG

The conserved oligomeric Golgi (COG) complex is an eight subunit (COG1-8) peripheral Golgi membrane hetero-oligomeric protein complex. It is organized into lobes A (COG2-4) and B (COG5-7) with COG1 and COG8 bridging these lobes. This complex is thought to play a critical role in vesicle tethering processes involving retrograde Golgi transport of resident proteins responsible for glycan biosynthesis. Defects have been reported in COG1 (COG1-CDG ; *CDG-IIg*), COG7 (COG7-CDG ; *CDG-IIe*), COG8 (COG8-CDG ;*CDG-IIh*), COG4 (COG4-CDG; *CDG-IIj*), COG5 (COG5-CDG), and COG6 (COG6-CDG). Common features were feeding problems, growth retardation, microcephaly, dysmorphy, hypotonia, and cerebral atrophy. Two of the 3 COG1-deficient patients showed a cerebrocostomandibular-like syndrome. The seven reported COG7-deficient patients were all of North African origin. They had a lethal disorder (six of them died in their first year of life) and showed in addition hyperthermia, ventricular/atrial septum defect, and cholestatic liver disease. They showed a type 2 pattern on serum transferrin IEF and an abnormal pattern of serum apolipoprotein C-III IEF. Studies of fibroblast glycoproteins showed a partial *N*- and *O*-glycosylation defect caused by a decreased transport of CMP-sialic acid and UDP-galactose into the Golgi, and a reduced activity of two glycosyltransferases involved in the galactosylation and sialylation of *O*-glycans. The six patients with early lethality were homozygous for the same intronic mutation in *COG7*.[35] Since the COG complex is most probably not only involved in glycosylation but also in other cellular functions, we propose to call the COG defects "CDG-plus."

ATP6V0A2-CDG

In a subgroup of patients with autosomal recessive cutis laxa type 2 and of patients with wrinkly skin syndrome defects have been demonstrated in the a2 subunit of the vesicular ATPase H^+-pump. These patients have generalized cutis laxa at birth that becomes less obvious with age. They also present with increased joint laxity, ophtalmological abnormalities (mainly strabismus, myopia or amblyopia, and sometimes corneal dystrophy), microcephaly, and delayed motor development that improves with age. Tropoelastin aggregates are found in their Golgi apparatus.[36] This seems to be another "CDG-plus."

SEC23B-CDG

Congenital dyserythropoietic anemia type II (CDAII) has recently been shown to be due to mutations in the gene coding for SEC23B, a COPII component. One reason why this disorder is limited to the erythroblastic lineage might be that in the other tissues a SEC23A isoform compensates for the SEC23B deficiency.[37–39]

Diagnostic approaches in CDG

CDG should be looked for in any unexplained syndrome. Serum transferrin IEF remains the golden standard for the diagnosis of CDG due to an *N*-glycosylation defect. In case of an abnormal result, an artifact, a transferrin protein variant, and a secondary CDG (mainly galactosemia and hereditary fructose intolerance), should be excluded. There are two types of patterns due to a CDG; the type 1 pattern points to an assembly defect (in the cytosol or ER; CDG-Ix) and the type 2 pattern to a processing defect (in the ER or Golgi; CDG-IIx).

The first step in CDG-Ix is measurement of the enzymatic activity of phosphomannomutase and phosphomannose isomerase in fibroblasts or leukocytes. When PMM2-CDG and MPI-CDG are excluded, the next step is dolichol-linked oligosaccharide (LLO) analysis in fibroblasts. An increase of one or more glycan intermediates, permits to formulate a diagnostic hypothesis that can be tested by enzymatic analysis and/or mutation analysis. If there is no specific abnormal pattern, a defect in the dolichol-phosphate pathway should be considered.

CDG-IIx should be further investigated by two different analyses: transferrin glycan analysis, and apolipoprotein C-III IEF to look for core 1 mucin-type glycosylation defects. In a small minority of cases the transferrin glycan analysis reveals a specific pattern with an accumulation of (a) Golgi glycan intermediate(s). This permits to formulate a hypothesis. In most cases however, an aspecific pattern will be obtained (usually some degree of hyposialylation and/or hypogalactosylation). The further approach should then be guided by the clinical presentation. In case of a cutis laxa syndrome, mutational analysis of the ATP6V0A2 gene is indicated. If not, we recommended to perform mutation analysis of the COG subunit genes.

Treatment of CDG

CDG is still a poorly treatable family of disorders. Only in three CDG is a treatment option available.

(1) MPI-CDG: this is the only efficiently treatable CDG. The treatment consists of oral mannose (1 g/kg body weight per day, divided in 4–6 doses). The rationale for this treatment is that hexokinases phosphorylate mannose to mannose 6-phosphate, thus bypassing the defect.
(2) SLC35C1-CDG: some patients with GDP-fucose transporter deficiency respond to fucose, depending on the nature of the mutation. The treatment is only effective with regard to the typical recurrent infections with hyperleukocytosis.
(3) PIGM-CDG: the mutation causes histone hypoacetylation at the *PIGM* promoter, and this feature has been exploited as the basis for a treatment with butyrate, a histone deacetylase inhibitor. Butyrate markedly increases *PIGM* transcription, and is able to control the seizures.

Conclusions and perspectives

The field of CDG continues to grow at a surprisingly rapid rate. Forty-five CDG have been unraveled in the course of 30 years. Never before have so many disorders from the same family been identified in this time lapse. Major developments are to be expected in the following areas: defects in dolichol metabolism, defects in lipid glycosylation, defects in organ-specific glycosylation, and defects in other multifunctional proteins, that is, proteins that are not only involved in glycosylation but also in other functions. Advances in treatment are eagerly awaited since an efficient treatment is available for only one CDG (MPI-CDG). Efforts should be directed in particular toward a therapy for the important and often devastating disorder, PMM2-CDG. Since about 250 genes are considered to be involved in glycosylation, it is evident that we know only a small minority of existing CDG. The search for novel CDG will at the same time greatly boost our knowledge of the many unknown aspects of this most important protein and lipid modification.

Conflicts of interest

The author declares no conflicts of interest.

References

1. Jaeken, J., M. Vanderschueren-Lodeweyckx, P. Casaer, *et al.* 1980. Familial psychomotor retardation with markedly fluctuating serum proteins, FSH and GH levels, partial TBG-deficiency, increased serum arylsulphatase A and increased CSF protein: a new syndrome? *Pediatr. Res.* **14:** 179.
2. Freeze, H.H. 2006. Genetic defects in the human glycome. *Nat. Rev. Genet.* **7:** 537–551.
3. Jaeken, J. & G. Matthijs. 2007. Congenital disorders of glycosylation. A rapidly expanding disease family. *Annu. Rev. Genomics. Hum. Genet.* **8:** 261–278.
4. Jaeken, J., T. Hennet, H.H. Freeze & G. Matthijs. 2008. On the nomenclature of congenital disorders of glycosylation. *J. Inherit. Metab. Dis.* **31:** 669–672.
5. Jaeken, J., T. Hennet, G. Matthijs & H. Freeze. 2009. CDG nomenclature: time for a change! *Biochim Biophys Acta* **1792:** 825–826.
6. Jaeken, J., H.G. Van Eijk, C. van der Heul, *et al.* 1984. Sialic acid-deficient serum and cerebrospinal fluid transferrin in a

newly recognized genetic syndrome. *Clin. Chim. Acta.* **144:** 245–247.
7. Carchon, H.A., R. Chevigné, J.B. Falmagne & J. Jaeken. 2004. Diagnosis of congenital disorders of glycosylation by capillary zone electrophoresis of serum transferrin. *Clin. Chem.* **50:** 101–111.
8. Grünewald, S. 2009. The clinical spectrum of phosphomannomutase 2 deficiency (CDG-Ia). *Biochim. Biophys. Acta.* **1792:** 827–834.
9. de Lonlay, P. & N. Seta. 2009. The clinical spectrum of phosphomannose isomerase deficiency, with an evaluation of mannose treatment for CDG-Ib. *Biochim. Biophys. Acta.* **1792:** 841–843.
10. Al-Owain, M., S. Mohamed, N. Kaya, *et al.* 2010. A novel mutation and first report of dilated cardiomyopathy in ALG6-CDG (CDG-Ic): a case report. *Orphanet. J. Rare. Dis.* **5:** 7.
11. Kranz, C., L. Sun, E.A. Eklund, *et al.* 2007. CDG-Id in two siblings with partially different phenotypes. *Am. J. Med. Genet. A.* **143A:** 1414–1420.
12. Kranz, C., A.A. Basinger, Gülsavas-Calicoglu, *et al.* 2007. Expanding spectrum of congenital disorder of glycosylation Ig (CDG-Ig): sibs with a unique skeletal dysplasia, hypogammaglobulinemia, cardiomyopathy, genital malformations, and early lethality. *Am. J. Med. Genet. A.* **143A:** 1371–1378.
13. Stölting, T., H. Omran, A. Erlekotte, *et al.* 2009. Novel ALG8 mutations expand the clinical spectrum of congenital disorders of glycosylation type Ih. *Mol. Genet. Metab.* **98:** 305–309.
14. Schwarz, M., C. Thiel, J. Lubbehusen, *et al.* 2004. Deficiency of GDP-Man: GlcNAc2-PP-dolichol mannosyltransferase causes congenital disorder of glycosylation type Ik. *Am. J. Hum. Genet.* **74:** 472–481.
15. Jaeken, J., W. Vleugels, L. Régal, *et al.* 2009. RFT1-CDG: deafness as a novel feature of congenital disorders of glycosylation. *J. Inherit. Metab. Dis.* **32:** 756–757.
16. Garshasbi, M., V. Hadavi, H. Habibi, *et al.* 2008. A defect in the TUSC3 gene is associated with autosomal recessive mental retardation. *Am. J. Hum. Genet.* **82:** 1158–1164.
17. Molinari, F., F. Foulquier, P.S. Tarpey, *et al.* 2008. Oligosaccharyltransferase-subunit mutations in syndromic mental retardation. *Am. J. Hum. Genet.* **82:** 1150–1157.
18. de Cock, P. & J. Jaeken. 2009. MGAT2 deficiency (CDG-IIa): the life of J. *Biochim. Biophys. Acta.* **1792:** 844–846.
19. Heinritz, W., V. Hüffmeier, S. Strenge, *et al.* 2009. New mutations of EXT1 and EXT2 genes in German patients with multiple osteochondromas. *Ann. Hum. Genet.* **73:** 283–291.
20. Faiyaz-Ul-Haqve, M., S.H. Zaidi, M. Al-Ali, *et al.* 2004. A novel missense mutation in the galactosyltransferase-I (B4GALT7) gene in a family exhibiting facioskeletal anomalies and Ehlers-Danlos syndrome resembling the progeroid type. *Am. J. Med. Genet. A.* **128A:** 39–45.
21. Chefetz, I. & E. Sprecher. 2009. Familial tumoral calcinosis and the role of O-glycosylation in the maintenance of phosphate homeostasis. *Biochim. Biophys. Acta.* **1792:** 847–852.
22. Hiraoka, S., T. Furuishi, G. Nishimura, *et al.* 2007. Nucleotide-sugar transporter SLC35D1 is critical to chondroitin sulfate synthesis in cartilage and skeletal development in mouse and human. *Nat. Med.* **13:** 1363–1367.
23. Hewitt, J.E. 2009. Abnormal glycosylation of dystroglycan in human genetic disease. *Biochim. Biophys. Acta.* **1792:** 853–861.
24. Sparrow, D.B., G. Chapman, M.A. Wouters, *et al.* 2006. Mutation of the lunatic fringe gene in humans causes spondylocostal dysostosis with a severe vertebral phenotype. *Am. J. Hum. Genet.* **78:** 28–37.
25. Hess, D., J.J. Keusch, S.A.L. Oberstein, *et al.* 2008. Peters Plus syndrome is a new congenital disorder of glycosylation and involves defective O-glycosylation of thrombospondin type 1 repeats. *J. Biol. Chem.* **283:** 7354–7360.
26. Simpson, M.A., H. Cross, C. Proukakis, *et al.* 2004. Infantile-onset symptomatic epilepsy syndrome caused by a homozygous loss-of-function mutation of GM3 synthase. *Nature. Genet.* **36:** 1225–1229.
27. Almeida, A.M., Y. Murakami, D.M. Layton, *et al.* 2006. Hypomorphic promoter mutation in PIGM causes inherited glycosylphosphatidylinositol deficiency. *Nature. Med.* **12:** 846–851.
28. Krawitz, P.M., M.R. Schweiger, C. Rödelsperger, *et al.* 2010. Identity-by-descent filtering of exome sequence data identifies *PIGV* mutations in hyperphosphatasia mental retardation syndrome. *Nat. Genet.* **42:** 827–829.
29. Imbach, T., B. Schenk, E. Schollen, *et al.* 2000. Deficiency of dolichol-phosphate-mannose synthase-1 causes congenital disorder of glycosylation type Ie. *J. Clin. Invest.* **105:** 233–239.
30. Schenk, B., T. Imbach, C.G. Frank, *et al.* 2001. MPDU1 mutations underlie a novel human congenital disorder of glycosylation, designated type If. *J. Clin. Invest.* **108:** 1687–1695 (Note: Erratum: J. Clin. Invest.: 111: 925 only).
31. Huizing, M. & D.M. Krasnewich. 2009. Hereditary inclusion body myopathy: a decade of progress. *Biochim. Biophys. Acta.* **1792:** 881–887.
32. Hellbusch, C.C., M. Sperandio, D. Frommhold, *et al.* 2007. Golgi GDP-fucose transporter-deficient mice mimic congenital disorder of glycosylation IIc/leukocyte adhesion deficiency II. *J. Biol. Chem.* **282:** 10762–10772.
33. Kranz, C., C. Jungeblut, J. Denecke, *et al.* 2007. A defect in dolichol phosphate biosynthesis causes a new inherited disorder with death in early infancy. *Am. J. Hum. Genet.* **80:** 433–440.
34. Cantagrel, V., D.J. Lefeber, B.G. Ng, *et al.* 2010. *SRD5A3* is required for converting polyprenol to dolichol and is mutated in a congenital disorder of glycosylation. *Cell* **142:** 1–15.
35. Foulquier, F. 2009. COG defects, birth and rise! *Biochim. Biophys. Acta.* **1792:** 896–902.
36. Guillard, M., A. Dimopoulou, B. Fischer, *et al.* 2009. Vacuolar H+-ATPase meets glycosylation in patients with cutis laxa. *Biochim. Biophys. Acta.* **1792:** 903–914.
37. Denecke, J. & T. Marquardt. 2009. Congenital dyserythropoietic anemia type II (CDAII/HEMPAS): where are we now? *Biochim. Biophys. Acta.* **1792:** 915–920.
38. Bianchi, P., E.Fermo, C. Vercellati, *et al.* 2009. Congenital dyserythropoietic anemia type II (CDAII) is caused by mutations in the SEC23B gene. *Hum. Mutat.* **30:** 1292–1298.
39. Schwarz, K., A. Iolascon, F.Verissimo, *et al.* 2009. Mutations affecting the secretory COPII coat component SEC23B cause congenital dyserythropoietic anemia type II. *Nat. Genet.* **41:** 936–940.

Ann. N.Y. Acad. Sci. ISSN 0077-8923

ANNALS OF THE NEW YORK ACADEMY OF SCIENCES
Issue: *The Year in Human and Medical Genetics*

New insights in gene-derived therapy: the example of Duchenne muscular dystrophy

Annemieke Aartsma-Rus, Johan T. den Dunnen, and Gert-Jan B van Ommen
Department of Human Genetics, Leiden University Medical Center, Leiden, the Netherlands

Address for correspondence: Annemieke Aartsma-Rus, Leiden University Medical Center, Department of Human Genetics, Postzone S4-P, PO Box 9600, 2300 RC, Leiden, the Netherlands. a.m.rus@lumc.nl

The two therapeutic approaches currently most advanced in clinical trials for Duchenne muscular dystrophy are antisense-mediated exon skipping and forced read-through of premature stop codons. Interestingly, these approaches target the gene product rather than the gene itself. This review will explain the rationale and current state of affairs of these approaches and will then discuss how these gene-derived therapies might also be applicable to other diseases.

Keywords: monogenic diseases; Duchenne muscular dystrophy; forced read-through; exon skipping; personalized medicine

Introduction

Duchenne muscular dystrophy (DMD) is a severe, progressive neuromuscular disorder.[1] Patients present with first symptoms before the age of 5, generally lose ambulation by the age of 12, and die of respiratory or heart failure in the second to fourth decade of life. The incidence is around 1 in 3,500 newborn boys, with a *de novo* frequency of 1 in 3, while the worldwide prevalence is estimated to be 1 in 30,000. The causative gene was named *DMD* and was one of the first genes to be identified by reverse genetics (also called "positional cloning").[2] This was both because the disease was obviously X-chromosomal linked and because it had a relatively high frequency, so there were many patients available for genotyping, including some patients with large deletions that were microscopically visible.[3,4] This facilitated the hunt for the gene, as a probe derived from a patient with a large deletion could confirm that the same area was deleted in many other DMD patients.[5] The causative gene happened to be the largest gene in the human genome, spanning no less than 2.2 Mb on the short arm of the X chromosome. The associated protein "dystrophin" was predicted to have actin-binding sites at the N-terminus and a C-terminal cysteine rich domain that could interact with a transmembrane complex in the sarcolemma.[6,7] These two essential domains are joined by a central rod domain consisting of 24 spectrin-like repeats joined by 4 hinge regions.

Surprisingly, the probes deleted in DMD patients were also found to be deleted in another disease, Becker muscular dystrophy (BMD).[8] These patients typically presented with a much less severe disease phenotype and a milder progression than DMD.[1] In addition, between BMD patients there is a lot of variation in the age at which the disease is diagnosed and ambulation is lost. Assisted ventilation and scoliosis surgery are required for the majority of DMD patients but are rarely needed for BMD patients, while life expectancy for BMD patients is slightly reduced compared to normal. Initially, it was hypothesized that the two diseases were caused by two different genes, located close to each other. However, once dystrophin cDNAs were identified and these were used as probes, it turned out that both diseases were mainly caused by large deletions or duplications within the same gene. This puzzled the field for a while, until Monaco *et al.* presented the reading frame rule to explain the discrepancy between DMD and BMD, proposing that in DMD the reading frame of the mRNA is disrupted by the deletion or duplication of one or multiple exons, resulting in a protein that lacks the essential C-terminal

doi: 10.1111/j.1749-6632.2010.05836.x

domain and therefore is not functional.[9] By contrast, BMD patients carry deletions or duplications that do not affect the open reading frame and yield dystrophins that, while internally truncated, contain the essential interacting domains and so are partially functional. This rule indeed holds true for the vast majority (>90%) of DMD and BMD patients.[10] Unexpectedly, a number of BMD patients carried premature nonsense mutations. These mutations are mainly located in in-frame exons and disrupt exonic motives needed for exon recognition, causing the mutated exons to be skipped, thus bypassing the stop mutation while maintaining the reading frame.[10,11]

Although the gene was discovered in 1986, there is no therapy for the disease and treatment is only symptomatic, providing assisted ventilation and applying corticosteroids, which were found to slow disease progression. Corticosteroids probably act by suppressing the immune response in muscles of DMD patients due to chronic muscle damage. The immune cells exacerbate the pathology by inducing fibrosis through production of cytokines.[12] It has to be noted that due to improved care, the life expectancy of patients has increased from less than 20 to sometimes over 40 years of age in the last 20 years. Nevertheless, there is a clear need for a therapy for this disabling disease. Initial research focused mainly on cell and gene therapy. Although these approaches were quite successful in animal models (mainly mouse), neither has yet been successful in patients as many more hurdles have to be overcome than were initially anticipated. Cell therapy is problematic because muscle is a postmitotic tissue, and while the satellite cells that repair muscle upon damage can be isolated and expanded *ex vivo*, transplanting them is not very effective because these cells do not migrate from the bloodstream into muscle tissue and poorly migrate throughout the tissue when injected in muscle.[13] There are pluripotent nonmuscle-derived cells that can migrate to muscle from blood. However, the efficiency of this migration is currently too low to be of any benefit for patients.[14] The sheer size of the dystrophin gene and its coding region (11 kb) hampers gene therapy, as the capacity of the only virus able to efficiently infect muscle, the adeno-associated virus (AAV), is limited to ~4.5 kb.[15] Furthermore, gene therapy has some inherent safety issues, such as insertion mutagenesis, nonreversibility, and an immune response to the vector. Another complication for DMD cell and gene therapy in general is the abundance of muscle tissue. Roughly 30–40% of the human body consists of muscle tissue, and therefore the number of cells or viral particles required for effective treatment is well beyond the current biotechnological capacity.

In face of these challenges it is probably not surprising that the approaches currently closest to clinical application target the gene product rather than aiming to replace the gene: drug-induced read-through of premature stop codons and antisense-mediated exon skipping. In this review we will explain the rationale of these approaches, describe the current state of affairs, and discuss the hurdles. Finally, we will give examples of how these approaches that were developed for DMD, have been applied, and might be applied to other monogenic and complex diseases.

Read-through of premature nonsense mutations

Approximately 25% of DMD patients carry small mutations within an exon.[16] In roughly 15% of all patients these are substitution mutations causing a premature stop codon.[16] It was discovered that antibiotics of the aminoglycoside class could force cells to ignore both regular and aberrant stop codons during protein translation[17] (Fig. 1). This system appeared to be more efficient for mutation-induced stop codons, probably because these stop codons were not optimal due to, for example, their location (in the middle of the coding sequence) and their sequence context. Most research has been performed with gentamicin, which has shown varying results in cultured cells, mouse models, and patients.[18–20] This variation was attributed to the different gentamicin isomers. Every isomer has a different potency, and the relative amount of isomers differed between different batches.[18] In addition, the read-through response was partly determined by the type of stop codon (TGA, TAA, or TAG) and immediate sequence context. A recently published clinical trial reported that a 6-month treatment with gentamicin resulted in a restoration in dystrophin levels of 13–15% in 3 out of 16 treated DMD patients.[21] As chronic gentamicin treatment can lead to renal and ototoxicity, PTC therapeutics performed a high throughput screening in a cell-based system to identify drugs that are more potent in inducing

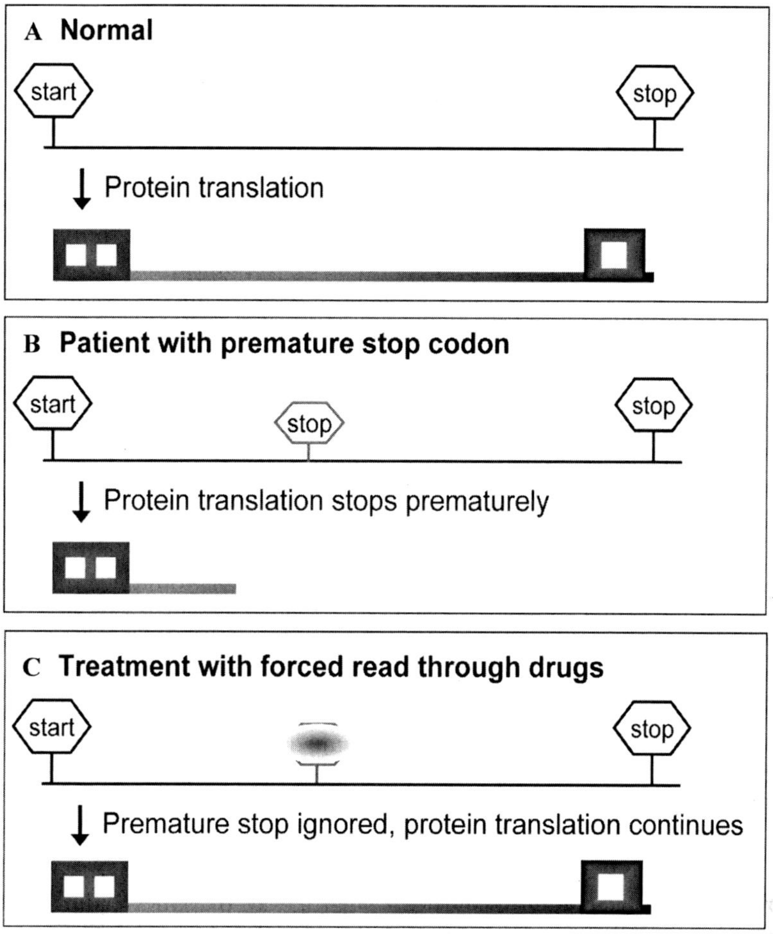

Figure 1. Forced read-through of premature truncation codons. (A) Normal mRNAs have a start and stop codon to allow translation into a protein. (B) When patients have a premature truncation codon, protein translation is stopped prematurely, even though these premature stop codons are generally less potent than regular stop codons. (C) Drugs that can force read-through of protein translation make the cell ignore preferentially mutated stop codons. Instead of a stop codon, an amino acid is built into the protein, and translation further continues until the end of the protein.

forced read-through and less toxic than gentamicin, using a luciferase construct containing a stop codon to prevent expression in the absence of a drug that induces read-through. The compound PTC124 (currently named ataluren®) was identified as being very efficient and was selected for further development. Preclinical studies showed dystrophin restoration in cultured cells and the *mdx* mouse model (which carries a nonsense mutation in exon 23) in the absence of toxicity.[22] The safety and tolerability of the drug was confirmed in a first clinical trial in healthy volunteers[23] followed by a trial in DMD patients (see www.clinicaltrials.gov). As the latter trial lasted for 4 weeks, an effect on muscle function was not found or anticipated. A longer trial of 48 weeks in 174 DMD patients has recently been completed. The trial contained a placebo group, a high PTC124 dose group, and a low-PTC124 dose group. The primary outcome was in muscle function, which was assessed by the 6-min walk test. Unfortunately, no difference could be observed between the high-dose and placebo group, while the improvement in the low-dose group compared to the placebo group was not significant (http://www.duchenne.nl/976_resultsataluren.pdf). The trial has been put on hold, and the analysis of other parameters (e.g., dystrophin levels) is pending.

Two recent publications partly explain why PTC124 performs less well than anticipated from the luciferase screening assay.[24,25] Some of the PTC124 derivatives generated during the screening process can stabilize the luciferase enzyme, resulting in enhanced efficiency in the screening assay. However, the additional preclinical tests in cell and animal models confirmed that PTC124 has at least some read-through potential,[22,26] so it is possible that in the clinical trial PTC124 treatment resulted in some dystrophin restoration, but that the levels of dystrophin were simply not sufficient to result in a significant difference in muscle function. For PTC124 to be effective in targeting the *DMD* gene product, it is essential that there is muscle tissue present. Because the treatment aims to stabilize a slowly progressive disease, it is challenging to perform functional analysis. Furthermore, it is currently not known how much dystrophin is needed to slow down disease progression. Therefore, dystrophin analysis of muscle biopsies will be essential to determine if long term PTC124 treatment results in dystrophin restoration and, if so, at which levels.

A factor that might have influenced the efficiency of PTC124 treatment is a process known as nonsense-mediated mRNA decay (NMD). This mechanism specifically breaks down mRNAs that carry premature stop codons. However, its efficiency varies between individuals, for different stop codons, location within the mRNA, and sequence context. When this process is very active, the levels of mutated dystrophin mRNA will be accordingly low, and as such the amount of protein that can be produced after PTC124 treatment is also expected to be low. In fact, using gentamicin to treat cystic fibrosis patient-derived cells *in vitro* indeed showed that cells from some patients resulted in restoration of CTFR protein, while cells from other patients did not respond to treatment.[27] Analysis of a subset of these patient-derived cells revealed that on average the nonresponders had a more active NMD than responders. Furthermore, when NMD was blocked in cultured cells from the nonresponders, these cells became more responsive to gentamicin treatment.[27] While this work is preliminary, it might be possible in future to prescreen cultured patient cells (e.g., from a skin biopsy) to identify likely responders and nonresponders.

Meanwhile, screens to identify additional, more potent read-through compounds are ongoing. A luciferase independent high-throughput screening method has recently been published.[28] This involves plasmids encoding myc-tagged transcripts with different premature stop mutation in an enzyme-linked immunosorbent assay (ELISA)-based system (using antibodies against the myc tags). One of the leading compounds identified in this screen (RTC13) induced read-through leading to significant dystrophin production in cultured *mdx* cells and *mdx* mice *in vivo*[28] (Dr. Carmen Bertoni, personal communication, 2010). The compound was less efficient than gentamicin but also less toxic. There is probably a fine balance between efficiency and toxicity, as it can be anticipated that compounds that lead to very efficient read-through of the premature stop codons in the *DMD* gene, also induce read-through of genuine stop codons. Further research will have to assess whether the newly identified compounds are able to induce sufficient restoration of dystrophin levels in the absence of side effects.

Antisense-mediated exon skipping

DMD is generally caused by mutations that result in prematurely truncated, nonfunctional dystrophins that are unstable, while BMD is caused by internally deleted, partly functional proteins.[9] In the late 1990s, the idea arose to manipulate the splicing of the DMD transcript in patients in such a way that the reading frame was restored through the skipping of one or more exons, to allow for the production of a BMD-like protein and conversion into a milder phenotype (Fig. 2). This indeed appeared to be feasible using antisense oligonucleotides (AONs) that specifically bind to their target sequence, hide an exon from the splicing machinery, and cause it to be skipped.[29–32] Initial tests were performed in patient-derived cell cultures carrying deletions,[30,31,33,34] duplications,[35] and point mutations[33,34,36,37] in exons and introns[38] and in cultured cells derived from the *mdx* mouse model.[32] In each case, skipping of the targeted exon(s) was achieved, resulting in restoration of varying levels of internally deleted dystrophin. In mouse models, intramuscular injections confirmed exon skipping and dystrophin restoration *in vivo*, and this was accompanied by a restoration in muscle function.[39] These encouraging results were followed by initial clinical trials where patients received intramuscular injections of AONs consisting of different chemistries (2′-O-methyl RNA

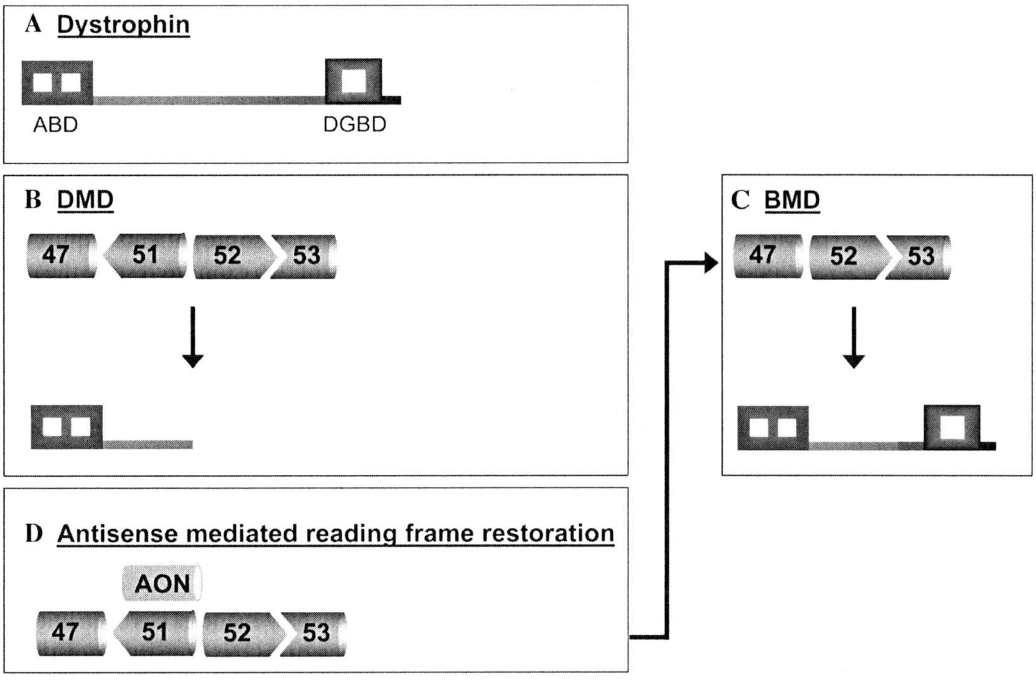

Figure 2. Antisense-mediated exon skipping. (A) The dystrophin protein contains an N-terminal actin-binding domain (ABD) and a dystroglycan-binding domain (DGBD), which are crucial for protein function. (B) In DMD, the open reading frame is disrupted (through a deletion of exon 48–50 in this example), leading to a prematurely truncated, nonfunctional dystrophin protein. (C) In BMD, mutations maintain the open reading frame (a deletion of exon 48–51 in this example), leading to an internally deleted protein that is partially functional, as the crucial domains are present. (D) Antisense-mediated reading frame restoration aims to manipulate the splicing so that an exon is skipped to restore the open reading frame (exon 51 in this example). This would allow the production of a BMD-like protein and conversion of a severe into a milder phenotype.

with a phosphorothioate backbone [2OMePS] and phosphorodiamidate morpholino oligomers [PMO].[40,41] Both resulted in exon skipping and dystrophin restoration in levels of up to 35%. While very promising as proof-of-principle, intramuscular administration is not a viable therapeutic strategy, since muscle tissue makes up 30–40% of the human body, making efficient systemic delivery of AONs essential. It is known that the AON chemistries used in the first trials primarily lead to uptake into liver and kidney and hardly into muscle when injected intravenously or subcutaneously in healthy animals and humans. Surprisingly, when the 2OMePS AONs were injected in the *mdx* mouse model, uptake in muscle was 10-fold higher than in healthy mice.[42,43] This increased uptake is thought to arise from the muscle pathology that leads to a defective muscle fiber membrane causing muscle enzymes to leak out into the bloodstream, but apparently also allowing AONs to enter the damaged muscle fibers. Systemic delivery of AONs has been further optimized in the *mdx* mouse model,[43–45] leading to two systemic Phase I/II trials where 2OMePS and PMO AONs are injected subcutaneously and intravenously, respectively. The preliminary results are promising in that dystrophin restoration was confirmed in both trials (http://prosensa.eu/press-room/press-releases/2009-09-14-PRO051-shows-favourable-results.php and http://www.avibio.com/news_detail.php?newsId=0068).

As the majority of patients die due to respiratory or heart failure, targeting the diaphragm and heart is crucial. For 2OMePS AONs, exon skipping levels in *mdx* mice are generally higher in diaphragm when compared to skeletal muscles, while skipping levels in heart are lower.[43,45,46] For PMO AONs, skeletal muscles show highest skipping levels, while dystrophin levels are slightly lower in diaphragm and again much lower in heart.[44,45] The low response in heart is likely due to the different structure and pathology of the heart muscle, which is made up of individual cardiomyocytes and thus does not have the "advantage" of large muscle fibers that become leaky in DMD. Indeed, higher doses of AONs show better exon skipping and dystrophin restoration in heart.[47] Alternatively, it was shown that it is possible to improve uptake of PMOs in all tissues using peptide conjugates (pPMO).[48,49] These pPMOs lead to very efficient exon skipping at reduced AON doses, even allowing phenotypic restoration of the very severely affected dystrophin–utrophin double knockout mouse model.[50] However, further optimization of these peptide conjugates is required, as toxicity has been observed for some pPMOs in primates.[51] These peptide-conjugated AONs have the disadvantage that the peptide part may elicit an immune response, while humeral or cellular immune responses have never been found after unconjugated 2OMePS and PMO AON treatment in animal models or clinical trials.[52]

Toward broad application of AON treatment for DMD

There are several hurdles that impede broad clinical application of the AON treatment approach for every DMD patient. Exon skipping is a mutation-specific approach (different exons need to be skipped to correct different mutations) and will not work for mutations that involve all essential actin-binding domains and/or the cysteine rich domain and/or a deletion of more than 75% of the central domain.[53] Fortunately, these mutations are rare (less than 10% of all patients). Two out of three patients carry deletions of one or more exons, and 75% of these deletions involve a hotspot between exon 45 and exon 55.[10] Therefore, the skipping of a few well-chosen exons already applies to a relatively large group of patients. The largest of these (13% of all patients) would benefit from exon 51 skipping, the exon currently furthest in clinical development.[53] However, the benefit of this approach gradually diminishes for additional exons and to target larger groups of patients, AONs targeting multiple exons need to be developed. To increase the applicability of this approach, it has been postulated to generate a "common" skip that would be beneficial for larger groups of patients, for example, exon 45–55 skipping,[54] but unfortunately, skipping larger stretches of exons has not yet been successful.[55]

Currently, each AON is considered a new drug and therefore has to go through all stages of clinical development. Though cost and labor intensive, this is feasible for exons targeting larger groups of patients but will be very problematic for exons applicable to increasingly smaller groups of patients; there may not even be sufficient patients to perform the clinical trials. Also, some patients require simultaneous skipping of two exons to restore the reading frame, which was shown to be feasible *in vitro* and *in vivo* in animal models.[34,35,56–58] It is plausible that AONs targeting individual exons would benefit other patients when used in combination (e.g., exon 51 and exon 45 skipping for a deletion of 46–50). These combinations, when individually approved, may not need to be tested in an equally rigorous trial. To ensure that exon skipping is developed for as many patients as possible in a safe but also an effective manner, a dialogue between industry, academics, and regulatory agencies has been initiated.[59] As the properties of AONs consisting of the same chemistry are very similar and in the vast majority sequence independent, there are opportunities for reduced toxicity tests to confirm the new AON of the same chemistry behaves as the rest of their class and to identify the occasional bad actor. This all depends on the outcome of placebo-controlled trials with exon 51 skipping drugs that are currently planned for the 2OMePS chemistry (GSK/Prosensa) and for the PMO chemistry (AVI-Biopharma).

It is currently not yet known which levels of dystrophin are needed to elicit a beneficial effect. Based on work done in *mdx* mice and occasional case reports of mildly affected BMD patients, the most common estimation is ~20–30% of normal dystrophin levels.[39,60] However, even exon 51 skipping induces different internally truncated dystrophins for different mutations (e.g., lacking exon 45–51, 47–51, 48–51, 49–51, 50–51, or 51–52) and it is likely that these will have different functionality and stability. Furthermore, variation in dystrophin levels is observed between BMD patients within the same families,[61] making it likely that even Duchenne patients with the same mutation will respond differently to the exon-skipping approach.

Due to AON turnover and clearance, repeated treatment will be required. As AONs have thus far been used to treat acute conditions such as cancer and inflammatory disorders, nothing is known about the long-term effects of AON treatment (longer than 1 year). In addition, repeated treatment may pose a burden for patients. Subcutaneous injections (2OMePS) might be given at home, but it may be necessary in the early phase to monitor adverse events at the injection site when administered over a long period of time. For intravenous infusion (PMO), patients would need to visit their doctor regularly. A method to deliver the antisense sequences through gene therapy is under development. This involves small nuclear ribonucleoproteins (snRNPs) U1 and U7, which have been modified so the original antisense sequence targeting the donor splice site and histones, respectively, are replaced with antisense sequences of choice.[62,63] The U1 and U7 snRNPs genes are very small and therefore fit in the AAV vector. The company AMT (Amsterdam, the Netherlands) is developing the U1 snRNP approach for clinical application. The main problem to this snRNP approach, besides the gene therapy hurdles mentioned earlier, is the optimization of good manufacturing practice (GMP) production of viral particles to have sufficient amounts to treat all muscles of the body. Alternatively, one could harvest or induce autologous pluripotent stem cells from patients, expand and transduce them *ex vivo* with appropriate U1 or U7 snRNP genes, and then transplant them back into patients.[64] However, this approach will need further optimization, since the efficiency of stem cell transplantation is currently too low.

Finally, it should be noted that the dystrophin transcript is produced in muscle tissue. In patients, however, muscle tissue is lost and replaced by adipose and fibrotic tissue, which do not express dystrophin. Thus, for the DMD exon skipping approach to be successful, it is essential that there is still muscle tissue present because exon skipping will not bring back muscle tissue once it is lost. Therefore, therapeutic approaches are being developed to increase the amount of muscle tissue to compensate for the loss of muscle. In the future, these approaches might be combined with approaches that target dystrophin transcripts. Examples include myostatin inhibition, activin-receptor type IIB blockers (e.g., ACE-031, which is currently tested in a phase 2 clinical trial in DMD patients), and IGF1 treatment.

Genetic therapies for other diseases

An overview of how forced read-through and antisense-mediated modulation of splicing is used as a research or therapeutic tool for other diseases is given in Table 1. PTC124 targets only genetic diseases and is furthest advanced for cystic fibrosis.[65] One might envisage how it could also be useful for cancer predispositions due to nonsense mutations in tumor suppressor genes.

Table 1A. Overview of ataluren trials

Disease	Phase	Status
Cystic fibrosis	3	Recruiting
DMD ambulatory	2B extension	Terminated
DMD nonambulatory	2	Suspended
Hemophilia A and B	2	Recruiting

Note: Information from www.clinicaltrials.gov.

Exon skipping could be used for any disease caused by the activation of aberrant splice sites. Since RNA is rarely analyzed, deep intronic mutations that activate cryptic exons and exonic variants that affect splicing by creating new splice sites or by activating/destroying exonic splice enhancers or silencer sites are generally missed. Nevertheless, for some diseases some of the most frequently occurring mutations are those that affect splicing, such as cystic fibrosis. The modulation of splicing through AONs to counteract the consequences of a mutation can be applied in many ways. Most obvious is the application to correct for mutations that induce aberrant intronic splice sites. For other mutations, mostly deletions and point mutations, it depends on the characteristics of the encoded protein whether or not exon skipping can be useful. The application is obvious for large structural proteins that contain (partly) dispensable repeated domains, such as dystrophin and some collagens.[53,66] Exon skipping can then be used to re-open a truncated reading frame and/or to remove partial domains that interfere with folding or function.

Rather than skipping exons, AONs can also be used to enable exon inclusion. The prime example is spinal muscular atrophy (SMA) caused by deletion of the functional *SMN1* gene. *SMN1* and *SMN2* are almost identical genes involved in the survival of motor neurons and are located antiparallel and close together on chromosome 5q. *SMN2* has a single synonymous nucleotide difference in exon 7, causing 90% of the SMN2 exon 7 to be skipped, which is translated into an instable protein. Reduced levels of SMN protein are produced from the 10% of

Table 1B. Overview of antisense-mediated exon-skipping applications

Disease	Disease-causing gene	Results	References[a]
Reading frame restoration			
Duchenne muscular dystrophy	DMD	Dystrophin was restored in patient-derived cell cultures and a functional improvement in the *mdx* mouse model was observed	31
Dystrophic epidermolysis bullosa	*COL7A1*	Allele-specific exon 70 skipping, which leads to internally deleted protein with near-normal functionality	66
Limb girdle muscular dystrophy type 2B	*DYSF*	Exon 32 skipping bypasses a truncating mutation in patient-derived cells	71
Mendelian susceptibility to mycobacterial disease	*IL-12RB1*	Skipping of nonsense codon containing exon 2 resulted in partially functional IL-12R production in patient-derived cultured cells.	72
Spinal muscular atrophy	*SMN2*	Antisense-mediated exon 7 inclusion leading to increased SMN levels *in vitro* and *in vivo*	67
Restoration of aberrant splicing			
β-thalassemia	β-*globin*	Restoration of β-globin and significant hemoglobin levels were observed in erythrocytes of a thalassemic mouse model carrying a common human mutation	73
Afibrinogemia	*FGB*	Antisense-mediated skipping of a cryptic exon was induced by an intronic mutation	74
Ataxia telangiectasia	*ATM*	Restoration of functional ATM in three patient-derived cell lines and normalization of radiosensitive phenotype	75
Breast cancer	*BRCA1*	Restoration of normal splicing in a pcDNA3.0 plasmid minigene construct containing a BRCA1 mutation transfected in HeLa cells	76
Congenital disorder of glycosylation type 1A	*PMM2*	AONs restored cryptic splicing in patient-derived cell cultures resulting in functionally active PMM protein	77
Cystic fibrosis	*CFTR*	Restoration of CFTR protein levels in cells stably expressing a *CFTR* cDNA construct that mimicked a common aberrant splice mutation was observed	78
Growth hormone insensitivity	*GHR*	AONs targeting the donor or acceptor splice site induced skipping of pseudoexon 6ψ in a minigene construct.	79
Iron-sulphur cluster deficiency myopathy	*ISCU*	Complete restoration of normal splicing pattern in fibroblasts from three patients carrying intronic mutations	80
Menkes disease model	*ATP7A*	*In vivo* correction of disrupted splicing in a zebrafish model for Menkes disease. AON treatment resulted in normal protein and completely correct copper-deficient defects	81

Continued

Table 1B. *Continued*

Disease	Disease-causing gene	Results	References[a]
Methylmalonic acidemia	*MUT*	Restoration of *MUT*-encoded MCM enzyme activity in patient-derived fibroblasts to up to 100% of control levels was observed	82
Neurofribro-matosis type 1	*NF1*	AONs restored cryptic splicing for three deep intronic mutations, leading for normalized NF GTPase activity	83
Nevoid basal cell carcinoma syndrome	*PTCH1*	Restoration of normal splicing in a pcDNA3.0 plasmid minigene construct containing a PTCH1 mutation transfected in HeLa cells was observed	76
Niemann-Pick type C disease	*NPC1*	AONs restored normal splicing in fibroblasts from a patient with an intronic mutation leading to inclusion of a pseudoexon	84
Ocular albinism type 1	*OA1*	Restoration of OA protein levels in patient-derived melanocytes was observed	85
Propionic acidemia	*PCCA*		
	PCCB	Restoration of PCCA/B-encoded PCC enzyme activity in patient-derived fibroblasts to close to 100% of control levels was observed	82
Alternative Splicing			
Arthritis and hepatitis	*TNFR2*	Antisense- mediated exon 7 skipping resulted in a switch from a transmembrane to a soluble isoform, which was accompanied by improved phenotypes in TNFα-induced hepatitis and collagen-induced arthritis mouse models.	86
Atherosclerosis	*APOB*	Exon skipping disrupts only the ORF of the APOB100 isoform, leading to decreased levels of APOB100 protein in liver cell cultures	87
Basal cell carcinoma	*Mcl-1*	Switching of the anti-apoptotic Mcl-1L to the proapoptotic Mcl-1S leads to apoptosis of cultured basal cell carcinoma and gastric adenocarcinoma cell lines	88
Breast cancer	*SRA1*	Restoration of the ratio of fully spliced and intron 1 retained SRA mRNAs in breast cancer cells might influence their growth and invasive properties	89
Breast cancer	*ErbB-2*	AON treatment resulted in exon 19 skipping, leading to a truncated, inactive Her-2 protein in cultured cells	90
Cancer	*Bcl-X*	Switching from anti-apoptotic Bcl-xL to proapoptotic Bcl-xS isoform leads to spontaneous apoptosis in various treated cancer cell lines and sensitization to chemotherapy of all treated cancer cell lines were observed	91

Continued

Table 1B. *Continued*

Disease	Disease-causing gene	Results	References[a]
Cancer	*tNOX*	AON treatment resulted in a shift from an invasive detrimental cell surface isoform to an isoform that is not surface expressed and induced apoptosis in HELA cells	92
FTDP-17[b]	*MAPT*	AONs restore normal splicing patterns, leading to decreased levels of the 4R tau isoforms, accompanied by a change in morphology of cultured rat cells	93
Inflammatory diseases	*IL-5Rα*	Exon skipping resulted in a decrease in the transmembrane and an increase in the soluble protein isoform	69
Inflammatory diseases	*MyD88*	Switching from immune stimulatory $MyD88_L$ isoform to the antistimulatory $MyD88_S$ isoform resulted in diminished proinflammatory IL-1 receptor signaling in cultured cells and in mouse liver *in vivo*	70
Leukemia	*c-myc*	Antisense-mediated disruption of c-myc ORF leads to lowered C-myc protein levels in various leukemic and lymphoma cell lines was observed	94
Leukemia	*WT1*	Decreased cell survival of treated cultured leukemic cells due to switching to a proapoptotic isoform	95
Myotonic dystrophy	*ClC-1*	AONs restored normal splicing in a myotonic mouse model, eliminating the myotonic pathology	96
Preterm labor	*COX-2*	Decreased levels of enzymatic activity in amnion-WISH and myometrial cells were observed due to skipping of an exon encoding the functional domain	97
Prostate cancer	*FOLH1*	Exon skipping leads to a switch from transmembrane FOLH1 to proapoptotic, enzymatically inactive intracellular isoform in prostate cancer cell cultures	98

[a]Due to space limitations, only one reference per application is included.
[b]Frontotemporal dementia and Parkinsonism linked to chromosome 17.

SMN2 transcripts that has exon 7 included. This can compensate for loss of *SMN1* to some extent, and when there are more (2–6) copies SMN2 it increasingly reduces SMA severity. Through the blocking of intron sequences that inhibit exon 7 inclusion, the amount of exon 7 containing SMN2 mRNA has been shown to increase, leading to higher SMN protein levels and a less severe phenotype in mouse models.[67]

Alternative splicing is a process that naturally increases protein diversity by either the inclusion or exclusion of one or more exons in the mRNA. As common diseases are thought to be typically due to more subtle deregulation than rare Mendelian diseases, it is plausible to consider that next to genetically or epigenetically caused gene-level alterations, differences in splicing or an imbalance in alternative splicing may be important causes of common disease.[68] If true, this would open a large new application for splice modulation. Moreover, even when there is no obvious disbalance, the subtle manipulation of the positioning or composition of cellular proteins may hold therapeutic benefit. Often, alternative isoforms of a protein have opposite functions, for example, pro- and anti-apoptotic, or pro- and anti-inflammatory. For example, in the case of the transmembrane domain-containing exon of IL-5Rα, exon skipping resulted in a decrease of the transmembrane isoform and increase of the soluble protein isoform.[69] This is expected to result in a reduction of proinflammatory signaling. The same concept of disrupting inflammatory signaling underlies the removal of exon 2 in MyD88, yielding an isoform incapable of coupling IRAK-1 to NF-κB.[70]

For dominant negative diseases caused by mutations in proteins that form complexes (e.g., dimers), one might reduce the pathology occurring from dominant-negative dimerization of a mutated and wild-type proteins by targeting either the mutated or the normal allele to increase the amount of dimerized proteins containing two identical proteins. Similarly, removal of specific, mutated domains from proteins might help to reduce (dominant) negative consequences. Exon skipping can also be used to deliberately target a frame-shifting exon simply to induce knock down of a protein either overall or for the mutated allele only. AON-mediated exon skipping is often not 100% efficient, but for some targets this might be an advantage, notably when complete knockdown would be deleterious but partial reduction of expression levels beneficial. General transcription reduction can probably also be achieved by targeting the 3′-terminal exon or the poly(A)-addition signal or poly(A)-addition site. Regarding the latter, when the normal poly(A)-addition signal is blocked an alternative polyA-site might be found, which most probably will affect the overall yield and/or stability of the transcript. In addition to destabilization, this could also stabilize RNA, counteracting mutations that reduce transcript levels.

In conclusion, while exon skipping was initially developed to treat rare Mendelian genetic diseases, the insights gained so far, as well as information from the ongoing clinical trials with respect to administration, dosage, toxicity, and regulatory aspects will offer great potential for complex diseases, too. This makes a strong case for abolishing the economic and social gap between translational research efforts for rare diseases and common diseases.

Acknowledgments

Mark Einderhand and Willeke van Roon-Mom are acknowledged for critically reading the manuscript. The Department of Human Genetics of the Leiden University Medical Center participates in the Center for Medical System Biology (CMSB) and the Center for Biomedical Genetics (CBG), and is a partner in the FP6 TREAT-NMD network of excellence (LSHM-CT-2006–036825) and the FP7 BIO-NMD project (HEALTH-F2–2009–241665).

Conflicts of interest

The authors report being employees of Leiden University Medical Center and coinventors on patent applications for antisense sequences and exon-skipping technology. Leiden University Medical Center has licensed the rights to part of these patents exclusively to Prosensa Therapeutics. The inventors specified on the patents (including the authors) are jointly entitled to a share of any future royalties paid to Leiden University Medical Center, should the therapy eventually be brought to the market.

References

1. Emery, A.E. 2002. The muscular dystrophies. *Lancet* **359:** 687–695.
2. Koenig, M., E.P. Hoffman, C.J. Bertelson, *et al.* 1987. Complete cloning of the Duchenne muscular dystrophy (DMD) cDNA and preliminary genomic organization of the DMD gene in normal and affected individuals. *Cell* **50:** 509–517.
3. Francke, U., H.D. Ochs, B.de Martinville, *et al.* 1985. Minor Xp21 chromosome deletion in a male associated with expression of Duchenne muscular dystrophy, chronic granulomatous disease, retinitis pigmentosa, and McLeod syndrome. *Am. J. Hum. Genet.* **37:** 250–267.
4. Greenstein, R., M. Reardon & C. TS. 1977. An X-autosomal translocation in a girl with Duchenne muscular dystrophy, evidence for DMD gen localisation. *Pediatr. Res.* **11:** 475A.
5. Monaco, A.P., C.J. Bertelson, W. Middlesworth, *et al.* 1985. Detection of deletions spanning the Duchenne muscular dystrophy locus using a tightly linked DNA segment. *Nature* **316:** 842–845.
6. Hoffman, E.P., R.H. Brown, Jr. & L.M. Kunkel. 1987. Dystrophin: the protein product of the Duchenne muscular dystrophy locus. *Cell* **51:** 919–928.
7. Koenig, M., A.P. Monaco & L.M. Kunkel. 1988. The complete sequence of dystrophin predicts a rod-shaped cytoskeletal protein. *Cell* **53:** 219–226.

8. Kingston, H.M., M. Sarfarazi, N.S. Thomas, *et al.* 1984. Localisation of the Becker muscular dystrophy gene on the short arm of the X chromosome by linkage to cloned DNA sequences. *Hum. Genet.* **67:** 6–17.
9. Monaco, A.P., C.J. Bertelson, S. Liechti-Gallati, *et al.* 1988. An explanation for the phenotypic differences between patients bearing partial deletions of the DMD locus. *Genomics* **2:** 90–95.
10. Aartsma-Rus, A., J.C. van Deutekom, I.F. Fokkema, *et al.* 2006. Entries in the Leiden Duchenne muscular dystrophy mutation database: an overview of mutation types and paradoxical cases that confirm the reading-frame rule. *Muscle Nerve* **34:** 135–144.
11. Flanigan, K.M., D.M. Dunn, N.A. von *et al.* 2009. Mutational spectrum of DMD mutations in dystrophinopathy patients: application of modern diagnostic techniques to a large cohort. *Hum. Mutat.* **30:** 1657–1666.
12. Deconinck, N. & B. Dan. 2007. Pathophysiology of duchenne muscular dystrophy: current hypotheses. *Pediatr. Neurol.* **36:** 1–7.
13. Skuk, D., B. Roy, M. Goulet, *et al.* 2004. Dystrophin expression in myofibers of Duchenne muscular dystrophy patients following intramuscular injections of normal myogenic cells. *Mol. Ther.* **9:** 475–482.
14. Farini, A., P. Razini, S. Erratico, *et al.* 2009. Cell based therapy for Duchenne muscular dystrophy. *J. Cell Physiol.* **221:** 526–534.
15. van Deutekom, J.C. & G.J. van Ommen. 2003. Advances in Duchenne muscular dystrophy gene therapy. *Nat. Rev. Genet.* **4:** 774–783.
16. Aartsma-Rus, A., W.E. Kaman, R. Weij, *et al.* 2006. Exploring the frontiers of therapeutic exon skipping for Duchenne muscular dystrophy by double targeting within one or multiple exons. *Mol. Ther.* **14:** 401–407.
17. Barton-Davis, E.R., L. Cordier, D.I. Shoturma, *et al.* 1999. Aminoglycoside antibiotics restore dystrophin function to skeletal muscles of mdx mice. *J. Clin. Invest.* **104:** 375–381.
18. Barton E.R. 2003. Constituents of gentamicin promote nonsense suppression and restoration of dystrophin function in mdx mice. Montpellier (France). *J. Muscle Res. Cell Motil.* **24:** 361–362.
19. Dunant, P., M.C. Walter, G. Karpati, *et al.* 2003. Gentamicin fails to increase dystrophin expression in dystrophin-deficient muscle. *Muscle Nerve* **27:** 624–627.
20. Wagner, K.R., S. Hamed, D.W. Hadley, *et al.* 2001. Gentamicin treatment of Duchenne and Becker muscular dystrophy due to nonsense mutations. *Ann. Neurol.* **49:** 706–711.
21. Malik, V., L.R. Rodino-Klapsac, L. Viollet, *et al.* 2010. Gentamicin-induced readthrough of stop codons in duchenne muscular dystrophy. *Ann. Neurol.* **67:** 771–780.
22. Welch, E.M., E.R. Barton, J. Zhuo, *et al.* 2007. PTC124 targets genetic disorders caused by nonsense mutations. *Nature* **447:** 87–91.
23. Hirawat, S., E.M. Welch, G.L. Elfring, et al. 2007. Safety, tolerability, and pharmacokinetics of PTC124, a nonaminoglycoside nonsense mutation suppressor, following single- and multiple-dose administration to healthy male and female adult volunteers. *J. Clin. Pharmacol.* **47:** 430–444.
24. Auld, D.S., N. Thorne, W.F. Maguire, *et al.* 2009. Mechanism of PTC124 activity in cell-based luciferase assays of nonsense codon suppression. *Proc. Natl. Acad. Sci. USA* **106:** 3585–3590.
25. Auld, D.S., S. Lovell, N. Thorne, *et al.* 2010. Molecular basis for the high-affinity binding and stabilization of firefly luciferase by PTC124. *Proc. Natl. Acad. Sci. USA* **107:** 4878–4883.
26. Du M., X. Liu, E.M. Welch, *et al.* 2008. PTC124 is an orally bioavailable compound that promotes suppression of the human CFTR-G542X nonsense allele in a CF mouse model. *Proc. Natl. Acad. Sci. USA* **105:** 2064–2069.
27. Linde, L., S. Boelz, M. Nissim-Rafinia, *et al.* 2007. Nonsense-mediated mRNA decay affects nonsense transcript levels and governs response of cystic fibrosis patients to gentamicin. *J. Clin. Invest.* **117:** 683–692.
28. Nichols, W.C., D.K. Kissell, N. Pankratz, *et al.* 2009. Variation in GIGYF2 is not associated with Parkinson disease. *Neurology* **72:** 1886–1892.
29. Dunckley, M.G., M. Manoharan, P. Villiet, *et al.* 1998. Modification of splicing in the dystrophin gene in cultured Mdx muscle cells by antisense oligoribonucleotides. *Hum. Mol. Genet.* **7:** 1083–1090.
30. Takeshima, Y., H. Wada, M. Yagi, *et al.* 2001. Oligonucleotides against a splicing enhancer sequence led to dystrophin production in muscle cells from a Duchenne muscular dystrophy patient. *Brain Dev.* **23:** 788–790.
31. van Deutekom, J.C., M. Bremmer-Bout, A.A. Janson, *et al.* 2001. Antisense-induced exon skipping restores dystrophin expression in DMD patient derived muscle cells. *Hum. Mol. Genet.* **10:** 1547–1554.
32. Wilton S.D., F. Lloyd, K. Carville, *et al.* 1999. Specific removal of the nonsense mutation from the mdx dystrophin mRNA using antisense oligonucleotides. *Neuromuscul. Disord.* **9:** 330–338.
33. Aartsma-Rus, A., A.A. Janson, W.E. Kaman, *et al.* 2003. Therapeutic antisense-induced exon skipping in cultured muscle cells from six different DMD patients. *Hum. Mol. Genet.* **12:** 907–914.
34. Aartsma-Rus, A., A.A. Janson, W.E. Kaman, *et al.* 2004. Antisense-induced multiexon skipping for duchenne muscular dystrophy makes more sense. *Am. J. Hum. Genet.* **74:** 83–92.
35. Aartsma-Rus, A., A.A. Janson, G.J. van Ommen, *et al.* 2007. Antisense-induced exon skipping for duplications in Duchenne muscular dystrophy. *BMC Med. Genet.* **8:** 43.
36. Spitali, P., P. Rimessi, M. Fabris, *et al.* 2009. Exon skipping-mediated dystrophin reading frame restoration for small mutations. *Hum. Mutat.* **30:** 1527–1534.
37. Surono, A., T. Van Khanh, Y. Takeshima, *et al.* 2004. Chimeric RNA/ethylene-bridged nucleic acids promote dystrophin expression in myocytes of duchenne muscular dystrophy by inducing skipping of the nonsense mutation-encoding exon. *Hum. Gene. Ther.* **15:** 749–757.
38. Gurvich, O.L., T.M. Tuohy, M.T. Howard, *et al.* 2008. DMD pseudoexon mutations: splicing efficiency, phenotype, and potential therapy. *Ann. Neurol.* **63:** 81–89.
39. Lu, Q.L., C.J. Mann, F. Lou, *et al.* 2003. Functional amounts of dystrophin produced by skipping the mutated exon in the mdx dystrophic mouse. *Nat. Med.* **8:** 1009–1014.

40. Kinali, M., V. rechavala-Gomeza, L. Feng, *et al.* 2009. Local restoration of dystrophin expression with the morpholino oligomer AVI-4658 in Duchenne muscular dystrophy: a single-blind, placebo-controlled, dose-escalation, proof-of-concept study. *Lancet Neurol.* **8:** 918–928.
41. van Deutekom, J.C., A.A. Janson, I.B. Ginjaar, *et al.* 2007. Local dystrophin restoration with antisense oligonucleotide PRO051. *N. Engl. J. Med.* **357:** 2677–2686.
42. Bremmer-Bout, M., A. Aartsma-Rus, E.J. de Meijer, *et al.* 2004. Targeted exon skipping in transgenic hDMD mice: a model for direct pre-clinical screening of human-specific antisense oligonucleotides. *Mol. Ther.* **10:** 232–240.
43. Heemskerk, H., W.C. de, K.P. van *et al.* 2010. Preclinical PK and PD studies on 2′-O-methyl-phosphorothioate RNA antisense oligonucleotides in the mdx mouse model. *Mol. Ther.* **18:** 1210–1217.
44. Alter, J., F. Lou, A. Rabinowitz, *et al.* 2006. Systemic delivery of morpholino oligonucleotide restores dystrophin expression bodywide and improves dystrophic pathology. *Nat. Med.* **12:** 175–177.
45. Heemskerk, H.A., C.L. De Winter, S.J. de Kimpe, *et al.* 2009. In vivo comparison of 2′-O-methyl phosphorothioate and morpholino antisense oligonucleotides for Duchenne muscular dystrophy exon skipping. *J. Gene. Med.* **11:** 257–266.
46. Heemskerk, H., C.L. De Winter, G.J. van Ommen, *et al.* 2009. Development of antisense-mediated exon skipping as a treatment for duchenne muscular dystrophy. *Ann. N.Y. Acad. Sci.* **1175:** 71–79.
47. Wu, B., P. Lu, E. Benrashid, *et al.* 2010. Dose-dependent restoration of dystrophin expression in cardiac muscle of dystrophic mice by systemically delivered morpholino. *Gene. Ther.* **17:** 132–140.
48. Wu, B., H.M. Moulton, P.L. Iversen, *et al.* 2008. Effective rescue of dystrophin improves cardiac function in dystrophin-deficient mice by a modified morpholino oligomer. *Proc. Natl. Acad. Sci. USA* **105:** 14814–14819.
49. Yin, H., H.M. Moulton, Y. Seow, *et al.* 2008. Cell-penetrating peptide-conjugated antisense oligonucleotides restore systemic muscle and cardiac dystrophin expression and function. *Hum. Mol. Genet.* **17:** 3909–3918.
50. Goyenvalle, A., A. Babbs, D. Powell, *et al.* 2010. Prevention of dystrophic pathology in severely affected dystrophin/utrophin-deficient mice by morpholino-oligomer-mediated exon-skipping. *Mol. Ther.* **18:** 198–205.
51. Moulton, H.M. & J.D. Moulton. 2010. Morpholinos and their peptide conjugates: therapeutic promise and challenge for Duchenne muscular dystrophy. *Biochim. Biophys. Acta* **1798:** 2296–2303.
52. Crook, S.T. (Ed.). 2001. *Antisense drug technology: Principles, strategies, and applications.* Marcel Dekker Inc. New York.
53. Aartsma-Rus, A., I. Fokkema, J. Verschuuren, *et al.* 2009. Theoretic applicability of antisense-mediated exon skipping for Duchenne muscular dystrophy mutations. *Hum. Mutat.* **30:** 293–299.
54. Beroud, C., S. Tuffery-Giraud, M. Matsuo, *et al.* 2007. Multiexon skipping leading to an artificial DMD protein lacking amino acids from exons 45 through 55 could rescue up to 63% of patients with Duchenne muscular dystrophy. *Hum. Mutat.* **28:** 196–202.
55. van Vliet, L., C.L. De Winter, J.C. van Deutekom, *et al.* 2008. Assessment of the feasibility of exon 45–55 multiexon skipping for duchenne muscular dystrophy. *BMC. Med. Genet.* **9:** 105.
56. McClorey, G., H.M. Moulton, P.L. Iversen, *et al.* 2006. Antisense oligonucleotide-induced exon skipping restores dystrophin expression in vitro in a canine model of DMD. *Gene Ther.* **13:** 1373–1381.
57. Mitrpant, C., S. Fletcher, P.L. Iversen, *et al.* 2009. By-passing the nonsense mutation in the 4 CV mouse model of muscular dystrophy by induced exon skipping. *J. Gene Med.* **11:** 46–56.
58. Yokota, T., Q.L. Lu, T. Partridge, *et al.* 2009. Efficacy of systemic morpholino exon-skipping in Duchenne dystrophy dogs. *Ann. Neurol.* **65:** 667–676.
59. Muntoni F. 2010. The development of antisense oligonucleotide therapies for Duchenne muscular dystrophy: report on a TREAT-NMD workshop hosted by the European Medicines Agency (EMA), on September 25th 2009. *Neuromuscul. Disord* **20:** 355–362.
60. Neri, M., S. Torelli, S. Brown, *et al.* 2007. Dystrophin levels as low as 30% are sufficient to avoid muscular dystrophy in the human. *Neuromuscul. Disord.* **17:** 913–918.
61. Nordenskjold, M., L. Nicholson, L. Edstrom, *et al.* 1990. A normal male with an inherited deletion of one exon within the DMD gene. *Hum. Genet.* **84:** 207–209.
62. Denti, M.A., T. Incitti, O. Sthandier, *et al.* 2008. Long-term benefit of adeno-associated virus/antisense-mediated exon skipping in dystrophic mice. *Hum. Gene Ther.* **19:** 601–608.
63. Goyenvalle, A., A. Vulin, F. Fougerousse, *et al.* 2004. Rescue of dystrophic muscle through U7 snRNA-mediated exon skipping. *Science* **306:** 1796–1799.
64. Meregalli, M., A. Farini & Y. Torrente. 2008. Combining stem cells and exon skipping strategy to treat muscular dystrophy. *Expert. Opin. Biol. Ther.* **8:** 1051–1061.
65. Kerem, E., S. Hirawat, S. Armoni, *et al.* 2008. Effectiveness of PTC124 treatment of cystic fibrosis caused by nonsense mutations: a prospective phase II trial. *Lancet* **372:** 719–727.
66. Goto, M., D. Sawamura, W. Nishie, *et al.* 2006. Targeted skipping of a single exon harboring a premature termination codon mutation: implications and potential for gene correction therapy for selective dystrophic epidermolysis bullosa patients. *J. Invest. Dermatol.* **126:** 2614–2620.
67. Hua, Y., T.A. Vickers, H.L. Okunola, *et al.* 2008. Antisense masking of an hnRNP A1/A2 intronic splicing silencer corrects SMN2 splicing in transgenic mice. *Am. J. Hum. Genet.* **82:** 834–848.
68. Thorsen, K., K.D. Sorensen, A.S. Brems-Eskildsen, *et al.* 2008. Alternative splicing in colon, bladder, and prostate cancer identified by exon array analysis. *Mol. Cell Proteomics.* **7:** 1214–1224.
69. Karras, J.G., M.A. Maier, T. Lu, *et al.* 2001. Peptide nucleic acids are potent modulators of endogenous pre-mRNA splicing of the murine interleukin-5 receptor-alpha chain. *Biochemistry* **40:** 7853–7859.
70. Vickers, T.A., H. Zhang, M.J. Graham, *et al.* 2006. Modification of MyD88 mRNA splicing and inhibition of

IL-1beta signaling in cell culture and in mice with a 2′-O-methoxyethyl-modified oligonucleotide. *J. Immunol.* **176:** 3652–3661.
71. Wein, N., A. Avril, M. Bartoli, *et al.* 2010. Efficient bypass of mutations in dysferlin deficient patient cells by antisense-induced exon skipping. *Hum. Mutat.* **31:** 136–142.
72. Van de Vosse, E., E.M. Verhar, *et al.* 2009. Antisense-mediated exon skipping to correct IL-12Rbeta1 deficiency in T cells. *Blood* **113:** 4548–4555.
73. Svasti, S., T. Suwanmanee, S. Fucharoen, *et al.* 2009. RNA repair restores hemoglobin expression in IVS2–654 thalassemic mice. *Proc. Natl. Acad. Sci. USA* **106:** 1205–1210.
74. Davis, R.L., V.M. Homer, P.M. George, *et al.* 2009. A deep intronic mutation in FGB creates a consensus exonic splicing enhancer motif that results in afibrinogenemia caused by aberrant mRNA splicing, which can be corrected in vitro with antisense oligonucleotide treatment. *Hum. Mutat.* **30:** 221–227.
75. Du, L., J.M. Pollard & R.A. Gatti. 2007. Correction of prototypic ATM splicing mutations and aberrant ATM function with antisense morpholino oligonucleotides. *Proc. Natl. Acad. Sci. USA* **104:** 6007–6012.
76. Uchikawa, H., K. Fujii, Y. Kohno, *et al.* 2007. U7 snRNA-mediated correction of aberrant splicing caused by activation of cryptic splice sites. *J. Hum. Genet.* **52:** 891–897.
77. Vega, A.I., C. Perez-Cerda, L.R. Desviat, *et al.* 2009. Functional analysis of three splicing mutations identified in the PMM2 gene: toward a new therapy for congenital disorder of glycosylation type Ia. *Hum. Mutat.* **30:** 795–803.
78. Friedman, K.J., J. Kole, J.A. Cohn, *et al.* 1999. Correction of aberrant splicing of the cystic fibrosis transmembrane conductance regulator (CFTR) gene by antisense oligonucleotides. *J. Biol. Chem.* **274:** 36193–36199.
79. David, A., U. Srirangalingam, L.A. Metherell, *et al.* 2010. Repair of aberrant splicing in growth hormone receptor by antisense oligonucleotides targeting the splice sites of a pseudoexon. *J. Clin. Endocrinol. Metab.* **95:** 3542–3546.
80. Kollberg, G. & E. Holme. 2009. Antisense oligonucleotide therapeutics for iron-sulphur cluster deficiency myopathy. *Neuromuscul. Disord.* **19:** 833–836.
81. Madsen, E.C., P.A. Morcos, B.A. Mendelsohn, *et al.* 2008. In vivo correction of a Menkes disease model using antisense oligonucleotides. *Proc. Natl. Acad. Sci. USA* **105:** 3909–3914.
82. Ugarte, M., C. Aguado, L.R. Desviat, *et al.* 2007. Propionic and methylmalonic acidemia: antisense therapeutics for intronic variations causing aberrantly spliced messenger RNA. *Am. J. Hum. Genet.* **81.**
83. Pros, E., J. Fernandez-Rodriguez, B. Canet, *et al.* 2009. Antisense therapeutics for neurofibromatosis type 1 caused by deep intronic mutations. *Hum. Mutat.* **30:** 454–462.
84. Rodriguez-Pascau, L., M.J. Coll, L. Vilageliu, *et al.* 2009. Antisense oligonucleotide treatment for a pseudoexon-generating mutation in the NPC1 gene causing Niemann-Pick type C disease. *Hum. Mutat.* **30:** E993–E1001.
85. Vetrini, F., R. Tammaro, S. Bondanza, *et al.* 2006. Aberrant splicing in the ocular albinism type 1 gene (OA1/GPR143) is corrected in vitro by morpholino antisense oligonucleotides. *Hum. Mutat.* **27:** 420–426.
86. Graziewicz, M.A., T.K. Tarrant, B. Buckley, *et al.* 2008. An endogenous TNF-alpha antagonist induced by splice-switching oligonucleotides reduces inflammation in hepatitis and arthritis mouse models. *Mol. Ther.* **16:** 1316–1322.
87. Khoo, B., X. Roca, S.L. Chew, *et al.* 2007. Antisense oligonucleotide-induced alternative splicing of the APOB mRNA generates a novel isoform of APOB. *BMC Mol. Biol.* **8:** 3.
88. Shieh, J.J., K.T. Liu, S.W. Huang, *et al.* 2009. Modification of alternative splicing of Mcl-1 pre-mRNA using antisense morpholino oligonucleotides induces apoptosis in basal cell carcinoma cells. *J. Invest. Dermatol.* **129:** 2497–2506.
89. Cooper, C., J. Guo, Y. Yan, *et al.* 2009. Increasing the relative expression of endogenous non-coding steroid receptor RNA activator (SRA) in human breast cancer cells using modified oligonucleotides. *Nucleic Acids Res.* **37:** 4518–4531.
90. Pankratova, S., B.N. Nielsen, T. Shiraishi, *et al.* 2010. PNA-mediated modulation and redirection of Her-2 pre-mRNA splicing: specific skipping of erbB-2 exon 19 coding for the ATP catalytic domain. *Int. J. Oncol.* **36:** 29–38.
91. Mercatante, D.R., J.L. Mohler & R. Kole. 2002. Cellular response to an antisense-mediated shift of Bcl-x pre-mRNA splicing and antineoplastic agents. *J. Biol. Chem.* **277:** 49374–49382.
92. Tang, X., D.J. Morre & D.M. Morre. 2007. Antisense experiments demonstrate an exon 4 minus splice variant mRNA as the basis for expression of tNOX, a cancer-specific cell surface protein. *Oncol. Res.* **16:** 557–567.
93. Kalbfuss, B., S.A. Mabon & T. Misteli. 2001. Correction of alternative splicing of tau in frontotemporal dementia and parkinsonism linked to chromosome 17. *J. Biol. Chem.* **276:** 42986–42993.
94. Giles, R.V., D.G. Spiller, R.E. Clark, *et al.* 1999. Antisense morpholino oligonucleotide analog induces missplicing of C-myc mRNA. *Antisense Nucleic Acid Drug Dev.* **9:** 213–220.
95. Renshaw, J., R.M. Orr, M.I. Walton, *et al.* 2004. Disruption of WT1 gene expression and exon 5 splicing following cytotoxic drug treatment: antisense down-regulation of exon 5 alters target gene expression and inhibits cell survival. *Mol. Cancer Ther.* **3:** 1467–1484.
96. Wheeler, T.M., J.D. Lueck, M.S. Swanson, *et al.* 2007. Correction of ClC-1 splicing eliminates chloride channelopathy and myotonia in mouse models of myotonic dystrophy. *J. Clin. Invest* **117:** 3952–3957.
97. Tyson-Capper, A.J. & G.N. Europe-Finner. 2006. Novel targeting of cyclooxygenase-2 (COX-2) pre-mRNA using antisense morpholino oligonucleotides directed to the 3′ acceptor and 5’ donor splice sites of exon 4: suppression of COX-2 activity in human amnion-derived WISH and myometrial cells. *Mol. Pharmacol.* **69:** 796–804.
98. Williams, T. & R. Kole. 2006. Analysis of prostate-specific membrane antigen splice variants in LNCap cells. *Oligonucleotides* **16:** 186–195.

Ann. N.Y. Acad. Sci. ISSN 0077-8923

Retraction

B.J. Yang, R.W. Smith, M. Saho, and M. Sadayappan. 2002. The Influence of Marangoni Flows on Crack Growth in Cast Metals. *Annals of the New York Academy of Sciences.* **974:** 110–123.

The above article, published in October 2002, has been retracted by agreement among the authors, the *Annals* director and executive editor, and Wiley Periodicals, Inc. The retraction has been agreed upon due to duplication with a previous publication.

References

Yang, B.J., R.W. Smith, M. Saho & M. Sadayappan. 2002. The Influence of Marangoni Flows on Crack Growth in Cast Metals. *Ann. N.Y. Acad. Sci.* **974:** 110–123.

doi: 10.1111/j.1749-6632.2002.tb05901.x

doi: 10.1111/j.1749-6632.2010.05916.x